NEUTRINO GEOPHYSICS:

PROCEEDINGS OF NEUTRINO SCIENCES 2005

NEUTRINO GEOPHYSICS: PROCEEDINGS OF NEUTRINO SCIENCES 2005

Edited by

STEPHEN T. DYE

Reprinted from *Earth Moon, and Planets*
Volume 99, Nos. 1–4, 2006

Springer

A C.I.P catalogue record for this book is available from the library of Congress

ISBN 978-0-387-70766-2

Published by Springer,
P.O. Box 17, 3300 AA, Dordrecht, The Netherlands

www.springer.com

Printed on acid-free paper

Printed in the Netherlands

Table of Contents

Preface

These pages present a collection of recent papers primarily documenting the nascent science of neutrino geophysics. Most of the papers followed from talks given at Neutrino Sciences 2005: Neutrino Geophysics[1] held at the University of Hawaii in December 2005. Several papers were solicited later in an effort to make the collection as comprehensive as possible. Every paper was scrutinized by an external reviewer to assure the quality of scientific content. These reviewers are thanked for lending their scientific expertise through their many thoughtful comments and suggestions. All authors are commended for providing excellent manuscripts of their important work while maintaining a spirit of cooperative collaboration throughout. Although every attempt was made to produce a thoroughly accurate volume, it is the accepted responsibility of the associate editor for any mistakes, errors, or omissions in the presented material. The recommendations, advice, and wisdom of John Learned and Sandip Pakvasa were indispensable in organizing and completing this project. Production charges were generously provided by the University of Hawaii. The support of Hawaii Pacific University, which contributed teaching release time, is gratefully acknowledged.

Stephen T. Dye
Associate Editor
November 28, 2006

Stephen T. Dye is an associate professor of physics at Hawaii Pacific University and an affiliate to the graduate faculty of the Department of Physics and Astronomy at the University of Hawaii at Manoa.

[1]See www.phys.hawaii.edu/~sdye/hnsc.html

Earth, Moon, and Planets (2006) 99:1–15
DOI 10.1007/s11038-006-9117-3

Neutrino Geophysics Conference Introduction

JOHN G. LEARNED, STEPHEN T. DYE and SANDIP PAKVASA
Department of Physics and Astronomy, University of Hawaii, Manoa, Honolulu, HI, USA
(E-mail: jgl@phys.hawaii.edu)

(Received 3 July 2006; Accepted 11 August 2006)

Abstract. Long distance detection of electron anti-neutrinos from reactors at distances of order 200 km has been achieved with the 1000 ton liquid scintillator-based KamLAND instrument in Japan. In summer 2005 the KamLAND group reported the first detection of anti-neutrinos from the natural radioactivity of the earth. These measurements are due to uranium and thorium decays dominantly from the nearby crust in Japan, and are expected to have only a small contribution from the earth's mantle (and core). Several new detectors are under consideration around the world for measurements which when taken together can reveal the location of these heavy elements, which are expected to contribute a major share of the internal earth's heating via their radioactivity. This heating is of course associated with providing the power to drive the geomagnetic field and plate tectonics. Geologists have only indirect evidence about the deep earth, mostly from seismic wave velocity and inferences from a few meteorites. Anti-neutrino detection, on the other hand, yields direct information about earth's interior. The location and magnitude of the earth's uranium and thorium are crucial to understanding the origin and evolution of the earth and present day activity.

Keywords: Neutrino, anti-neutrino, uranium, thorium, rare metal abundance, earth heat balance, geomagnetic field, plate tectonics

1. Welcome

This meeting provides for most of us the opportunity to meet with colleagues from different disciplines on an exciting topic of mutual interest: the study of neutrinos emanating from the earth, and what we can learn about the deep and inaccessible interior from these ghostly messengers. We share the unusual chance to meet in a small and informal venue where one can ask those "dumb questions" (the ones that are usually hard to answer), and we can all benefit from new viewpoints. In fact, writing in retrospect, this did indeed take place, and we had a wonderful three days of very active information exchange.

2. Neutrino Science Blossoming

Many of the physicists at this meeting have participated in the great advances made in neutrino physics in the last decade. Neutrinos have been largely a curiosity in the study of elementary particle physics due to the extreme difficulty of their detection. This is mainly due to their only interaction being via the weak force, and hence their traveling right though detectors and even the earth and sun, almost unimpeded. The cross-section between a proton and a geo-neutrino (an electron anti-neutrino of a few MeV) is a mere 10^{-42} cm^2, or about 1/100 billionth the cross section for a photon striking a proton. Hence, huge instruments are necessary to get into the game, and these have been developed over the last twenty-five years. Some were built mainly to search for the decay of ordinary matter (proton decay), which has yet to be found. Yet, by (open eyed) serendipity three major events have occurred.

First, a burst of neutrinos was observed arriving from a collapsing star at a distance of 150,000 light years, on 23 February 1987. This provided evidence for the end of life of large stars, and gave confirmation to many things suspected of neutrinos. In fact neutrinos are invoked as vital participants driving the type-II supernova explosion shock wave responsible for the production of the heavy elements under study at this meeting: the general belief is that the only significant source of heavy elements in our universe is during the explosion of the envelope surrounding the collapsing neutron core of an exhausted star.

The second major event in neutrino studies was the reported observation by the Super-Kamiokande collaboration in 1998 of "oscillations" of neutrinos, and in this instance the morphing in flight of one of the three flavors of neutrinos (electron, muon and tau neutrinos) from muon neutrino to tau neutrino (and back). This observation, now multiply confirmed, indicates the peculiarity that the muon neutrinos and tau neutrinos are essentially as mixed as they can be, transforming almost entirely, with a frequency of about 10^3 Hz for energies of about 1 GeV, back and forth between flavors. This unique oscillation phenomenon tells us that neutrinos do have some rest mass, albeit much less than an electron (by about a factor of 20 million). These observations were not anticipated by model builders, however the phenomenon of neutrino oscillations had been discussed for many years. Non-zero neutrino masses require physics beyond the Standard Model of particle physics.

The third major step occurred as a concluding series of experiments elucidated the origin of the 35 year "solar neutrino puzzle." This began with the measurement by Ray Davis and collaborators of solar neutrino interactions in a radiochemical experiment in the Homestake gold mine in 1968. His experiment and other later experiments found a deficit of neutrinos from the sun, as compared to well developed predictive models of solar burning. Much speculation and even acrimony over whether the solar models were wrong or the experiments were wrong, was dismissed by results from two experiments in the last four years. First, the SNO experiment in a deep nickel mine in Canada, using heavy water as (an expensive) target was able to measure not only the flux of electron neutrinos, as had several other experiments (including Super-Kamiokande), but most importantly they were able to detect a process called the "neutral current interaction" which responds to any type of neutrino, and hence measure the total flux of neutrinos arriving from the sun, not just electron neutrinos. Indeed their results indicated that there were no missing neutrinos, but that the number of electron neutrinos was greatly depleted at earth.

Finally, the KamLAND experiment measuring electron anti-neutrinos (which should show the same behavior as electron neutrinos) was able to measure the flux of electron anti-neutrinos due to all the power reactors operating around Japan, at a typical distance of 200 km. Not only did KamLAND measure the decrease in rate as expected from solar neutrinos, but the energy distribution of the anti-neutrinos as observed showed the unique oscillatory signature (see Figure 1). This eliminated the last competing hypotheses.

This latter experiment, KamLAND has the sensitivity to see neutrinos all the way down in energy to around 1 MeV, where one begins to find the flux of neutrinos from radioactive decays throughout the earth, and hence the present meeting is made possible.

It should be said that neutrinos, while presenting great challenges to detection, on the flip-side, present the possibility for peering inside luminous objects from throughout the universe. Not only can we see inside the earth and into the core of the sun where the fusion furnace burns, but into exploding stars, and we hope in the near future into the cores of galaxies, other of the most energetic objects in the universe, and ultimately back to the time of the Big Bang.

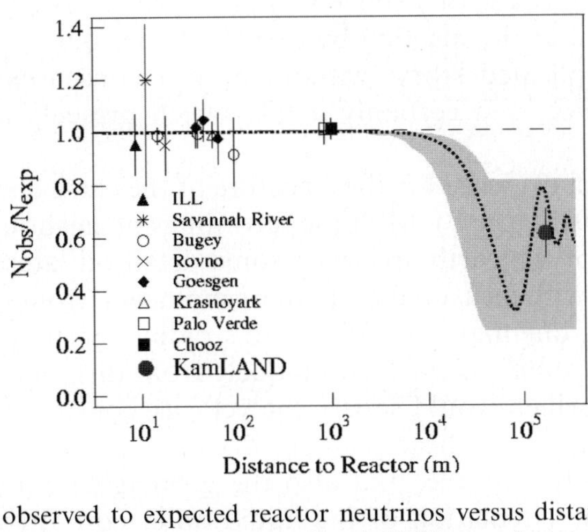

Figure 1. Ratio of observed to expected reactor neutrinos versus distance from the reactor from KamLAND and previous experiments. KamLAND was first to see a deficit due to neutrino oscillations.

On the terrestrial front, experiments and neutrino beams are being planned and even constructed to study details of the newly recognized oscillation phenomenon, in hopes of elucidating some of the most peculiar aspects of particle physics. Some of the heavier quarks slightly violate a symmetry called CP invariance. It is of great interest to know if neutrinos share this peculiar property, and this will be tested in some difficult experiments in the next several decades.

Practical applications are also being studied, from nearby reactor monitoring with truck-sized detectors, to dreams of a world network of (as yet impractical) instruments to monitor all the world's reactors as well as keep track of clandestine nuclear weapons testing. The field is indeed blossoming and this is the right time to start dreaming of doing neutrino tomography of the earth.

3. Where is the Uranium and Thorium in the Earth?

What do we know about the inner earth (see Figure 2)? The answer is, unfortunately, not very much and perhaps less than what we know about the inside of the sun. We can only guess the composition by analogy, using spectroscopy of the outer sun and direct measurements of meteorites. In fact only three Carbonaceous Chondrites are generally taken to provide the template for terrestrial composition. Of course we can directly sample only the material at or near the earth's surface. There are expected differences in this composition and the proto-earth abundances, due to early heating which drove off light elements, and due to chemical combinations of some elements which may be shallower or deeper within the earth. All this will be (was) discussed in detail at the meeting by experts. From our viewpoint as physicists it is a complicated story, without even a consensus upon the earth formation sequence, and certainly a tale which presents multiple possible scenarios.

The most direct evidence for the structure of the earth comes from seismic measurements (see Figure 3). Multiple recordings of earthquakes yield sound velocity profiles of the earth and even some detail on lateral heterogeneity. Combining these with measured earth mass moments (from satellites), and an equation of state, one may infer the earth's radial density profile. However, the composition cannot be inferred uniquely from this; one can only posit a possible mixture which would satisfy the velocity constraints.

The internal terrestrial heat gradient drives circulation producing continental drift from the mantle, and also the geomagnetic field is thought to originate from the circulation of the liquid outer core. From the frozen-in magnetic fields of dated rocks on the surface, we know that geo-magnetic

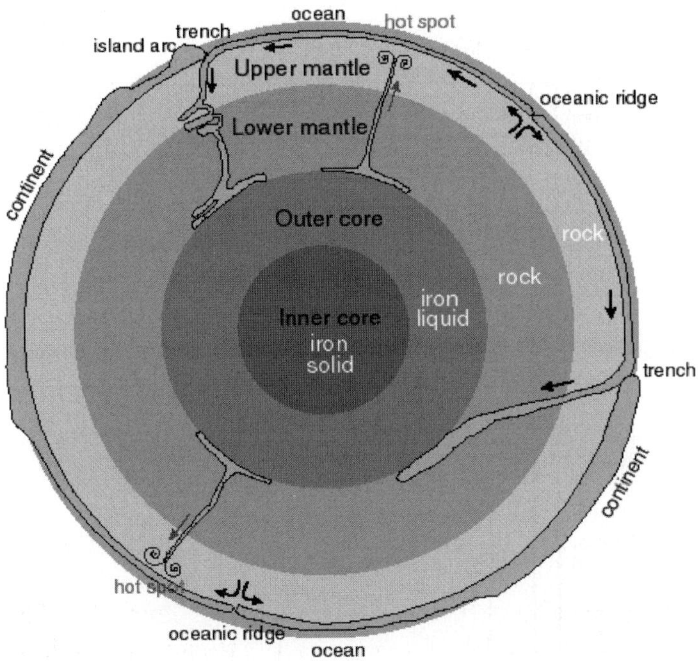

Figure 2. Sketch of earth in cross section (http://133.5.170.64/Museum/Museum-e/Part5-e/P51-e/earth_struct-e.htm).

fields have been around for billions of years, and though fluctuating in direction have had reasonably constant magnitude.

Earth surface heat flow is tiny compared to solar irradiance and measurements are difficult, particularly from the ocean floor. Data with model-dependent interpolation gives ranges from 30 to 45 TW for the total heat emission. Given expectations on the U/Th content, the U/Th radiogenic heat may be in the range of 20 TW, but could be twice that. Many other possible sources of heat have been suggested, but radioactivity is thought to be dominant, though the heat budget remains uncertain to a factor of two, see Table I.

The big question is not how much U/Th but whether it has mostly floated like slag up under the crust (as most experts believe), or remains in solution in the mantle, or has sunk onto the core-mantle boundary, or even into the core (and combinations of all of the above). One controversial model by Herndon (Herndon, 1996; Hollenbach and Herndon, 2001), has enough U in the inner core to power a natural breeder reactor providing 1–10 TW from the inner core. (This geo-reactor, if it exists, will be easy to detect in the new experiments discussed below.) While most geologists do not accept this geo-reactor model, it is not at all certain where the U/Th resides in the earth. Where the U/Th delivers the radiogenic heating makes a big difference, even without a

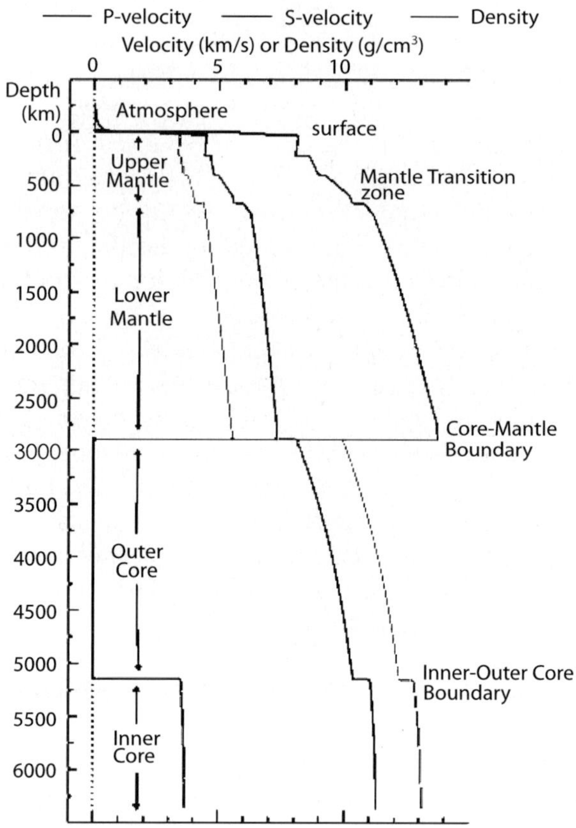

Figure 3. Plot of radial velocity and density profiles from Preliminary Reference Earth Model (Dziewonski and Anderson 1981).

TABLE I
Overview of earth heat flow balance (Verhoogen, 1980)

Element/Source	Abundance (ppm)	Earth Total Heat (TW)	
		Calc.	Meas.
Potassium (K)	170	3.7 ± 50%	
Uranium (U)	0.018	10.0 ± 50%	
Thorium (Th)	0.065	10.5 ± 50%?	
Total Radioactive		24.2 ± 50%?	
Other Sources		< 10?	
Geo-Reactor		0–10 ?	< 20
Total Heat Flow		30–50	30–45

geo-reactor, since presumably the circulation of liquid outer core is the region of origin of the geomagnetic field. It would seem that one would need a fire under the pot to drive the presumed geo-dynamo. And, one would imagine that the mantle would do well with the heat from below, though there are some who argue that the circulation can originate in dropping cooled flows.

Another issue has to do with the content of potassium, in particular the radioactive potassium-40. The earth seems to be somewhat depleted in potassium, relative to external reference abundances, and models have been made suggesting that it may have disappeared due to volatility. However, the inner earth core does seem to have slightly less density (based upon seismic velocity) than from the expected nickel-iron mix. Some speculate that the decreased density is due to potassium in solution, which might then allow for potassium heating to be the real pot-boiler from below.

Sadly the potassium-40 neutrinos are of very low energy, and do not make the signature inverse beta decay reaction (not enough energy to promote a proton to a neutron plus positron). Particle physicists have not found any viable plans as yet to measure the potassium-40 neutrinos, though one new idea is put forward at this meeting by Mark Chen.

4. How Can We Detect Geo-Neutrinos?

4.1. Inverse beta decay and history

The neutrino was posited to exist in the 1930's but was thought to be undetectable due to its miniscule probability for interaction with matter. Fred Reines and Clyde Cowan did in fact observe electron anti-neutrinos from large military nuclear reactors in the mid 1950's, at a distance of around 10 m from the reactor core. Since that time various groups have made further measurements at ever greater distances, concluding in the successful detection of reactors from all around Japan at ranges of typically 200 km, in the KamLAND experiment in 2002 (Eguchi et al., 2003).

The detection mechanism takes advantage of the "inverse beta decay" process, whereby a proton is transmuted to a neutron and the anti-neutrino turns into a positron (see Figure 4). The positron immediately annihilates with an electron in the medium, giving off two (back-to-back) gamma rays. If the medium is some scintillating material (typically an organic liquid which converts ionization by particles into nanosecond pulses of blue light), this results in a flash of light whose energy is proportional to the initial neutrino energy, less a portion (0.8 MeV) of the 1.8 MeV of energy needed to make the reaction. The neutron is given a small forward momentum from the incoming neutrino, but slows rapidly, walking randomly in direction, and captures on a proton in the medium in a time of about 200 μs, making a

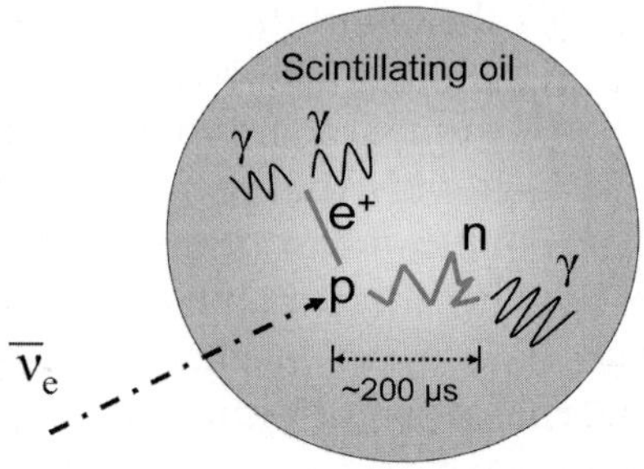

Figure 4. Schematic of electron anti-neutrino detection, showing production and prompt annihilation of a positron, followed by a 2.2 MeV emission from the neutron capturing on a proton (making a deuteron).

deuteron. The deuteron binding energy of 2.2 MeV is then released as a gamma ray, and this causes a second flash in the scintillating liquid. The combination of the two sets of light flashes, one of known energy, and both close in space and time, provides a powerful filter for extracting the electron anti-neutrino interactions from the wealth of other single flashes (due to radioactivity, solar neutrinos, electronic noise, cosmic rays, etc.).

In KamLAND (see Figure 5) the liquid scintillator consists of hydrocarbons, so the target is hydrogen and the neutron capture is also accomplished by hydrogen. (Neutrino interaction on the protons or neutrons in the carbon nucleus requires too much energy because of the nuclear binding energy to be of use here.) However, some detectors (including the original Reines–Cowan instrument) use a scintillating liquid "loaded" with elements which have much larger neutron capture cross sections, and higher energy emissions (eg. Gd, Cd). Such liquids can be expensive and some have proved unstable, but

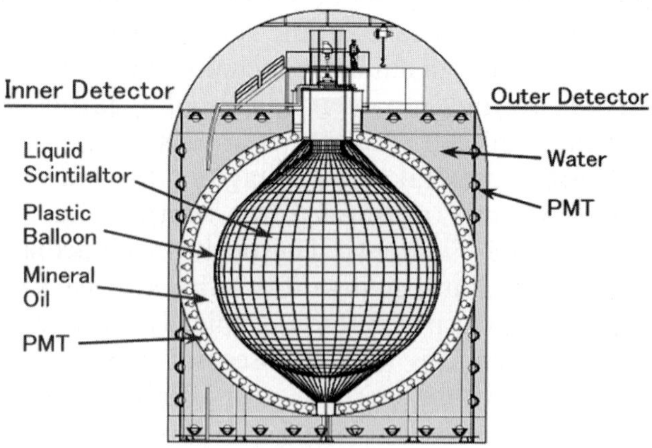

Figure 5. Cross section of the 1000 ton liquid scintillator-based KamLAND detector located in a deep mine in Japan, and able to measure electron anti-neutrinos arriving from nuclear power reactors at distances of hundreds of kilometers around Japan.

may be worth pursuing to improve the directional signature since they can shorten the neutron capture distance and time by a large factor (> 10).

4.2. Natural neutrino spectra

As illustrated below in Figure 6, the dominant fraction of the reactor signal as observed by KamLAND, is in an energy region between about 2.0 and 7.0 MeV neutrino energy, corresponding to 1.2 – 6.2 MeV in the observed first pulse energy in the detector. The decay energies attributable to uranium-238 decay chain and to thorium-232 decay chain are all below 3.4 MeV (see Figure 7). There is an additional background shown in Figure 6, due to a contamination of the KamLAND detector by radon and a reaction of alpha particles with carbon-13. This background is avoidable and will not be a factor in later measurements, though it was a nuisance in the initial KamLAND attempt at measuring the U/Th neutrinos, as reported in the cover issue of *Nature* in July 2005 (Araki et al., 2005). Note that the reactor fission products fortunately do make higher energy neutrinos than the natural radioactive decay neutrinos, so we can clearly distinguish the two sources by their characteristic spectra, as illustrated in Figure 6.

One may also note that the spectra from U and Th differ significantly, so with adequate statistics we can measure the ratio of U/Th as well as observe the total flux and hence amount of U and Th. Observing the total rate does not correspond to the total abundance of U and Th however, even in a uniformly layered earth (it is not the same as for electrical charges and Gauss' law). Moreover there are surely great lateral heterogeneities due to the varying crustal composition. Most U/Th is expected to be in or near the crust, so discerning the amount distributed throughout the mantle and core is

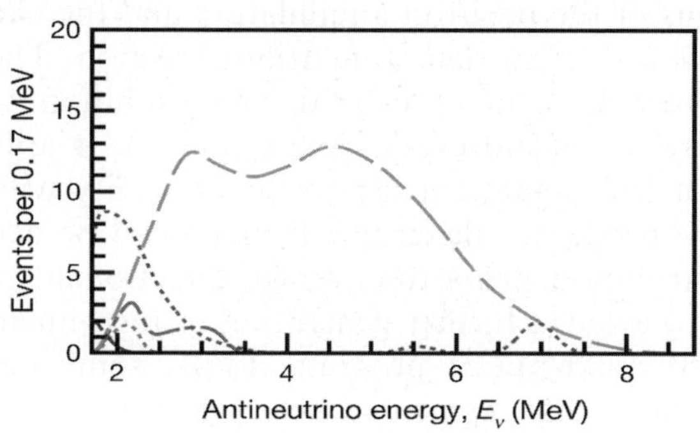

Figure 6. Spectrum of electron anti-neutrinos from reactors at KamLAND (dashed curve) and several backgrounds, including those due to natural radioactive decay of U and Th at energies below 3.4 MeV (dot-dashed curve) (Araki *et al.*, 2005).

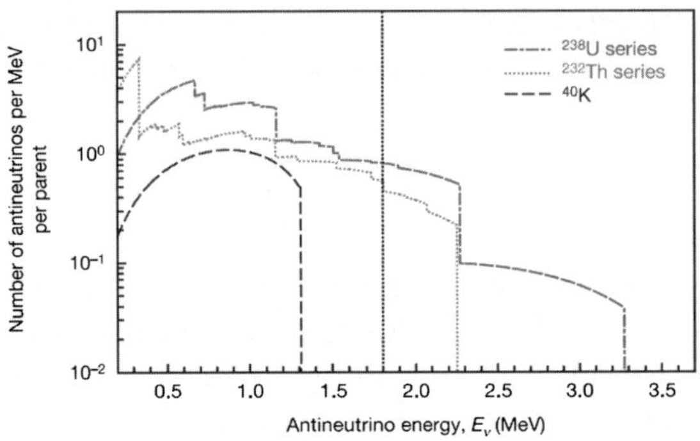

Figure 7. ^{238}U and ^{232}Th series decay anti-neutrinos extend to energies of 3.3 MeV, while ^{40}K anti-neutrinos are all below 1.4 MeV, and not detectable by inverse beta decay.

very difficult. For example, only 27% of the flux from U/Th decay at KamLAND is expected to be from the mantle and core, and most due to the local mountains and deeper plate. Oceanic crust is younger and thinner and expected to have typically only 1/10 as much U/Th as that when measuring from a continental location, and hence the crucial issue of how much of the terrestrial radioactivity is in the mantle/core will need to be measured from an oceanic location.

4.3. Can we measure more than rate?

Of course we would like to measure the arrival directions for the neutrinos and hence map out the origin in direct fashion. However, directional measurement is very hard at these energies and particularly in a scintillating material with its isotropic light emission. A small handle can be had from the net momentum transmitted to the neutron by the neutrino, which statistically biases the locations of the positron annihilation and the neutron capture to be slightly aligned with the original neutrino direction. The Chooz reactor experiment in France did achieve an 18 degree resolution from their nearby reactor with several thousand events (see Figure 8). It appears that in the next generation of instruments, neutrino directionality will be of marginal utility. Studies are needed to determine if much can be deduced about the internal earth distribution, from the average directionality of U/Th neutrinos. It will have to await a further generation of instruments with neutrino tracking ability to make much progress. Ideas, some mentioned at this meeting, range from highly segmented detectors with high electron density and shorter neutron capture distance, to coherent neutrino scattering. None are close to implementation on a large and affordable scale (and coherent scattering has not been observed at all as yet).

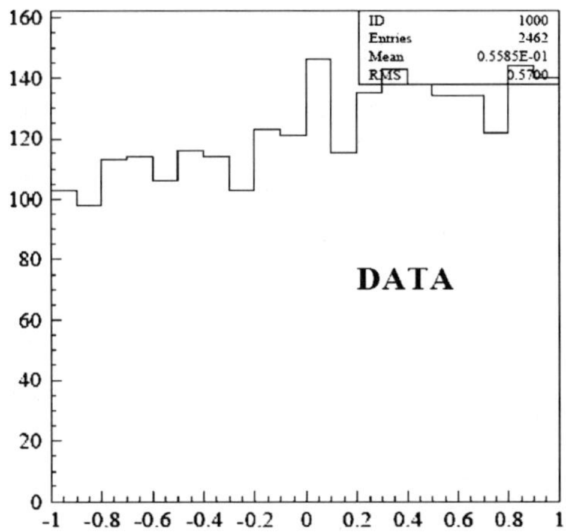

Figure 8. Chooz experiment data exhibiting weak directional dependence of vector between positron annihilation and neutron capture locations (Appolonio *et al.*, 2000).

A similarly difficult situation obtains for measuring the neutrinos from potassium 40. Unfortunately the highest potassium energy is 1.3 MeV, which is below the inverse beta threshold. There is also a weak electron neutrino line from potassium 40 due to electron capture.

5. Synergy in Multiple Observations around the World

Various groups have proposed or built experiments which can contribute to the study of natural electron anti-neutrinos (see Table II). The KamLAND detector in Japan has reported results, as discussed above. The long delayed and somewhat small Borexino detector in Italy will hopefully come into operation late in 2006. The SNO detector, located in Canada, which made such excellent measurements of solar neutrinos using a heavy water target will end its run in 2006, and is being considered for conversion to a scintillating liquid detector (SNO+). Proposals have also been put forward for other instruments; in Baksan (Caucasus, Russia), Homestake Mine (USA), Finland (LENA), by a Dutch group for EARTH, drilling below the island of Curacao, and by the Hanohano group for deep ocean emplacement near Hawaii. See discussions of these projects in these proceedings, and see map of locations in Figure 9. All proposed projects except EARTH and Hanohano would be dominantly measuring anti-neutrinos from U and Th decay from the local crust and to a lesser degree from the mantle. The ratio of neutrinos from the local crust to neutrinos from the mantle (and core) measured at KamLAND is expected by calculation to be about 3:1. In a deep ocean location, due to the thin and not so radioactive oceanic crust, the situation is

TABLE II
Some proposed geo-neutrino instruments, location, size and status

Project	KamLAND	Borexino	SNO+	Hanohano
Location	Japan	Italy	Canada	Hawaii
Crust	Continental	Continental	Continental	Oceanic
Current status or Start date	Operating	2007	2008	Planning
Depth (meters water equivalent)	2700	3700	6000	4500
Target (10^{32} free protons)	0.35	0.18	0.57	8.7
Geo-neutrinos per year- Total	13	8	30	110
Geo-neutrinos per year- Mantle	4	2	5	81
Reactor neutrinos per year	39	6	32	12

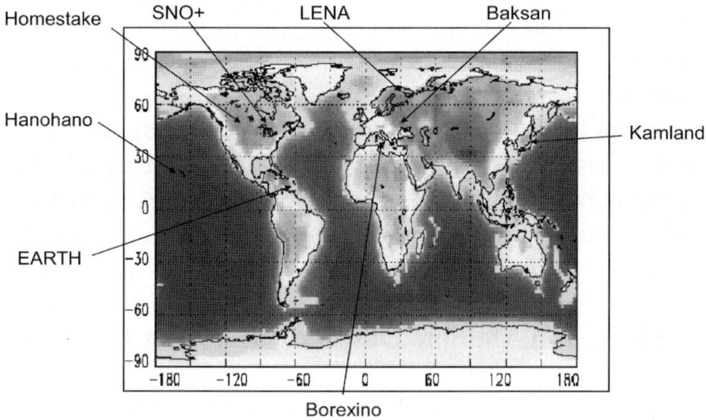

Figure 9. Map of proposed geo-neutrino detector locations.

expected to be reversed (3:1 mantle and core to crust). Thus, given at least a 20% uncertainty in the calculations of expected fluxes for locations dominated by the local crust, one cannot extract the mantle (and core) contribution from continental measuring locations alone. There is nice synergy however with measurements both from an (one or better more) oceanic location and continental locations from which the mantle (and core) contribution can be subtracted.

Discussions are underway about the possibility of making measurements at multiple ocean locations and what can be learned and sensitivity required to detect expected contrasts (such as between ordinary mantle and that near large upwellings). The Hanohano design target volume (now set at 10 kilotons) is being aimed at this prospect, given the mobility of this instrument. We envisage a one year run at several locations. Hanohano can also be located off shore from a nuclear reactor complex, and may provide an opportunity for a precision measurement of the solar mixing angle (θ_{12}) and atmospheric mass-squared differences leading to determination of neutrino mass hierarchy.

6. Other Applications of Future Large, Low-Energy Neutrino Detectors

In the future we can anticipate many uses of neutrinos both for fundamental science in particle physics and astrophysics, and in applications as probes due to their unique penetrating ability. For some time now people have written papers suggesting some far-out possibilities, such as active earth tomography with accelerator-produced neutrino beams and perhaps natural neutrinos, using neutrino beams to search for oil, measure heterogeneities, determine earth core properties in ways unrivaled, and even as carrier beam for an ultimate galactic time standard.

In the shorter term we can begin to think seriously about using neutrinos to monitor nuclear reactors, both for checking on use of the reactor and reactor performance. This can only reasonably be carried out from close-in (10–100m) and with cooperative facilities. For locations which may not be cooperative, one can stand away distances of hundreds or even thousands of kilometers. However, the price for range is great since the flux falls with distance squared and moreover one starts to have competing signals from other reactors (see Figure 10). Since there are about 500 reactors in the world, one can imagine a network of roughly that number of detectors which can monitor all the world's reactors, and can subtract the known contributions from cooperative sites, revealing clandestine operations. While there are other means to search for rogue reactors (eg. thermal signatures), one cannot shield the neutrinos. And, the synergistic application of multiple monitoring techniques may yield more powerful constraints.

Another application in this line, which comes for free with remote (close-in detectors would not have the sensitivity) reactor monitoring is the detection of clandestine nuclear weapons testing. There are many mechanisms in place for detecting such activity, although there have been cases of both false positives and false negatives. The neutrino signature cannot be faked or masked, and is a definitive measure of the weapons fission yield. Studies show that a large array for reactor monitoring as above, could detect weapons down to the one kiloton level anywhere in the world.

Science applications of future huge low energy neutrino detectors are also very exciting. For example, a one gigaton instrument (or collection totaling that effective mass) could detect supernovae from throughout our galactic supercluster, recording perhaps one per week. Such would have many associated studies ranging from stellar evolution to general relativity and particle physics. The measurement of the sum of electron anti-neutrinos

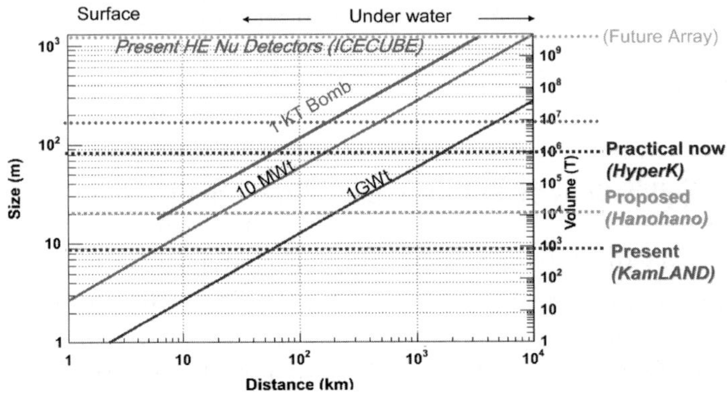

Figure 10. Size requirements for an anti-neutrino detector to measure a reactor flux to 25% uncertainty in 6 months. Upper is for a 10 MWt reactor, lower for a 1 GWt reactor. Background is not considered. Practical volumes now are up to about 1–10 MT, and in future may go to 1 GT. Detectors beyond a few km must be underground or underwater for cosmic ray shielding.

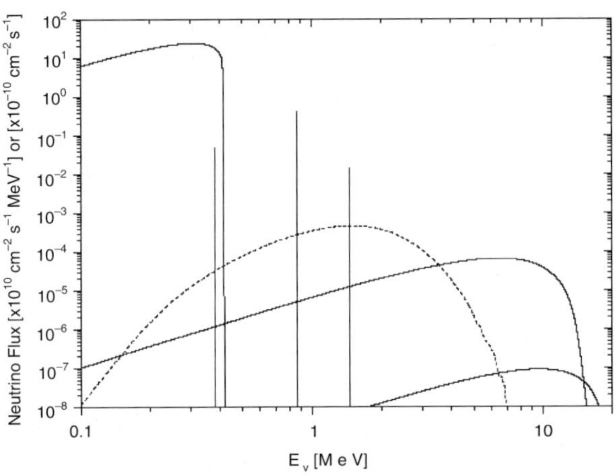

Figure 11. Neutrino spectra, illustrating the flux expected from a 1 kiloparsec distant massive star during the last few days of silicon burning. Notice that this (electron neutrino and anti-neutrino) flux rises by about two orders of magnitude above the solar electron neutrino flux in the energy range between roughly 1 and 2 MeV (Odrzywolek, Misiaszek, and Kutschera, 2004).

from all previous supernovae throughout the universe would yield much interesting information on stellar formation rates and cosmology. On a more local level, increased thermal neutrino output of a star within our galaxy during the last few days of burning prior to implosion may be registered with large instruments, giving earth a supernova early warning system (see Figure 11).

7. Challenge and Outlook

- Geologists: tell us what you need to know most and what are the range of possibilities one may find. Also, please tell us about potential for unorthodox phenomena, such as the hypothetical geo-reactor.
- Physicists: tell us what can be done now, and in a few years. How well can we measure the U/Th content? Can we measure the K content? Can radial distributions be measured?
- All: What are the larger implications of these measurements for life on earth and in other circumstances?
- Outlook: lots of fun, as we witness here perhaps the birth of a new and important area of science.

Acknowledgements

In retrospect, we want to thank our participants for making a great meeting, and we look forward to the ventures we hope will flow from this event. We want to thank the University of Hawaii for support for this meeting, particularly Vice President James Gaines and Dean Charles Hayes, and our hosts at the East-West Center.

References

Appolonio, M. et al.: 2000, *Phys. Rev. D* **61**, 012001.
Araki, T. et al.: 2005, *Nature* **436**, 499.
Dziewonski, A. M. and Anderson, D. L.: 1981, *Phys. Earth Plan. Int.* **25**, 297.
Eguchi, K. et al.: 2003, *Phys. Rev. Lett.* **90**, 021802.
Herndon, J. M.: 1996, *Proc. Nat. Acad. Sci.* **93**, 646.
Hollenbach, D. F. and Herndon, J. M.: 2001, *Proc. Nat. Acad. Sci.* **98**, 11085.
Odrzywolek, A., Misiaszek, M., and Kutschera, M.: 2004, *Astropart. Phys.* **21**, 303.
Verhoogen, J.: 1980. *Energetics of Earth*, National Academy of Sciences, Washington, DC.

Earth, Moon, and Planets (2006) 99:17–22
DOI 10.1007/s11038-006-9102-x

Can the Future of Neutrino Physics Compare with its Past?

SHELDON L. GLASHOW
Physics, Boston University, 590 Commonwealth Ave., Boston, MA 02215, USA
(E-mail: slg@bu.edu)

(Received 6 February 2006; Accepted 28 June 2006)

I shall summarize the marvellous accomplishments in neutrino physics of the past decade, very briefly sketch our current understanding of these elusive particles and provide a personal list of the most important challenges that remain in this discipline, given in (my) order of priority. Because the following discussion is more provocative than novel, I must apologize in advance for the inevitable omissions, distortions, inequities and iniquities resulting from my ignorance or infirmitude, inattention or ineptitude and unintended arrogance.

The neutrino story began with the desperate remedy Pauli proposed in 1931 and later may have regretted: "I have done something terrible," he wrote, "I have predicted an undetectable particle." A quarter of a century later, Pauli was proven right about the neutrino but wrong about its unobservability. Since then, we have learned a great deal more about neutrinos: that they come in three species, that they oscillate in identity, that they have mass, and much more. Indeed, Pauli's desperate remedy soon became a powerful tool which helped reveal the quark structure of hadrons, and through the discovery of neutral currents, established the validity of the electroweak theory. Furthermore, neutrino observations have confirmed our understanding the nature of both supernovæ and the sun. This very conference shows that neutrinos may someday help us to understand the inner structure of our planet as well. Consider, if you will, some of the many remarkable advances in neutrino physics of just the past decade:

1994: The surprising and anomalous indications of the LSND experiment should *not* be included in this list for two reasons. The experiment did not take place in the past decade and its result has not been confirmed.

1998: Super-Kamiokande dramatically resolved an apparent atmospheric neutrino anomaly that was first noted by IMB and Kamiokande. The announcement at Takayama of decisive evidence for neutrino mass and atmospheric oscillations was met by a standing ovation.

1999: CHOOZ set the current upper bound to the subdominant mixing angle θ_{13} associated with 3-family neutrino oscillations.
2000: DONUT (at Fermilab) was the first experiment to provide *direct* evidence for the existence of tau neutrinos.
2003: SNO decisively solved the solar neutrino problem by measuring neutral current interactions of all the neutrinos coming from the sun.
2003: KamLAND observed the disappearance of electron anti-neutrinos from relatively distant nuclear reactors.
2005: KamLAND reported the detection of antineutrinos from natural sources within the Earth.

These, among other, of the many recent accomplishments in neutrino physics have led to numerous prizes and medals. I have selected ten recent awards to list below:

1995: Fred Reines wins the Nobel Prize for his detection of the neutrino (with Clyde Cowan, since deceased) for work done 40 years earlier.
2000: Ray Davis and Masatoshi Koshiba share the Wolf Prize for their observations of neutrinos of astrophysical origin.
2001: Ray Davis is awarded the National Medal of Science for his lifelong contributions to neutrino physics.
2002: Nick Samios wins the Pontecorvo Prize for discovering, via neutrino interactions, the Ω^- particle and the first charmed baryon.
2003: John Bahcall wins the Dan David Prize for his contributions to neutrino astrophysics.
2002: Ray Davis and Masatoshi Koshiba share the Nobel Prize for the detection of cosmic neutrinos.
2002: Masatoshi Koshiba, Yoji Totsuka and Tokaaki Kajita share the Panofsky Prize for their compelling evidence for neutrino oscillations.
2003: John Bahcall and Ray Davis share the Fermi Prize for work leading to a revolution in the understanding of the properties of neutrinos.
2003: Art McDonald is awarded the Hertzberg Medal for developing and exploiting the Sudbury Neutrino Observatory.
2003: Tom Bowles, Vladimir Gavrin and Vadim Kuzmin share the Markov Prize for their radiochemical research on the solar neutrino problem.

It is clear from all these awards that neutrino physics has become a recognized, well-regarded, and exciting discipline. So, what have we learned (or not learned) from all this research? An executive summary follows in the form of a very short list:

- There seem to be exactly three active (i.e., weak doublet) neutrino states, at least two of which have mass.
- The masses and mixings of neutrinos seem to be described by a minimal model: a 3×3 matrix that involves precisely six observable parameters, all but one of them either roughly measured or constrained. These consist of three angles: θ_{12} (solar), θ_{23} (atmospheric), and θ_{13} (subdominant), one CP-violating phase δ, and two squared-mass differences: Δ_{solar} and Δ_{atm}.
- No experiment (with the exception of LSND) suggests or requires the mixing of active neutrinos with sterile light singlet states.
- Neutrino masses may just as well be 'Majorana' (lepton-number violating) or 'Dirac' (lepton-number conserving). Intermediate possibilities, such as would require sterile neutrinos, are disfavored.
- Despite hundreds of published theoretical papers, little is known about the origin of neutrino masses, and nothing at all about why the various parameters are what they are. The latter question is quite analogous to the mystery of quark masses and mixings.
- Neutrino experiments have enabled the most sensitive searches for flavor-dependent Lorentz violation in the neutrino sector. The relevant dimensionless parameters are known to be less than 10^{-25}.

I conclude with yet one more list addressing the question of what are the future challenges for neutrino physicists, in order of importance as I see it:

(1) First on my list, *pour rendre une politesse*, must be *The Study of Geoneutrinos*: antineutrinos produced by naturally radioactive elements within our planet's crust, mantle and core. Why on Earth should this be interesting?

- Because their precise measurement is a daunting and difficult challenge for experimenters who thrive on doing the near impossible.
- Because theorists have written (and will write) lots of papers about geoneutrinos, such as Eder and Marx in the 1960s and at least 13 others (including me) in the 1980s.
- To determine how uranium and thorium are distributed within the Earth, and (optimistically) to see if potassium has accumulated in the core, thus helping power the geomagnetic dynamo.
- To test the speculative suggestion that there is a natural nuclear reactor within Earth's core.
- Additional motivations to investigate geoneutrinos are likely to appear in the Proceedings to this conference.

(2) *The Quest for Neutrinoless Double Beta Decay*

- If this process is decisively observed, we will *know* that lepton number is not conserved. This would be a truly important discovery.
- The existence of neutrinoless $\beta\beta$ decay would show that neutrinos are likely to have Majorana masses. Conversely, if it be shown that the process is absent at an appropriate level, then neutrino masses are likely to be Dirac, although there are more exotic possibilities.
- If the process is both seen and measured, we will have a quantitative test of our simple 3×3 matrix description of neutrino masses and mixings, and as well, an estimate of the absolute size of neutrino masses.
- In 2001, Klapdor et al. claimed 'Evidence for No-Neutrino Double Beta Deacay' [Mod. Phys. Lett. A16, 2409]. However, Aalseth et al. responded that 'there is no basis' to this claim [Ibid. A17, 1475]. And so?
- Although one experiment may suffice in principle to establish the existence of no-ν $\beta\beta$ decay, several positive results will be necessary to clinch the case. Great discoveries demand decisive proof! Furthermore, several experiments will be needed to pin down the neutrino parameter M_{ee}, because the relevant nuclear matrix elements are poorly known.

(3) *Setting the Parameters of the Minimal Model*

- The subdominant angle θ_{13} must be measured (and must not be too small) if we are to detect CP violation in the neutrino sector. The present (CHOOZ) limit is roughly $\sin^2 2\theta_{13} < 0.2$, where I say roughly because the constraint strongly depends on Δ_{atm}. We must do much better. Minos-Numi may achieve 0.06; Double-CHOOZ will aim for 0.03; while the (not yet funded) Daya Bay project will target 0.01.
- Aside from the question of how far θ_{13} departs from zero, we must know how close are θ_{12} to $\pi/6$ and θ_{23} to $\pi/4$.
- The squared-mass differences are roughly known: $\Delta_{atm} \simeq 3 \pm 1 \times 10^{-3}$ eV^2 and $\Delta_{sol} \simeq 8 \pm 1 \times 10^{-5}$ eV^2. More accuracy is required, especially for the atmospheric difference. Furthermore, it is essential to determine whether the neutrino mass hierarchy is normal or inverted... a question that could be addressed by long baseline off-axis experiments.
- It would be wonderful, but difficult, to demonstrate CP violation in the neutrino sector. JPARC may accomplish this feat, if θ_{13} is at least a few degrees and δ is nearly maximal. Again, more sensitive experiments are needed.
- Finally, the big question: Are all aspects of neutrino propagation determined by the six parameters of the minimal model? If not, great! We will be challenged. But if so, what more can experimenters do? Will they have accomplished their appointed task?

(4) *What about LSND?* This experiment alleges to see $\bar{\nu}_\mu \to \bar{\nu}_e$ with a small mixing angle and large (0.1–10 eV^2) squared-mass difference. This result is unique in that it cannot be explained in the otherwise triumphant minimal (3-state) model. Hundreds of theorists are in hot pursuit, but I am not yet persuaded to join the chase. It is the intent of the long-running MiniBoone experiment to confirm or refute the LSND experiment, but its result has not yet been announced. While a decisive refutation would put this issue to rest, I fear that theorists will be permitted to play their games for years to come.

(5) *To Catch a Supernova* We observed a total of 18 events from supernova 1987a at IMB and Kamioka, thereby confirming our understanding of core-collapse supernovæ. How wonderful it would be to detect and measure thousands of neutrinos from the next relatively nearby supernova... in a year or a century from now. So much the better if our ever-ready detector could (like SNO) distinguish between charged-current and neutral-current events. Such a device (with some cooperation from nature) could tell us a lot about both neutrinos and supernovæ. I would assign a high priority to the indefinite maintenance of such a facility.

(6) *Absolute Neutrino Masses* Oscillation phenomena depend only upon differences of squared masses, and not upon their absolute masses. At present, we have only a weak lower bound to their size from oscillation experiments:

$$\sum m_i > 0.05\,\mathrm{eV}/c^2,$$

and a surprisingly strong upper bound from our more astrophysical colleagues:

$$\sum m_i < 1\,\mathrm{eV}/c^2.$$

With further work, the constraints may come much closer together, and it is just possible that the mass scale will be set through the observation of neutrinoless double beta decay.

Searches for neutrino-mass effects on the endpoints of beta spectra may also be relevant to this issue. At present, studies of the tritium endpoint constrain the 'ν_e mass' to be less than 2.2 eV, which is compatible with, but no stronger than, the cosmological claim. Can we anticipate improvement? Of course, better *direct* limits on the masses of the muon and tau neutrinos can and will be set, but these are unlikely to be relevant.

(7) *Cosmic Ray Neutrinos* We have seen and studied neutrinos of extra-terrestrial origin: from the sun, from a supernova and as tertiary products of cosmic-ray impacts. But what about neutrinos as primary cosmic rays?

- As point sources in the sky?
- As transient sources such as GRBs?
- As ultra-high energy neutrinos?
- From '*W*-bursts' via the Glashow resonance?
- From WIMP anihillation within the Earth or sun?

In short, will there ever be a true science of neutrino astronomy?

(8) *Surprises?* Who knows what Nature has in store for us?

Sterile neutrinos,
Mass-Varying Neutrinos,
Lorentz Violation,
Effects due to Extra Dimensions,
Neutrino with Magnetic Moments,
Decaying Neutrinos,
Departures from Flavor Universality,
What about $N_\nu = 2.984 \pm 0.008$, a 2-σ discrepency?
Or will there be something even more interesting, Because as always, we must allow Nature to call the shots.

Research supported in part by the National Science Foundation under grant number NSF-PHY 0099529.

Earth, Moon, and Planets (2006) 99:23–32
DOI 10.1007/s11038-006-9109-3

Radioactivity of the Earth and the Case for Potassium in the Earth's Core

V. R. MURTHY
Department of Geology and Geophysics, University of Minnesota, Minneapolis, MN 55455, USA

Institute of Meteoritics, Department of Earth and Planetary Sciences, University of New Mexico, Albuquerque, NM 87031, USA
(E-mail: vrmurthy@umn.edu)

(Received 28 February 2006; Accepted 11 July 2006)

Abstract. The radioactivity of the earth is an important parameter in understanding the dynamics of the planet and the evolution of the crust–mantle–core system but geochemical and geophysical approaches have had only a limited success in defining it. The opportunity of a direct estimate of the radioactivity of the earth by measurement of the geoneutrino flux takes on an added significance in this context. Such an independent new measurement will help resolve and/or clarify a number of questions about global scale processes in the earth and will help advance earth sciences.

Keywords: Geochemical models, geonuetrino, global dynamics, heat flux, potassium, radioactivity of core, radioactivity of the earth, thorium, uranium

"— *In science, conventional wisdom is difficult to overturn*"

—Richard L. Armstrong, 1991.

1. Introduction

The radioactivity of the earth is an important parameter for understanding the earth's internal energy budget and the evolution and dynamics of the planet. As described below, attempts to estimate the overall radioactivity of the planet over the past few decades by using tools of geochemistry and geophysics have had only a partial success. This is not because that the tools are inadequate but because of some fundamental limitations in observing the earth. The net result is that at present we have only some model-dependent notions about the radioactive content of the earth's mantle and core, about 99% of the mass of the earth. The only parts of the earth for which we have some reliable estimates of the radioactive content are the crust and the upper mantle, because they are directly accessible for sampling. Since heat produced in the interior of the earth by decay of radioactive isotopes U, Th, and K is the driving force responsible for the dynamic activity of the earth, a precise estimate of the radioactivity of the earth will aid in clarifying the energy budget for a variety of global dynamic processes, such as the plate-tectonic motions and convection in the mantle, global surface heat flow, production and maintenance of the geomagnetic field for ~4 billion years, and the crystallization and evolution of the inner core. It is fortunate that a possibility to directly measure the earth's radioactivity seems to be at hand now. Radioactive decay in the earth of elements such as U and Th results in the production of geoneutrinos. Recent progress in the detection and measurement of this geoneutrino flux has advanced to a level that a new independent tool for measuring the radioactivity of the earth due to U and Th is available now (Mantovani et al., 2004; Fiorentini et al., 2005; and several articles in this volume). Present geoneutrino detection techniques do not permit measurement of K, due to the lower energy of K-neutrinos, relative to those from U and Th decay, and further development in the technique is needed for K-determination. I review here the current situation with respect to our understanding of the radioactivity of the earth. The prospect to look forward to is that we have an opportunity to examine many of the present notions of the radioactivity of the planet by this new tool offered by neutrino physics.

2. Radioactivity of the crust and the mantle

At present, information about the radioactivity of the earth is obtained in three approaches: (1) Direct measurement of K, U and Th in rocks from the crust and the outer few hundred kilometers thick layer of the mantle are now available. These provide a reasonably consistent estimate of the radioactivity of the crust, in a variety of crustal models employed by geoscientists. Beyond that, a fundamental difficulty exists in estimating the overall radioactivity of the entire mantle or the earth. The lower mantle, with a mass nearly half of that of the entire earth is not accessible for a direct measurement of its radioactivity. Seismic velocity measurements or mineral physics data cannot provide useful constraints on the trace and radioactive element content of the lower mantle (e.g., Mattern et al. 2005). Secondly, it is generally considered that the core with nearly one-third the mass of the earth is devoid of any radioactivity. This is being debated at present, however. Thus, in essence we are left with considerable uncertainty with respect to the radioactivity of about 80% of the mass of the earth in the lower mantle and the core. Nevertheless, several geochemical models have estimated the composition and the radioactive content of the mantle and the total earth based on data from the crust and the upper mantle samples (Jagoutz et al., 1979; Wanke et al., 1984; Hart and Zindler, 1986). One of the goals in these studies is to understand the differentiation history of the earth into a primitive silicate fraction (often referred to as the Bulk Silicate Earth – BSE) with a metallic core, and the further differentiation of the BSE into the crust and the mantle. These models consider the lower mantle to be similar in composition to the upper mantle. However, whether or not the mantle is compositionally layered is by no means a settled question at present (see for example Anderson 1989a, b; 2005; T. Lubetskaya and J. Korenaga, submitted).

(2) Geochemical information provided by meteorites combined with the terrestrial data has provided an enlarged framework for constructing geochemical models for the radioactivity of the earth. It has been known for a long time that the relative abundances of the lithophile refractory elements, for example Ca, Mg, Al, etc., in the earth's crust and upper mantle combined, match that of C1 carbonaceous chondrites, a class of volatile rich primitive meteorites with composition that closely matches the solar photosphere (Ringwood, 1966). In contrast, several volatile elements, such as the alkalis and halogens etc., appear depleted in the earth. Furthermore, the pattern of depletion of volatile elements is thought to follow closely the volatility of elements in the solar nebula during condensation processes. Since then it has become customary to use C1 chondrites as a reference initial composition to compare the terrestrial abundance estimates to understand the accretion processes and the differentiation history of the earth into the crust, mantle and the core. McDonough and Sun (1995) provide a detailed description of such a model with complete references to prior studies. Commonly used estimates of the radioactivity of BSE and the planet (Table I) are taken from this model.

(3) The terrestrial surface heat flow is an expression of heat production in the earth and can be an indicative measure of the internal radioactivity of the earth. We now have a large database of heat flow measurements from both the continents and the oceanic regions. Even with such information, the global average heat flux cannot be precisely determined; current estimates (Lee, 1970; Pollack et al., 1993; Stein, 1995; Hoffmeister and Criss, 2005) range from 30 to 44 TW (1 TW $= 10^{12}$ W). The higher value exceeds the heat production of chondritic model earth by nearly a factor of two (see the compilation by Lodders and Fegley, 1998). It is possible to rationalize this discrepancy by invoking a variety of possibilities such as a significant secular cooling delay of 1–2 Ga (1Ga $= 10^9$ years) which equates the present surface heat flux to radiogenic heat produced in the past, or by including residual primordial heat from accretion, or other sources (Anderson, 2005). No definitive statements can be made about these possibilities at present, but

TABLE I
Geochemical estimates of U, Th and K in the Bulk Silicate Earth – BSE)[a]

U	Th	K	Reference
0.021	–	231	Wanke et al., (1984)
0.018	0.064	180	Taylor and McClennan (1985)
0.0208	–	266	Hart and Zindler (1986)
0.0203	0.0813	258	Hofmann (1988)
0.0203	0.0795	240	McDonough and Sun (1995)
0.0202	0.0764	235	Van Schmus (1995)[b]

[a]Based on the assumption that the lower mantle is compositionally similar to the upper mantle. All values are in parts per million.
[b]Suggested average values.

one is left with the notion that the present heat flux is much greater than can be accounted for by the radioactive content of the earth shown in Table I.

The K/U ratio in rocks from the crust and upper mantle is ~1×10^4 but in chondrite meteorites is higher (see Lassiter, 2004 for an up-to-date discussion of K/U ratios in terrestrial materials and meteorites). The low K/U ratio in the earth and the low K abundance in the mantle is attributed to the loss of moderately volatile K from the earth. However, the volatile loss cannot be due to partial vaporization from the earth, because the isotopic composition of K in the earth is identical to that of meteorites (Humayun and Clayton, 1995). The volatile K-loss must have preceded planetary formation, implying incomplete condensation or loss of volatiles in the precursor materials that formed the earth. Since this type of volatile loss of alkalis is not shown by the chondritic meteorites across the entire range of composition from the most primitive volatile-rich, highly oxidized C1 meteorites to the highly reduced metal-rich E-chondrites (Lodders, 1995), one is forced to conclude that precursor material for the earth and the meteorites cannot be the same.

Terrestrial oxygen, the most abundant major element in the earth (> 30% by mass of the planet) is different in isotopic composition from that of C1 chondrites and ordinary chondrites. It is not possible to produce terrestrial type oxygen from chondritic type by any understandable volatility or fractionation processes. C1 chondrites also cannot provide for the abundance of ~30% Fe in the earth (Javoy, 1995). Drake and Righter (2002) have discussed in some detail the general difficulty of using chondrite meteorites for initial starting composition for the earth. It is difficult to avoid the conclusion that C1 and ordinary chondrites cannot be the precursor material for the earth, thereby diminishing their usefulness as reference material for understanding terrestrial fractionations.

The O-isotopic composition is a strong constraint in identifying earth precursor material, The only meteoritic material that closely matches the terrestrial O-isotope composition is that represented by the highly reduced metal-rich EH- enstatite chondrites. Based on this and the similarity of the oxidation states of the earth and the enstatite chondrites, some have suggested that EH condrites may be more appropriate precursors for the earth (Javoy and Pineau, 1983; Javoy et al., 1986; Javoy, 1995; Lodders, 1995, 2000). These types of models have the advantage of simultaneously satisfying the stringent requirement of the O- and other stable isotope characteristics, the high iron content of the earth, and the global heat flow data (Javoy, 1995; Hofmeister and Criss, 2005). Models of this type yield a significantly higher radioactive content in BSE than those listed in Table I.

There are other uncertainties as well. An important recent experimental work calls for a significant hidden reservoir of material enriched in radioactive elements in the deep mantle. Boyet and Carlson (2005) inferred from the ^{142}Nd evidence from the mantle that an early global differentiation of the mantle soon after the formation of the earth had segregated a radioactive rich layer deep in the earth. This enriched layer is isolated from the dynamic activity of the mantle and has been totally isolated since its inception in the early history of the earth. Boyet and Carlson (2005) estimate that such a layer could contain up to 43% of the earth's inventory of U, Th and K with a heat production of ~9 TW. A similar deep enriched reservoir has also been proposed to explain the rare gas evolution of the earth (Tolstikhin and Hofmann, 2005). If the mantle has such deep hidden reservoirs, estimating the radioactive content of the mantle from upper mantle samples would grossly underestimate the radioactivity of the mantle and therefore, of the planet.

It thus appears that our present knowledge of the radioactivity of the BSE, the bulk silicate portion of the earth (68% by mass of the planet) suffers from many fundamental uncertainties, such as to what its initial starting composition was, the question of whether the lower mantle is similar in composition to the upper mantle, and the possibility of hidden reservoirs of radioactivity in the deep interior. The radioactivity estimates in Table I are by no means uniquely constrained at present; rather, about all we have are some model-dependent estimates that need further evaluation.

3. Radioactivity of the Core

The presence of radioactive elements in the core has been a controversial issue for a long time. Conventional wisdom relegates the bulk inventory of K, U and Th in the earth exclusively to the silicate crust and the mantle (BSE), leaving the metallic core devoid of any radioactivity (see for example, McDonough, 1999). This is consistent with the familiar geochemical behavior of K, U and Th, which tend to partition preferentially and almost exclusively into silicates in the upper parts of the earth. However, in recent years it has become clear, both in theory and in experiment, that silicate-metal partitioning behavior of elements is not an invariant characteristic but is a function of pressure, temperature, composition of the metal, oxygen fugacity and other variables. It is is thus not useful to hold on to the rigid classification of elements as lihtophile, chalcophile or siderophile. Rather, to evaluate realistically whether or not there is radioactivity in the core, we need to investigate the geochemical behavior of these radioactive elements at the high P and T conditions relevant to core–mantle segregation in the earth. Such investigations are emerging just now.

Of the three main radioactive elements in the earth, K, U and Th, potassium is the only element for which there is some basis both in theory and experiment calling for its presence in the core. The suggestion that K can enter a sulfur-bearing metallic core of the earth was initially made by Lewis (1971) and Hall and Murthy (1971). Recent experiments have confirmed the solubility of K in Fe–S melts (Gessman and Wood, 2002; Murthy et al. 2003). If the earth's core formed by segregation of metallic liquids in the Fe–FeS system, the level of K-solubility observed in these experiments suggests that a significant amount of potassium can be present in the core and act as a radiogenic heat source in the core. The precise amounts cannot be estimated at present because the effect of pressure and temperature on the solubility is not yet completely determined.

Bukowinski (1976) showed by quantum mechanical calculations that a change in electronic structure (from 4d-like orbitals to 3s-like orbitals) makes K behave more like a transition element, thus allowing its entry into metallic Fe. Since then, other experimental and theoretical studies have found this change in chemical bonding of K at high pressure (Ito et al., 1993; Parker et al., 1996; Lee et al., 2004). Clear evidence that K can alloy with Fe at a pressure above >26 GPa and temperatures >2500 K, and at pressure corresponding to the core–mantle boundary (CMB) is provided by recent experimental works of Lee and Jeanloz (2003) and Hirao et al. (2006). These experiments uniquely show that K can also enter Fe-metal without the presence of O as suggested by Gessman and Wood (2002) or without the presence of sulfur (Hall and Murthy, 1971; Murthy et al., 2003). Thus it appears that significant K can be present and serve as a radioactive heat source in the core. Table II summarizes the estimates of K radioactivity in the core from various models, theoretical calculations and the recent experiments.

In spite of these experimental data from four different groups, McDonough (2003) has raised objections to the notion of the presence of K in the core. The objections are based on presumed behavior of elements like Ca and the rare-earth elements (REE) whose behavior at high P, T conditions relevant core formation is not known at present. For example, McDonough argues that Ca is also known to enter metallic liquid along with K (Gessman and Wood, 2002) so that conditions in the earth that would result in K

TABLE II

A comparison of inferred values of potassium abundance in the core from geochemical models, theoretical calculations, and recent experiments and the corresponding heat production in terawatts (TW) today

Method	Abundance (ppm)	Radiogenic heat production (TW)	Reference
Geochemical models	0	0	McDonough (1999), GERM database, McDonough (2003)
Theoretical calculations	550 ± 260	~4–5	Lodders (1995)
	200–400		Buffett (2003)
	250–750		Labrosse (2003)
	Up to 1420	9	Roberts et al. (2003)[a]
	400	~3	Nimmo et al. (2004)
Experimental	< 1	0.01	Chabot and Drake (1999)
	100–250	~0.8–2.0	Gessman and Wood (2002)
	60–130	0.4–0.8	Murthy et al. (2003)
	Up to 7000	Up to 45	Lee and Jeanloz (2003)[b]
	35	0.23	Hirao et al., submitted

[a]As quoted in Labrosse and Macouin, 2003.

[b]See discussion in Lee and Jeanloz (2003). This is an upper limit; for realistic conditions of core formation in the earth, the authors note that the value is likely to be much less, possibly around 1200 ppm.

extraction into the core should lead to subchondritic refractory element ratios such as Ca/Al, Ca /Sc etc., in the earth, contrary to observation. The argument is misleading in that a simple calculation using the D_{Ca} (abundance of Ca in metal/abundance of Ca in silicate) values from Gessman and Wood (2002) demonstrates that the effect of Ca extraction is so small and is entirely within the 10% error of the estimated abundance of Ca in the mantle. McDonough's (2003) arguments about the REE and some isotopic systems such as Sm/Nd and Lu/Hf suffer from the fact that nothing is known about D_{REE} (abundance of REE in metal/abundance of REE in silicate) in terrestrial sulfides at high pressure and temperature.

The CMB heat flux is estimated to be from 2–12 TW (Labrosse and Macouin, 2003, and references cited therein) and controls the rate of core cooling. Some recent studies of the energetics of the core, particularly the question of how best to reconcile the CMB heat flux with the size of inner core and the >3.5 Ga age of the magnetic field suggest the possibility of radioactivity in the core (Buffet, 2002; Nimmo et al. 2004). Using somewhat different parameters, similar conclusions have been reached by others

(Labrosse, 2003 and the references cited therein). In view of the recent experimental data it is reasonable to attribute this to the radioactivity of K in the core. About 400 ppm K is in best accord with the present size of the inner core and the power needs of a geomagnetic dynamo for the past 3.5 Ga.

There have been some suggestions for the presence of U (and Th) in the core. Furst et al. (1982) noted that under the highly reduced conditions relevant to core formation in the earth, U and Th tend to become chalcophile and thus might enter a sulfur-bearing core. Feber et al. (1984) noted the significant solubility of UO_2 in impure Fe-metal at 1 bar pressure and at temperature >3000 K and suggested that significant U may be present in the core to act as a heat source. Murrell and Burnett (1986) measured partitioning of K, U and Th between silicate and sulfide liquids to suggest that U and Th are more likely to be present in the core than K. A comparison of the recently determined Th/U ratios in chondrite meteorites to the terrestrial value, led Humayun (2003) to suggest the possibility of entry of U into the core.

In a series of papers Herndon (see for example, Herndon, 1980, 1998) has proposed a gravitational sinking of U to the core during the primordial differentiation of the earth and acts as an energy source in the core. The model requires U to be totally reduced to metallic form at the time earth formed. Such a low state of oxidation would have led to a reduction of all iron in the planet to metallic form, leaving no oxidized iron in the mantle. This is not supported by the observation that the mantle contains ~8% of oxidized iron. Direct modern experimental data to determine the presence of U in the core have been inconclusive and contradictory so far (Wheeler et al. 2004; Bao et al. 2004; Malvergne et al., 2005). Thus, whether U and Th are present in the core remains a moot question at present.

4. Conclusions

It thus appears that the radioactivity of the earth, either in the mantle or in the core is not well constrained by current geochemical and geophysical studies. There is a substantial uncertainty as to the precursor material that accreted to form the earth and the meteoritic models employed are plagued by many difficulties. While we continue to refine and expand these models and terrestrial observations, the fundamental difficulty that we have no way of directly measuring the radioactivity of a substantial fraction of the mass of the planet, i.e., the lower mantle and the core should be kept in mind. A way out of this dilemma is provided by the possibility of a direct measurement of the radioactivity of the mantle and the core by measurement of geoneutrino flux due to radioactivity in the earth.

References

Anderson, D.L.: 1989a, *Science* **243**, 367–370.

Anderson, D. L.: 1989b. *Theory of the Earth*. Blackwell Scientific Publications, Boston, http://www.resolver.caltech.edu/CaltechBOOK:1989.001.

Anderson, D. L.: 2005, *Earth's Deep Mantle: Structure, Composition and Evolution*, AGU Geophysical Monograph Ser., 160, pp. 165–186.

Bao, X., Secco, R. A., Gagnon, J. E. and Fryer, B. J.: 2004, 'Experimental Study of U, Th Solubility in Earth's Core: Toward a Solution of the Core Cooling Paradox. AGU Spring Meeting Abstracts, P41A., 02B.

Boyet, M. and Carlson, R. W.: 2005, *Science* **309**, 576–584.

Buffet, B. A.: 2002, Geophys. Res. Lett., **29**(12), doi: 10.1029/2001GL014649.

Buffett, B. A.: 2003, *Science* **299**, 1675–167.

Bukowinski, M. S. T.: 1976, *Geophys. Res. Lett.* **3**, 491–503.

Chabot, N. L. and Drake, M. J.: 1999, *Earth Planet. Sci. Lett.* **172**, 323–335.

Clayton, R. M.: 1993, *Ann. Rev. Earth Planet. Sci.* **21**, 115–149.

Drake, M. J. and Righter, K.: 2002, *Nature* **416**, 39–44.

Feber, R. C., Wallace, T. C. and Libby, L. M.: 1984, *Eos Trans. AGU*. **65**(44).

Furst, M. J., Stapanian, M. I. and Burnett, D. S.: 1982, *Geophys. Res. Lett.* **9**, 41–44.

GERM, A Geochemical Earth Reference Model. http://www.earthref.org/GERM/index.html.

Gessmann, C. K. and Wood, B. J.: 2002, *Earth Planet. Sci. Lett.* **200**, 63–78.

Hall, H. T. and Murthy, V. R.: 1971, *Earth Planet. Sci. Lett.* **11**, 239–244.

Hart, S. R. and Zindler, A.: 1986, *Chem. Geol.* **57**, 247–267.

Herndon, J. M.: 1980, *Proc. Roy. Soc. London, Series A* **372**, 149–154.

Herndon, J. M.: 1998, *Phys. Earth Planet. Inter.* **105**, 1–4.

Hirao, N., Ohtani, E., Kondo, T., Endo, N., Kuba, T., Suzuki, T. and Kikegawa, T.: 2006, *Geophys. Res. Lett.*, 33, L08303, doi: 10.1029/2005GL025324.

Hofmann, A. W.: 1988, *Earth Planet. Sci. Lett.* **90**, 297–314.

Hofmeister, A. M. and Criss, R. E.: 2005, *Tectonophysics* **395**, 159–177.

Humayun, M.: 2003, (Abstr.), *Eos. Trans. AGU* **84**(46).

Humayun, M. and Clayton, R. N.: 1995, *Acta* **59**, 2131–2148.

Ito, E., Morooka, K. and Ujike, O.: 1993, *Geophys. Res. Lett.* **20**, 1651–1654.

Jagoutz, E., Palme, H., Baddenhausen, K., Blum, M., Cendales, G., Dreibus, G., Spettel, V., Lorenz, V. and Wanke, H.: 1979, *Proc. Lunar Planet. Sci.*

Javoy, M.: 1995, *Geophys. Res. Lett.* **22**, 2219–2222.

Javoy, M. and Pineau, F.: 1983, *Meteoritics* **18**, 320–321.

Javoy, M., Pineau, F. and Delorme, H.: 1986, *Chem. Geol.* **57**, 41–62.

Labrosse, S.: 2003, *Phys. Earth Planet. Inter.* **140**, 127–143.

Labrosse, S. and Macouin, M.: 2003, *C. R. Geosci.* **335**, 37–50.

Lassiter, J. C.: 2004, *Geochem. Geophys. Geosyst.*, 5, Q 11012, doi: 10.1029/2004GC000711.

Lee, K. K. M. and Jeanloz, R.: 2003, *Geophys. Res. Lett.*, 30, 2312, doi: 10.1029/2003GL018515.

Lee, K. K. M., Steinle-Neumann, G. and Jeanloz, R.: 2004, *Geophys. Res. Lett.*, 31, doi: 10.1029/2004 GL 019839.

Lee, W. H. K.: 1970, *Phys. Earth Planet. Int.* **2**, 332–341.

Lewis, J. S.: 1971, *Earth. Planet. Sci. Lett.* **11**, 130–134.

Lodders, K.: 1995, *Meteoritics* **30**, 93–101.

Lodders, K.: 2000, *Space Sci. Rev.* **92**, 341–354.

Lodders, K. and Fegley, B. J., Jr.: 1998, *The Planetary Scientist's Companion*. Oxford University Press, Oxford.

Malvergne, V., Tarrida, M., Combes, R. and Bureau, H., 2005, *Lunar Planet. Sci.* **XXXVI**, 1823.

Mantovani, A., Garmignani, I., Fiorentini, G. and Lissia, M.: 2004, *Phys. Rev.* (D), **69**, 013001.

Mattern, E., Matas, J., Ricard, Y. and Bass, J.: 2005, *Geophys. J. Inter.* **160**, 973–990.

McDonough, W. F.: 2003, *Treatise on Geochemistry*, Vol. 2, Elsevier-Pergamon, Oxford, pp. 547–568.

McDonough, W. F.: 1999, *Encyclopedia of Geochemistry*, Kluwer Academic Publishers.

McDonough, W. F. and Sun, S.-S.: 1995, *Chem. Geol.* **120**, 223–253.

Murrell, M. T. and Burnett, D. S.: 1986, *J. Geophys. Res. B* **91**, 8126–8136.

Murthy, V. R., van Westrenen, W. and Fei, Y.: 2003, *Nature* **423**, 163–165.

Nimmo, F., Price, G. D., Brodholt, J. and Gubbins, D.: 2004, *Geophys. J. Inter.* **156**, 363–376.

Parker, L. J., Atou, T. and Badding, J. V.: 1996, *Science* **273**, 95–97.

Pollack, H. N., Hunter, S. J. and Johnson, J. R.: 1993, *Rev. Geophys.* **31**, 267–280.

Ringwood, A. E.: 1966, 'The Chemical Composition and Origin of the Earth, in *Advances in Earth Sciences*, MIT Press, Cambridge, MA, pp. 287–356.

Roberts, P. H., Jones, C. A. and Calderwood, C. A.: 2003, Earth's Core and Lower Mantle, Taylor & Francis, London.

Stein, C. A.: 1995, *A Handbook of Physical Constants: Global Earth Physics, AGU Reference Shelf 1*, American Geophysical Union, Washington DC, pp. 144–158.

Taylor, S. R. and McClennan, S. M.: 1985, *The Continental Crust: Its Composition and Evolution*. Blackwell Scientific Publications, Oxford, 312 pp.

Tolstikhin, I. and Hofmann, A. W.: 2005, *Phys. Earth Planet. Int.* **148**, 109–130.

Van Schmus, W. R.: 1995, *Global Earth Physics, A Handbook of Physical Constants, AGU Reference Shelf 1*, American Geophysical Union, Washington, DC, pp. 283–291.

Wanke, H. G., Dreibus, G. and Jagoutz, E.: 1984, *Archean Geochemistry*, Springer-Verlag, New York, pp. 1–24.

Wheeler, K., Walker, D., Fei, Y. and Minarik, W.: 2004, *Eos Trans. AGU*, **85**(47), MR43A–0879.

Earth, Moon, and Planets (2006) 99:33–49
DOI 10.1007/s11038-006-9108-4

Is there a Nuclear Reactor at the Center of the Earth?

R.D. SCHUILING
Inst. Geosciences, PO Box 80021, 3508 TA, Utrecht, The Netherlands
(E-mail: schuiling@geo.uu.nl)

(Received 24 February 2006; Accepted 11 July 2006)

Abstract. In this paper we discuss the Herndon hypothesis that a nuclear reactor is operating at the center of the Earth. Recent experimental evidence shows that some uranium can have partitioned into the core. There is no viable mechanism for the small amount of uranium that is dissolved in the molten metal to crystallize as a separate uranium phase (uranium metal or uranium sulfide) and migrate to the center of the core. There is no need for an extra heat source, as the total heat leaving the core can be easily provided by "classical" heat sources, which are also more than adequate to maintain the Earth's magnetic field. It is unlikely that nuclear georeactors (fast breeder reactors) are operating at the Earth's center.

Keywords: Earth's core, heat sources, nuclear georeactor, meteorites, uranium partitioning

1. Introduction

It is generally agreed that the core must produce a significant amount of energy, which is necessary to maintain convection in the outer core as well as the magnetic field of the Earth. It is claimed that a significant part of the heat production in the core is due to the presence of blobs of concentrated uranium that act as fast breeder reactors. The papers in which Herndon has developed this idea (see, e.g. Herndon, 1980, 1993, 2005, Hollenbach and Herndon, 2001) cover a period of more than 25 years. I will simplify the discussion, which involves the composition and redox state of the lower mantle and core, the distinction between an endo-earth (from 680 km downward) and an exo-earth (comprising the upper mantle and the crust), the composition of the inner core, stated to be Ni_2Si, and the nature of the boundary layer between core and mantle. Recently he has also proposed Whole Earth Decompression Dynamics Theory as a new concept to replace plate tectonics (Herndon and Edgerley, 2005).

I do not think that all these additional hypotheses are necessary to accept the possibility that a fraction of uranium under the assumed redox conditions can exist as U-sulfide or even as uranium metal and can find its way into the core. High-pressure experiments (Murthy et al., 2003) at core conditions have shown that potassium, which is also a lithophile element, can probably enter the core at concentrations between 60 and 250 ppm. Some experimental evidence for the behavior of uranium at high temperatures and pressures corresponding to core conditions has recently become available (Wheeler et al., 2006; Bao et al., 2005; Malavergne et al., 2005). This makes it easier to accept for the moment that part of the uranium could behave in a similar fashion as potassium, and reach concentrations between 1 and 6 ppb in the Earth's core, although Wheeler et al. (2006) note that "the transfer of U from metal sulfide to silicate under our experimental conditions is so complete that insufficient U would remain so as to be of any importance to the core's heat budget". The experiments of Bao et al. (2005) show that at high pressures several percentage of silicon enter the metal phase when liquid iron coexists with peridotite, and uranium concentrations in the metallic phase also increase with pressure. After accepting the possibility that uranium can partition into the core, we can focus on whether there is a plausible scenario for uranium to crystallize as a separate phase, and to concentrate into >100 kg blobs that can act as fast breeder reactors, and whether it is possible that the products of the reactor that poison the reactor process can be removed from time to time by diffusion.

2. Origin and Composition of the Earth

Most attempts at reconstructing the composition of the Earth are based on a particular choice of meteorites, usually chondritic, as ordinary chondrites are the most common type of meteorite. Herndon claims that this assumption is wrong, and that the Earth has formed mainly from a rare class of meteorites, the enstatite chondrites. Javoy (1995) and Wanke et al. (1984) have also argued that enstatite chondrites have played a major role in determining the composition of the Earth. In order to avoid discussions about the type of meteorites that have contributed to the composition of the Earth, I will derive the composition of the Earth in a different way by starting with average solar matter, and apply a condensation sequence to it (Schuiling, 1975; Schuiling et al., 1994). No primary evidence from meteorites will be used. This permits us to subsequently use the information obtained from meteorites for validation of the model.

If we start with solar matter, it will be necessary to use a yardstick to convert elemental ratios into elemental masses, because all solar abundances are relative to 10^{12} atoms of hydrogen, and all cosmic abundances, a more practical measure for condensed bodies, are expressed relative to 10^6 atoms of silicon. I have approached the problem in two steps. First I have made the arbitrary assumption that there is no silicon in the Earth's core, so the Earth's only inventory of silicon is the sum of the silicon in the mantle and in the crust. We will first treat the core as an unknown substance with a mass of 192×10^{25} g.

Contrary to Herndon's distinction between an endo-earth and an exo-earth, I have assumed that the mantle has a more or less homogenous composition throughout. There is no compelling geophysical evidence to the contrary, and seismic tomography shows that convection cells can pass unhindered through the 680 km boundary into the deeper mantle. Such convection cells would serve for some crude homogenization by mixing.

In the following we will often refer to an element as being lithophile, chalcophile or siderophile. Lithophile means that the element readily reacts with oxygen, and is commonly incorporated into silicate rocks. Chalcophile elements prefer to combine with sulphur as metal-sulphides, and siderophile elements are mostly found as native metals. When these substances melt, the lithophile elements and the metals form two immiscible liquids, like the slag and the metal in metallurgical processes, and the chalcophile elements are preferentially taken up by the metal melt.

Once the total mass of silicon in mantle + crust has been determined, we can calculate for each element the quantity of that element that should be present in the Earth relative to this calculated mass of silicon. We can then compare those calculated quantities to the quantities of each element actually found. We will define *apparent depletion* as the ratio between the mass of an element as found on Earth (in atmosphere, oceans, crust and mantle) and the mass of that element that should be present according to its cosmic abundance in the case of its complete condensation. So, apparent depletion of an element Z is

$$\text{Apparent depletion (Z)} = \frac{C_{Z(\text{earth})} \times \text{mass}_{(\text{earth})}}{\text{cosmic abundance}_Z \times \text{atomic wt.}_Z \times \text{atoms}_{\text{Si(earth)}} \times 10^{-6}}$$

Now we can predict the following outcome of these calculations:

- If an element is lithophile and refractory, its apparent depletion should have a value of 1.
- If an element is lithophile and volatile, its apparent depletion should be less than 1, and decrease with increasing volatility.
- If an element is siderophile (or chalcophile) and refractory, its apparent depletion should be less than 1; the missing part is then attributed to the core.
- If an element is siderophile and volatile, its apparent depletion should be less than 1. A value is assigned by comparison with its lithophile neighbors of similar volatility, and the remaining deficit is assigned to the core.

Figure 1 is the outcome of this exercise. It conforms in general to the predictions. All lithophile, refractory elements show an apparent depletion close to 1, and all siderophile refractory elements have a lower apparent depletion. At increasing volatility, the apparent depletions decrease. On closer look, there are a number of interesting minor features. First of all, the elements Li, Be and B show an apparent depletion well over 1. This is due to the fact that their solar inventory was higher at the origin of our solar system, but that these elements have since been used up in fusion reactions in the sun, whereas the Earth has preserved their abundances as they were 4.6 billion years ago. This discrepancy, of course, does not show up in Earth models based on chondrite compositions, as the meteorites have formed from the same primitive solar composition as the Earth. Argon and lead also show a relative excess, because part of the lead has formed in the Earth as daughter product of the decay of uranium and thorium, and part of the argon as a decay product of potassium.

Some more fundamental characteristics are as follows. All the refractory lithophile elements that condense at temperatures of 1200 K and above (Mg, Al, Ca, Sc, Ti, Sr, Y, Zr, Nb, Ba, La, Hf, Ta, Th and U) show apparent depletions not of 1.0 as predicted, but all slightly in excess of 1. The most logical explanation is that our yardstick is too small. By admitting about 7% of silicon in the core all these discrepancies disappear, because thereby the amounts of these elements that should be present in the Earth are all proportionally increased, and their apparent depletions are thereby reduced to values around 1. It also solves already part of the problem that according to geophysics the core should contain in the order of 10–15% of elements lighter than iron.

A second interesting point is the fact that the decrease of apparent depletions as a function of volatility is not a smooth curve, but that there seems to be a second level of apparent depletions around 0.1–0.15, comprising elements of different volatility that condense over a wide range of temperatures between 400 and 1200 K. The simplest explanation of these data is that the Earth consists of a mixture of two populations of condensed particles, one high temperature main fraction that completed its condensation around 1200 K, and a smaller second fraction (similar to carbonaceous chondrites?), including elements like Cs, Rb, Na and K, as well as the somewhat chalcophile elements Ag, Sn, Sb and As, that stopped condensation at temperatures around 400 K.

The data also permit the reconstruction of the composition of the core. By adding up all the missing masses of the siderophile elements, we can calculate the composition and the mass of the core, which is found indeed to consist mainly of Fe, Ni and S, with 7% Si, and has a calculated mass which closely corresponds to the observed mass of the core. We can now compare this calculated composition of the core, as an independent check, with the average

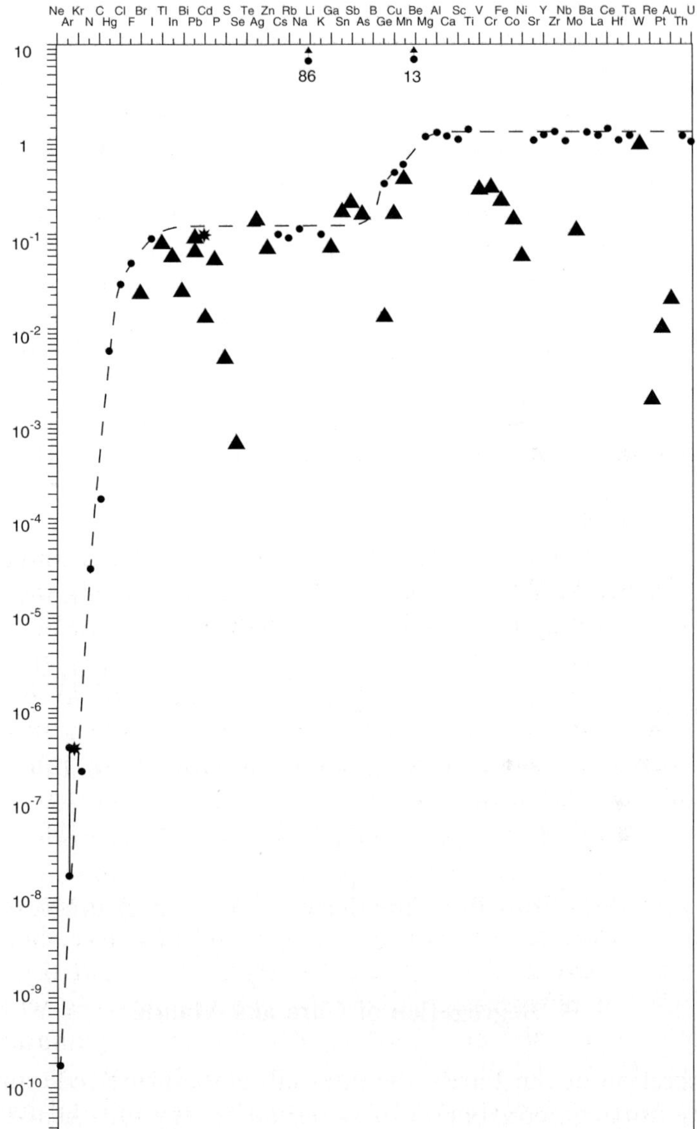

Figure 1. Apparent depletions of elements in the Earth. Triangles denote siderophile or chalcophile elements; dots denote lithophile (and some atmophile) elements.

composition of iron meteorites and find a surprising agreement (Figure 2). Remember that so far we have not used meteorite data in our calculation, so the agreement is an independent check. The fact that the calculated mass of the core also conforms closely to the observed mass is another independent outcome. Silicon and sulfur add up to 11%, in satisfactory agreement with

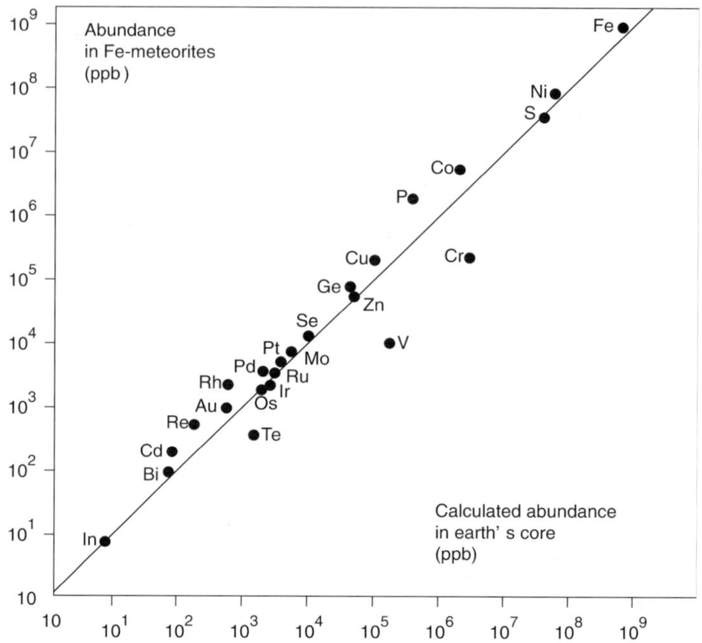

Figure 2. Average composition of iron meteorites compared to the calculated abundances of siderophile and chalcophile elements in the Earth's core.

the geophysical requirement of the presence of between 10 and 15% of elements lighter than iron in the core.

It should be noted that the internal consistency of the model is in itself a strong argument that the derived composition of the Earth can't be far from the truth. Large deviations from the derived abundances are very unlikely.

3. Segregation of Core and Mantle

After the accretion of the Earth, the internal temperature rose quickly (heat generated by impacts, conversion of potential energy into heat during self-compression, and the radioactive decay of short-lived isotopes). This led to widespread melting, and the segregation of a metal/metal sulfide melt and a silicate melt, that are immiscible. The heavier metal melt sank to the centre of the Earth. The segregation process itself contributed to heating, as the potential energy that was liberated by the process was transformed into heat. One can imagine, therefore, that once started, the segregation of core and mantle became a runaway process until it was completed. According to Birch (1965), the segregation event itself has caused an average heating of the entire Earth of about 1600 K. Silicate magma and metal melt remained in close contact at high temperatures (in excess of 4000 K) until they segregated. This must mean that they were maintaining thermodynamic distribution equilibrium until the moment of separation. As a result, lithophile elements must have maintained low, but measurable concentrations in the metal melt, and siderophile elements must have done likewise in the silicate melt. Even at the much more moderate temperatures of steel furnaces, the concentration of SiO_2 in the molten iron in equilibrium with a silicate slag is of the order of 0.5%. At the high temperatures near the core–mantle boundary one can expect that the equilibrium concentration of SiO_2 in molten iron was even higher.

3.1. Intermezzo: the behavior of elements in earth systems

Major elements in a system usually form their own compounds. Such natural crystalline compounds are called minerals. Trace elements rarely form their own minerals, because their low concentrations are normally accomodated by solid solution in the crystal lattice of the compounds of major elements. This process is known as isomorphic substitution, meaning that an atom of a trace element replaces a major element in its compounds, provided there is some similarity between the ionic radius and the charge of the major element and of the trace component. If the ionic radius of an element is much smaller or larger than those of "common" elements, the tendency to form their own compounds, even if they occur only in trace amounts, becomes larger. If the ionic charge is different, there is often the possibility to compensate this with so-called coupled substitutions or the creation of vacancies.

A case that may be relevant to the Herndon hypothesis concerns the fate of uranium. The ionic radius of four-valent uranium is 0.97 Å, very close to the ionic radius of two-valent Ca (0.99 Å). Under the normal reducing conditions in the Earth, uranium assumes a valence of four, and has a tendency to substitute for calcium in the lattice of a Ca-mineral.

4. The Cooling Stage of the Core

Obviously, the core, after its formation, must have been completely molten. The heat production of radioactive isotopes with short half-lives decreased rapidly, and the amount of heat from long-lived isotopes also decreased with time. Any superheat from the segregation event dissipated, so the inevitable outcome was that the core started a slow cooling. We know that the inner core is now solid. From the fact that the solidification has progressed to slightly over 1200 km it can be deduced that the whole core has cooled by about 170 K since the beginning of solidification. The moment at which the inner core started to solidify cannot be determined with any confidence.

Table I summarizes the main heating and cooling stages of the core in the course of the geological history.

From the limited knowledge we have about the properties of a metal melt at core conditions, it is likely that a solid nickel–iron will solidify from a melt that has a metal–metal sulfide composition. Herndon claims that the inner core consists of Ni_2Si, but this seems unlikely in view of the fact that Ni_2Si is lighter than Ni or Fe, and that its melting temperature is also lower, at least at low pressures. If any solid crystalline Ni_2Si would form in a cooling core, it will probably float to the top.

Earlier we have stated that the molten metal must have been in distribution equilibrium with the silicate or oxide phases in the mantle. These components cannot be incorporated in crystalline iron or nickel–iron, so if one part of the melt that was saturated with these compounds crystallizes, this automatically results in supersaturation of the remaining residual melt (the outer core) with these same lithophile phases, which will start to crystallize. Even the high-pressure equivalents of olivine are much lighter than the metal melt, so once they have formed crystals these will tend to rise in the melt. This additional (chemical) buoyancy helps to power the dynamo (Buffett et al., 2000). The crystals are trapped beneath the roof of the core (at the CMB, the core–mantle boundary). They continue to grow on their slow journey upward, and because they are free-floating and not disturbed by contact with other solids, they develop their own crystal shapes. This way a layer of silicate or oxide minerals floating in a matrix of molten metal is accumulating below the core–mantle boundary, and this layer will become thicker as a function of the crystallization of the inner core. This sequence of events is very similar to the formation history of a fairly rare class of meteorites, the so-called pallasites. Pallasites are stony-iron meteorites. If they are not too much deformed by later shock effects, they are composed of idiomorphic ("having their own crystalline shape") crystals of olivine, floating in a matrix of nickel–iron that displays a continuous Widmannstatten pattern (a subsolidus unmixing of nickel-rich metal lamellae from a nickel-poor matrix, indicating a very slow cooling history, in the range of 0.5–7.5 K per million years (Sears, 1978).

The core–mantle boundary layer must be forming in a similar way, only the olivine crystals will be substituted by their high-pressure equivalents, and the rate of cooling will be even lower than for the smaller planetesimals, in the order of 0.04–0.17 K per million years.

The solidification of the inner core releases a considerable amount of heat of crystallization. This will constitute a larger or smaller portion of the present heat flow from the core, depending on whether the inner core started to grow relatively late or relatively early in the Earth's history. It also makes assertions about the necessity of additional contributions to the heat flow from the core rather uncertain.

TABLE I
Main heating and cooling stages of the core during geological history

Main events in Earth history and their thermal consequences
Heating stages
1. Cooling of a solar nebula, and condensation of refractory elements. Final temperature of condensation around 1200 K, based on the condensation temperatures of refractory elements that have completely condensed and less refractory ("volatile") elements that have only partly condensed.
2. Accretion of condensed phases into a proto-earth. Temperatures inside the Earth rise by self-compaction, energy from impacting bodies and high levels of short-lived radiogenic elements. Temperatures rise rapidly to ~2000 K (melting point of iron at low pressures).
3. Iron starts to melt in the upper levels of the Earth (triggered locally by asteroid impacts?), and the metal magmas start to descend.
4. The lost potential energy is transformed into heat. At the completion of segregation this heating is equivalent to 1600 K for the whole Earth, but most of this heating takes place in the deeper levels of the Earth, where temperatures rise considerably more during the runaway process of segregation. The melting point of mantle material at pressures of the core–mantle boundary (CMB) is around 3800 K. It is highly probable that this temperature was considerably exceeded at the completion of core–mantle segregation.
Cooling stages
5. The superheat left after segregation is rapidly removed by whole mantle *liquid* convection, until the lower mantle has solidified.
6. This is followed by a period of slow cooling (but faster than at present) of the core by liquid convection, and of the mantle by heat conduction supplemented by whole mantle *solid* convection and/or mantle plumes rising from the CMB. Uncharted, but probably important heat sinks are the cold subducting plates and the endothermic reactions taking place in these plates like dehydration and decarbonation.
7. Temperatures of core and mantle drop to the point that a solid inner core starts to form. Initially the latent heat of crystallization contributes very little to the heat budget of the core. At this stage heat production of the core is mainly from (higher than present) levels of ^{40}K (and uranium?) and the loss of heat by cooling.
8. At present: heat production from ^{40}K has diminished considerably (more than 10 times since the formation of the Earth), and most of the heat produced at present is from the latent heat of core solidification. Temperatures throughout the whole core and the deeper mantle have dropped by about 170 K since the first formation of a solid core, and by more than 500 K since the completion of segregation.

5. Meteorites and the Earth

Most meteorites are so-called chondrites that never have been part of larger parent bodies. The other meteorites can bc divided into achondrites (crystalline stony meteorites), iron meteorites, and stony-iron meteorites. These are all believed to derive from larger planetesimals, and the iron meteorites and pallasites were part of planetesimals that had undergone segregation into a silicate mantle and a nickel–iron core. The achondrites are usually rich in olivine and pyroxene, and can be assimilated to mantle material. The iron meteorites, which consist of nickel–iron, usually contain some troilite. They show evidence of very slow cooling and are most likely similar to core material. The pallasite class of stony-iron meteorites, that also show evidence of very slow cooling, has probably formed as a boundary layer at the core–mantle boundary of planetesimals. Taking into account that in the Earth temperatures and pressures are much higher than they were in the smaller planetesimals, causing some qualitative differences, we can recognize the same fundamental classes of meteorites in the structure of the Earth as are reaching the Earth from outer space.

6. The Herndon Hypothesis. Behavior of Uranium in the Core

The total uranium inventory of the Earth amounts to approximately 9×10^{19} g, of which slightly over half is present in the crust. Uranium concentrations in the upper mantle are also fairly well constrained, and are around 10 ppb. This leaves a maximum of about 3×10^{19} g of uranium for the lower mantle + core (Herndon, 1993, assumes a total mass of U in the core + lower mantle of just over 1×10^{19} g). The whole mantle would have a U-concentration of ~10 ppb if this uranium is partitioned into the mantle, as is commonly believed on account of its lithophile character. If, however, the uranium of the endo-earth, in Herndon's terminology, would be very efficiently partitioned into the core, its average uranium concentration could reach a theoretical maximum of 15 ppb (1–6 ppb according to Malavergne et al., 2005). This is equivalent to 3×10^{13} ton of uranium, a staggering amount, but the problem is conceiving a mechanism to concentrate part of it in pure uranium blobs of >100 kg, which could act as fast breeder reactors.

In principle this can be treated as a problem of ore formation. Ore formation is the result of the sum of processes by which a low concentration of an element is extracted from a large volume of rock, transported and deposited in a small volume of rock with a high concentration (a high grade, in ore terms). Most ore forming processes are linked to steep gradients in chemical potential, pressure and temperature, the boiling of the solvent, the mixing of two different fluids that are out of equilibrium, or the cooling and crystallization of a melt. Crystals floating in a liquid will settle toward the bottom if they are heavier than the medium, or float when they are lighter. Sometimes the ore-forming process itself is preceded by an enrichment step leading to a protore, a rock volume that is already enriched in the component which later will form the ore deposit. It seems highly unlikely that any steep gradients in chemical gradient, pressure or temperature would persist in a fluid metallic core, and it is also hard to think of any separate immiscible fluids, except, of course, the immiscible silicate and metal melts that were responsible for the segregation process itself, and form the two reservoirs between which the uranium was distributed.

Is it possible that uranium, dissolved in a metal–metal sulfide melt at concentrations of 10 ppb or less could crystallize as a separate phase and sink to the center of the Earth? Although Herndon (1993) on the basis of the alleged higher temperature of melting of uranium compared to that of iron asserts that this is a straightforward proposition, it certainly is not. At low pressures, the melting point of uranium is much lower than that of iron, but even if we assume that the melting point of uranium rises much faster than that of iron as a function of pressure, the conclusion that uranium is the first metal to crystallize in a core fluid is not correct. If we look at the phase diagram of the system UO_2–Fe (Feber et al., 1984), we see that at the compositions to be expected in the core (~80% Fe, and 10 ppb U, a concentration ratio of 100 million) we are on the far right side of the diagram, which is the field of molten iron in which a few percentage of $UO_{2\text{-}x}$ are dissolved (It is strange to note that Feber et al. use a number for the mass of the core and the amount of iron in it that is almost three orders of magnitude less than the real value). According to their phase diagram (Figure 3), the solubility of uranium in this melt is a few percent, i.e., a million times larger than 15 ppb, so uranium at these extremely low concentrations will never crystallize out as a separate phase. One might also remark that even if the uranium would be the first to crystallize as a metal, there is no reason for it to settle toward the very center of the Earth, because the pull of gravity at the center of the Earth is zero.

We must conclude that the direct crystallization of uranium metal in the core is impossible. We will later come back to the possibility that the uranium collects in a separate mineral phase, and later is liberated from this enriched phase, but we will first discuss the second part of Herndon's hypothesis: Is there a plausible mechanism for the uranium, once formed, to collect in large blobs capable of maintaining a nuclear reactor? If we would accept for the moment that there were solid grains of uranium or uranium-sulfide floating around in the core, this is something that could indeed be envisaged, although it requires some wild speculation. One must then invoke convecting fluids that trap such uranium grains in irregularities at the outside of the inner core, much like the trapping of gold particles in a river bed. Although this remains pure speculation, it is at least a remote possibility.

The third step in the breeder concept, the removal of the reactor products by diffusion meets again serious obstacles. As uranium is heavier than nickel–iron,

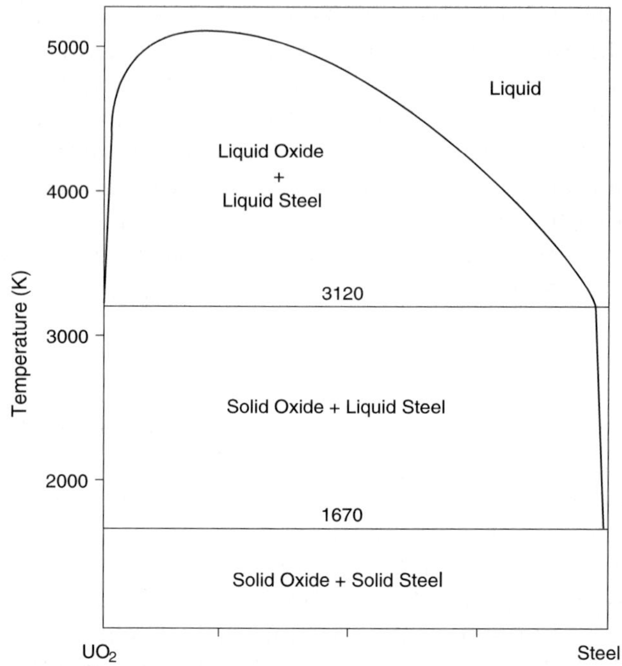

Figure 3. Phase diagram of the system UO_2–Fe.

any concentration of uranium that might form will settle on the growing inner core and soon become encapsulated by the next layer of solid nickel–iron. Diffusion rates of the reactor products (except maybe for helium) are extremely low through a solid medium, even at temperatures around 4000 K. So, unless an unknown efficient means of transport can be discovered, the self-poisoning of the breeder reactor remains a formidable obstacle.

From what has been said before, it seems that an efficient pre-concentrating step is an absolute requirement. Can we construct a viable mechanism by which uranium concentrations might form in the core, or in the CMB? As pointed out by Herndon, uranium in the Abee enstatite chondrite is mainly present in the mineral oldhamite (CaS) and in niningerite (Mg,Fe)S. Oldhamite is probably a carrier for trace amounts of uranium. It is not an uncommon mineral in enstatite chondrites and in aubrites (a type of achondrite). If, by any chance, the metal/metal sulfide melt would be saturated with CaS, this mineral would start to crystallize as a result of the oversaturation of the residual melt caused by the solidification of the inner core. Crystalline CaS is lighter than liquid nickel–iron, and will rise until it is trapped beneath the solid mantle. On its way up, it will have scavenged trace amounts of uranium. A mush of CaS crystals enriched in uranium can be considered as a protore, because concentrations may have gone up from 10 ppb in the homogenous melt to maybe 1–10 ppm, depending on the solubility of CaS in a metal melt under core conditions. If the oldhamite crystals are swept together over the downgoing limb of core convection cells, this can act as a first concentration step. It is unclear, however, how this uranium can subsequently be separated from its host mineral and become concentrated into small uranium blobs (uranium "ore bodies"). Concentrations of uranium-bearing oldhamite crystals, however, over the descending limbs of convection cells in the core may have interesting geophysical implications, because they will constitute local heat sources (from the radioactive decay of uranium), that may be the preferred birthplaces for mantle plumes.

7. Heat Production in the Core. Need for "Non-conventional" Heat Sources?

The following contributions to heat flow from the core can be distinguished:

- latent heat of solidification of inner core,
- cooling of the core,
- heat of crystallization of oxides/silicates in supersaturated outer core,
- potential energy from shrinking of the core during solidification and rise of lighter crystallizing solids to the CMB,
- heat production by decay of ^{40}K,
- heat production by decay of uranium,
- heat production by decay of ^{123}Te, ^{187}Re, ^{186}Os,
- breeder reactors?

In the table below (Table II) we have summarized these different contributions. It is clear that assumptions regarding the timing of the beginning of solidification, as well as regarding the potential amounts of K and U in the Earth's core lead to widely different heat-flows from the core, varying from a low value of 6.1 TW for a beginning of solidification of the inner core 4 billion years ago, and little potassium and no uranium to a high value of 27.3 TW for a late beginning of the onset of solidification and maximum allowable values for K- and U-concentration in the core. As the total heat flow from the core is estimated to be between 6 and 12 TW (Buffet, 2003), and the requirement for maintaining the Earth's magnetic field is only in the order of 1 TW (at a Carnot efficiency of 20% this would require a heat production of 5 TW), it is clear that there is no compelling reason to postulate the existence of additional "unconventional" heat sources like breeder reactors. One should realize that heat production from the core is not a fixed quantity, but is dependent on the "demand" of the overlying mantle. It is conceivable that when a cold subducting slab comes into contact with the

TABLE II
Heat sources and calculated heat-flow contributions in the Earth's core

Heat source	Present heat production if inner core started solidification 4 billion years ago	Present heat production if core started solidification 1 billion years ago
Heat of solidification of inner core	1.5 TW[a]	6.0 TW[a]
Cooling of the entire core	1.8 TW[a]	7.2 TW[a]
Heat of crystallization of oxides/silicates in core + shrinking of core and potential energy from rising crystals	2.4 TW (Melchior, 1986)[b]	9.6 TW[b]
^{40}K (60–250 ppm in core)	0.4–1.7 TW	0.4–1.7 TW
U decay (0–15 ppb in core)	0–2.8 TW	0–2.8 TW
Decay of ^{123}Te, ^{187}Re or ^{186}Os	Negligible?	Negligible?
Breeder reactors	0 to ?	0 to ?
Total energy production (conventional)	6.1–10.2 TW	23.2–27.3 TW

[a]Note that the relative contributions of the heat of solidification and the cooling of the entire core change with time. At the start of solidification the contribution of the latent heat of melting is minimal, but as the inner core grows, an ever increasing part of the heat flow is provided by inner core solidification, because for every degree of cooling an increasingly larger volume of iron solidifies. The heat of solidification of iron at core pressures is considerably larger than its latent heat of melting at 1 bar, on account of the large $P\Delta V$ term (Schuiling et al. 2005).
The data as given in the table represent a time-averaged contribution to the heat flow. Although outside the topic of this contribution, it can be noted that the data, when compared to the assumed range of heat flow from the core, seem to suggest that solidification of the inner core probably started well over 1 billion year ago.
[b]Melchior's figure is a minimum, as he has not taken into account the heat of crystallization and the potential energy lost by rising crystals of lithophile compounds, that crystallize because the liquid outer core gets supersaturated when a pure nickel–iron inner core forms.

CMB, the heat demand increases, to which the core responds by faster cooling and a faster growth of the inner core. It is conceivable that the Earth's magnetic field may also change in response to such variations in the thermal regime.

The heat of solidification of iron at core pressures is considerably larger than its latent heat of melting at 1 bar, on account of the large $P\Delta V$ term (Schuiling et al., 2005).

8. $^{3}He/^{4}He$ Isotope Geochemistry

For some time it seemed that the existence of marked helium isotope anomalies in rocks that were supposed to come from the deep mantle, or even to have originated at the CMB (mantle plumes) constituted a strong argument for the existence of nuclear reactors in the core. Observations of $^{3}He/^{4}He$ higher than ~10 times the atmospheric value were generally interpreted as evidence for a plume from the lower mantle, even in the absence of supporting data. This, however, is strictly an assumption. There is a growing body of observations that make a shallow, upper mantle origin for many helium anomalies likely (Meibom et al., 2003; Anderson et al., 2005), such as high $^{3}He/^{4}He$ in Samoan xenoliths that are known to be of upper mantle origin and in diamonds known to have been mined from pipes. High $^{3}He/^{4}He$ is also observed at Yellowstone, where extensive work has provided a strong case that the magmatic system there is lithospheric only. So, although high $^{3}He/^{4}He$ certainly fits in with Herndon's fast breeder concept, it is by no means compelling evidence.

9. Conclusions

Herndon's postulate that uranium is for a significant part partitioned into the core seems possible, and would automatically lead to an increased heat production in the core by radioactive decay. If this uranium is subsequently taken up by a mineral like oldhamite, this could lead to localized anomalies in heat production near the CMB, which might trigger the rise of mantle plumes.

There is no conceivable mechanism by which a uranium compound or uranium metal could crystallize from a metal–metal sulfide melt containing uranium at 15 ppb or less, and concentrate into large enough blobs to act as fast breeder reactors. *One should always be aware, though, that our inability to conceive such a mechanism should never be taken as absolute proof that it is impossible, and it remains interesting to see what geoneutrino fluxes would tell us.*

There is no need for the assumption of an additional heat source in the core. Conventional heat sources are more than adequate to provide the assumed heat flow from the core and the energy source for maintaining the Earth's magnetic field.

Acknowledgments

I wish to thank Valerie Malavergne for providing me with the latest information on uranium partitioning and critically reviewing my text. I thank the organizers of the geoneutrino session in Hawaii for inviting me to participate in this exciting and thought-provoking meeting.

References

Anderson, D.L., Foulger, G.R. and Meibom, A.: 2005, Helium: Fundamental Models. http://www.MantlePlumes.org.

Bao, X., Secco, R.A., Gagnon, J.E. and Fryer, B.J.: 2005, Experiments of U solubility in Earth's core. *AGU Abstract*.

Birch, F.: 1965, *J. Geophys. Res.* **70**, 6217–6221.

Buffett, B.A.: 2003, *Science* **299**, 1675–1676.

Buffett, B.A., Garnero, E.J. and Jeanloz, R.: 2000, *Science* **290**, 1338–1342.

Feber, R.C., Wallace, T.C. and Libby, L.M.: 1984, *EOS* **65**, 44.

Herndon, J.M.: 1980, *Proc. R. Soc. Lond. A* **368**, 495–500.

Herndon, J.M.: 1993, *J. Geomagn. Geoelectr.* **45**, 423–437.

Herndon, J.M.: 2005, Teaching earth dynamics: what's wrong with plate tectonics theory? http://www.arXiv.org/physics/0510090.

Herndon, J.M. and Edgerley, D.A.: 2005, Background for terrestrial antineutrino investigations: radionuclide distribution, georeactor fission events, and boundary conditions on fission power production. http://www.arXiv.org./hep-ph/0501216.

Hollenbach, D.F. and Herndon, J.M.: 2001, *Proc. Natl. Acad. Sci. USA*.

Javoy, M.: 1995, *Geophys. Res. Lett.* **22**, 2219–2222.

Malavergne, V., Tarrida, M., Combes, R. and Bureau, H.: 2005, *Proceedings Lunar and Planetary Science Conference*, XXXVI, p. 1823.

Meibom, A., Anderson, D.L., Sleep, N.H., Frei, R., Chamberlain, C.P., Hren, M.T. and Wooden, J.L.: 2003, *EPSL* **208**, 197–204.

Melchior, P.: 1986, The Physics of the Earth's Core. Pergamon Press.

Murthy, V.R., van Westrenen, W., and Fei, Y.: 2003, *Nature* **423**, 163–165.

Schuiling, R.D.: 1975, The earth as a cosmochemical body. Abstr. Ann. Meeting Meteoritics Soc., Tours, France.

Schuiling, R.D., Andriessen, P.A.M., Frapporti, G., Kreulen, R., de Leeuw, J.W., Poorter, R.P.E., de Smeth, J.B., Vergouwen, L., Vriend, S.P., Zuurdeeg, B.W. and Nijenhuis, I.: 1994, Introduction to Geochemistry (6th ed.), 350 pp. Universiteitsdrukkerij Technische Universiteit Delft.

Schuiling, R.D., Jacobs, M.H.G. and Oonk, H.A.J.: 2005, Heat from the Earth's core; delivery on demand. Unpublished notes.

Sears, D.W.: 1978, The Nature and Origin of Meteorites. Monographs on astronomical subjects 5, Adam Hilger Ltd., 187 pp.

Wanke, H., Dreibus, G. and Jagoutz, E.: 1984, Mantle Chemistry and Accretion History of the Earth. Archean Geochemistry, Springer-Verlag, Berlin, pp. 1–24.

Wheeler, K.T., Walker, D., Fei, Y., Minarik, W.G. and McDonough, W.F.: 2006, *Geochim. Cosmochim. Acta* **70**, 1537–1547.

Earth, Moon, and Planets (2006) 99:51
DOI 10.1007/s11038-006-9133-3

Comment on R. D. Schuiling's Paper

To appreciate the essential scientific reasons for the possible existence of a nuclear reactor at Earth's center, it is necessary to understand precisely the oxidation state of the deep interior of the Earth as well as the nature and probable circumstances of Earth's origin, which led to that state of oxidation. For example, in referring to the quote from Wheeler et al. (2006), "the transfer of U from metal sulfide to silicate under our experimental conditions is so complete that insufficient U would remain so as to be of any importance to the core's heat budget," Schuiling neglected to note that the silicate used in the laboratory experiment contained 8% FeO. A more highly reduced silicate – nearly devoid of FeO, such as $MgSiO_3$, consistent with the enstatite-chondritic deep interior of the Earth – would have yielded a significantly different laboratory result. Similarly, in referring to elemental behavior using Goldschmidt's term "chalcophile," Schuiling fails to mention that chalcophilicity is related to state of oxidation. Even making use of some condensation model, as Schuiling does, necessitates assuming a particular pressure, which leads to a particular range of oxygen fugacities. Schuiling adopts without reservation the so-called standard model of solar system formation, evidently without realizing that the resulting state of oxidation in that contemporary formation model would inevitably lead to Earth having an insufficiently massive core. And, he accepts the model-dependent idea that the inner core is partially crystallized iron metal, which produces energy by growing. But that thought for inner core composition was developed before data from the 1960s led to a different possibility that is the consequence of the highly reduced state of oxidation of the endo-Earth.

J. MARVIN HERNDON
Transdyne Corporation,
11044 Red Rock Drive,
San Diego,
CA, 92131, USA

Earth, Moon, and Planets (2006) 99:53–89
DOI 10.1007/s11038-006-9121-7

Solar System Processes Underlying Planetary Formation, Geodynamics, and the Georeactor

J. MARVIN HERNDON
Transdyne Corporation, 11044 Red Rock Drive, San Diego, CA 92131, USA
(E-mail: Herndon@UnderstandEarth.com)

(Received 9 February 2006; Accepted 22 August 2006)

Abstract. Only three processes, operant during the formation of the Solar System, are responsible for the diversity of matter in the Solar System and are directly responsible for planetary internal-structures, including planetocentric nuclear fission reactors, and for dynamical processes, including and especially, geodynamics. These processes are: (i) Low-pressure, low-temperature condensation from solar matter in the remote reaches of the Solar System or in the interstellar medium; (ii) High-pressure, high-temperature condensation from solar matter associated with planetary-formation by raining out from the interiors of giant-gaseous protoplanets, and; (iii) Stripping of the primordial volatile components from the inner portion of the Solar System by super-intense solar wind associated with T-Tauri phase mass-ejections, presumably during the thermonuclear ignition of the Sun. As described herein, these processes lead logically, in a causally related manner, to a coherent vision of planetary formation with profound implications including, but not limited to, (a) Earth formation as a giant gaseous Jupiter-like planet with vast amounts of stored energy of protoplanetary compression in its rock-plus-alloy kernel; (b) Removal of approximately 300 Earth-masses of primordial volatile gases from the Earth, which began Earth's decompression process, making available the stored energy of protoplanetary compression for driving geodynamic processes, which I have described by the new whole-Earth decompression dynamics and which is responsible for emplacing heat at the mantle-crust-interface at the base of the crust through the process I have described, called mantle decompression thermal-tsunami; and, (c) Uranium accumulations at the planetary centers capable of self-sustained nuclear fission chain reactions.

Keywords: crustal heat, Earth core, Earth structure, georeactor, geodynamics, geo-antineutrino, Solar System formation, thermal-tsunami, whole-Earth decompression dynamics

1. Introduction

Early in 1939, Hahn and Strassmann (1939) reported their discovery of neutron-induced nuclear fission. Just months later, Flügge (1939) speculated on the possibility that self-sustaining chain reactions might have taken place under natural conditions in uranium ore deposits. Kuroda (1956) used Fermi's nuclear reactor theory (Fermi, 1947) to demonstrate the feasibility that, two billion years ago or before, thick seams of uranium ore might have become critical and functioned as thermal neutron reactors moderated by ground water. Sixteen years passed before French scientists discovered in 1972 the first of several fossil remains of natural nuclear reactors at Oklo, in the Republic of Gabon, Africa (Bodu et al., 1972). These had operated about 1.8 billion years ago as thermal neutron reactors, in much the same manner as predicted by Kuroda (Maurette, 1976), and had also operated to some extent as fast neutron breeder reactors (Fréjacques et al., 1975; Hagemann et al., 1975).

There is evidence that certain planets contain internal energy sources. In 1969 astronomers discovered that Jupiter radiates to space more energy than it receives. Verification followed, indicating that not only Jupiter, but Saturn and Neptune as well each radiate approximately twice as much energy as they receive from the Sun (Aumann et al., 1969; Conrath et al., 1991). For two decades, planetary scientists could find no viable explanation for the internal energy sources in these planets and declared that "by default" (Stevenson, 1978) or "by elimination" (Hubbard, 1990) the observed energy must come from planetary formation about 4.5×10^9 years ago. In 1992, using Fermi's nuclear reactor theory, I demonstrated the feasibility for planetocentric nuclear fission reactors as the internal energy sources for the giant outer planets (Herndon, 1992). Initially, I considered only hydrogen-moderated thermal neutron reactors, but soon demonstrated the feasibility for fast neutron reactors as well, which admitted the possibility of planetocentric nuclear reactors in non-hydrogenous planets (Herndon, 1993, 1994, 1996).

It is known that the Earth has an internal energy source at or near the center of the planet that powers the mechanism that generates and sustains the geomagnetic field. In 1993, using Fermi's nuclear reactor theory, I demonstrated the feasibility of a planetocentric nuclear fission reactor as the energy source for the geomagnetic field (Herndon, 1993). Initially, I could only postulate that the georeactor, as it is called, would operate as a fast neutron breeder reactor over the lifetime of the Earth. Subsequent state-of-the-art numerical simulations, made at Oak Ridge National Laboratory, verified that the georeactor could indeed function over the lifetime of the Earth as a fast neutron breeder reactor and, significantly, would produce helium in the same range of isotopic compositions observed in oceanic basalts (Herndon, 2003; Hollenbach and Herndon, 2001).

Raghavan (2002) demonstrated the feasibility of using geo-antineutrinos as a means for verifying the existence of the georeactor. Why is verification extremely important? As noted by Domogatski et al. (2004), "Herndon's idea about georeactor located at the center of the Earth, if validated, will open a new era in planetary physics."

The purpose of this paper is to disclose the nature of Solar System processes that underlie planetary formation, geodynamics, and the georeactor. The processes revealed lead logically, in causally related ways, to planetary compositions, internal structures, and the basis for the georeactor. The processes disclosed also lead to a new vision of global dynamics, called *whole-Earth decompression dynamics* (Herndon, 2004c, 2005b, c), as well as to a new concept of heat transport within the Earth, called *mantle decompression thermal-tsunami*, which emplaces heat at the base of the crust. In a broader sense, the processes revealed lead to a fundamentally different view of planetary formation than considered over the past four decades and to a new understanding of the genesis of the matter that comprises the Solar System.

2. Nature and Origin of Planetary Matter

The constancy in isotopic compositions of most of the elements of the Earth, the Moon, and the meteorites indicates formation from primordial matter of common origin. Primordial elemental composition is yet manifest and determinable to a great extent in the photosphere of the Sun. The less volatile rock-forming elements, present in the outer regions of the Sun, occur in nearly the same relative proportions as in chondritic meteorites, the relative elemental abundances being related, not to chemical properties, but to nuclear properties.

Chondrites differ somewhat from one another in their respective proportions of major elements (Jarosewich, 1990; Wiik, 1969), in their states of oxidation (Herndon, 1996, Urey and Craig, 1953), mineral assemblages (Mason, 1962), and oxygen isotopic compositions (Clayton, 1993); accordingly, they are grouped into three distinct classes: *enstatite*, *carbonaceous* and *ordinary*. Virtually all approaches to whole-Earth composition are based upon the idea that the Earth is similar in composition to a chondrite meteorite. A major controversy within the Earth sciences began more than six decades ago with the choice of chondrite type as being representative of the Earth (Herndon, 2005a).

Only three major rock-forming elements, iron (Fe), magnesium (Mg) and silicon (Si), together with combined oxygen (O) and sulfur (S), comprise at least 95% of the mass of each chondrite and, by implication, each of the terrestrial planets. These five elements, because of their great relative abundances, act as a buffer assemblage. Minor and trace elements provide a great wealth of detail, but are slaves to that buffer system and are insufficiently abundant to alter conclusions derived from the major elements.

For decades, the abundances of major elements (E_i) in chondrites have been expressed in the literature as ratios, usually relative to silicon (E_i/Si) and occasionally relative to magnesium (E_i/Mg). By expressing Fe–Mg–Si elemental abundances as molar ratios relative to iron (E_i/Fe), as shown in Figure 1, I discovered a fundamental relationship bearing on the nature of chondrite matter that can be understood at different levels (Herndon, 2004b). In Figure 1, chondrite data points scatter about three distinct, well defined, least squares fit, straight lines, unique to their classes, despite mineralogical differences observed among members within a given class of chondrites.

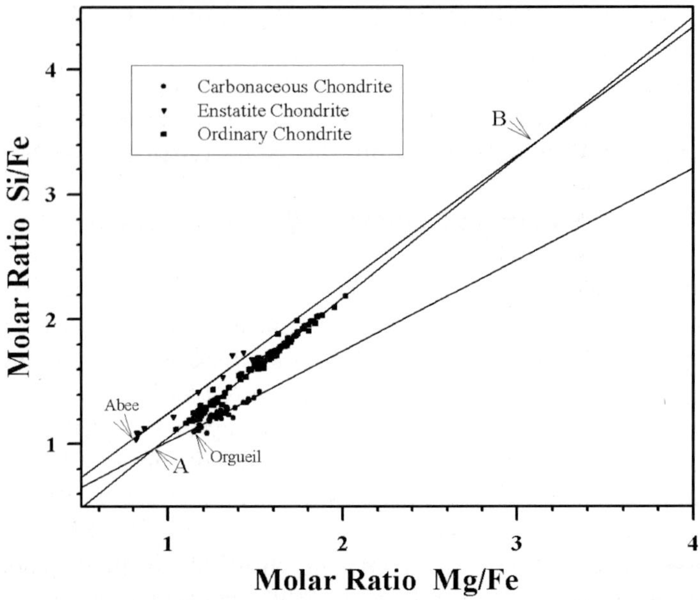

Figure 1. Molar (atom) ratios of Mg/Fe and Si/Fe from analytical data on 10 enstatite chondrites, 39 carbonaceous chondrites, and 157 ordinary chondrites. Data from Baedecker and Wasson (1975), Jarosewich (1990), Wiik (1969). Members of each chondrite class data set scatter about a unique, linear regression line. The locations of the volatile-rich Orgueil carbonaceous chondrite and the volatile-rich Abee enstatite chondrite are indicated. Line intersections A and B represent the compositions, respectively, of the primitive component and the partially differentiated-enstatite-chondrite-like component from which the ordinary chondrites appear to have formed.

At one level of understanding, Figure 1 means that the well-mixed primordial matter became, or evolved to become, only three distinct types of matter which still retain more-or-less the full complement of readily condensable elements and which became the building blocks of the terrestrial planets. At a deeper level, as discussed in reference (Herndon, 2004b), the relationship shown in Figure 1 admits the possibility of ordinary chondrites having been derived from mixtures of two components, representative of the other two types of matter, mixtures of a relatively undifferentiated carbonaceous-chondrite-like *primitive* component and a partially differentiated enstatite-chondrite-like *planetary* component.

The interest here is not simply to understand the origin of chondrite meteorites, but to understand the nature of the physical processes leading to the evolution of their components from the well-mixed primordial progenitor material. The components of chondrite meteorites are in a sense like the results of experiments made in a laboratory, but absent knowledge of exact experimental conditions. Making sense out of these data can lead to a broader understanding of what processes are possible and impossible in the medium from which the planets formed.

The Abee enstatite chondrite and the Orgueil carbonaceous chondrite typify the primitive (least differentiated) end members of their respective types of matter, as shown in Figure 1. In terms of their elemental compositions, including their respective complements of volatile trace elements, they are virtually identical meteorites, an indication of a relatively simple chemical progression from their essentially uniform, well-mixed primordial parent matter. But these two meteorites are strikingly different in terms of their states of oxidation, mineral compositions, evidence of thermal exposure, and formation-location in the Solar System.

There have long been mainly two ideas about how the planets of the Solar System formed. In the 1940s and 1950s, the idea was discussed about planets "raining out" from inside of giant-gaseous protoplanets with hydrogen gas pressures on the order of 10^2–10^3 bar (Eucken, 1944; Kuiper, 1951a; Urey, 1951). But, in the early 1960s, scientists instead began thinking of primordial matter, not forming dense protoplanets, but rather spread out into a very low-density "solar nebula" with hydrogen gas pressures on the order of 10^{-5} bar. The idea of low-density planetary formation, often referred to as the standard model, envisioned that dust would condense at fairly low temperatures, and then would gather into progressively larger grains, and become rocks, then planetesimals, and ultimately planets (Stevenson, 1982; Wetherill, 1980).

These two ideas about planetary formation embody fundamentally different condensation processes which, I submit, are the underlying cause for the two unique types of chondritic matter shown in Figure 1. The immediate implication is that both processes were operant during the formation of the Solar System. The relative extent and region of each process can be ascertained to some certitude from thermodynamic considerations together with planetary data. Even within present limitations, a consistent picture emerges that is quite unlike the standard model of Solar System formation.

3. Low-Temperature, Low-Pressure Condensation

Following the publication by Cameron (1963) of his diffuse solar nebula models at pressures of about 10^{-5} bar, confusion developed during the late 1960s and early 1970s about the nature of the products anticipated to result by condensation from an atmosphere of solar composition at such low pressures. The so-called "equilibrium condensation" model was contrived and widely promulgated (Larimer and Anders, 1970). That model was predicated upon the later refuted assumption (Herndon 1978; Herndon and Suess, 1977) that the mineral assemblage characteristic of ordinary chondrite meteorites formed as the condensate from a gas of solar composition at pressures of about 10^{-5} bar.

The great majority of chondrites observed falling to Earth are called ordinary chondrites, the name denoting their great frequency of occurrence, ~98%. In terms of the five major elements comprising about 95% of the mass of each ordinary chondrite, their mineral assemblage is quite simple, as shown in Table I. Silicon and magnesium occur combined with oxygen in the silicate minerals, olivine, $(MgO, FeO)_2SiO_2$, and pyroxene, $(MgO, FeO)SiO_2$. Some iron occurs combined with oxygen in the silicate minerals, some as iron metal, Fe, and some combined with sulfur as troilite, FeS. The minerals of ordinary chondrites are generally crystalline and typically show evidence of exposure to elevated temperatures.

Suess and I showed that the oxidized-iron content of ordinary-chondrite-silicate-minerals was consistent, not with their condensation from an atmosphere of solar composition, but from an atmosphere where hydrogen was about one-thousandth as abundant (Herndon and Suess, 1977). Subsequently,

TABLE I
The mineral assemblages characteristic of chondritic meteorites

Chondrite type	Major minerals
Hydrous chondrites	
Carbonaceous chondrites	Complex hydrous layer lattice silicate e.g. $(Mg, Fe)_6Si_4O_{10}(O, OH)_8$
	Epsomite, $MgSO_4 \cdot 7H_2O$
	Magnetite, Fe_3O_4
Anhydrous chondrites	
Carbonaceous chondrites	Olivine, $(Fe, Mg)_2SiO_4$
	Pyroxene, $(Fe, Mg)SiO_3$
	Pentlandite, $(Fe, Ni)_9S_8$
	Troilite, FeS
Ordinary chondrites	Olivine, $(Fe, Mg)_2SiO_4$
	Pyroxene, $(Fe, Mg)SiO_3$
	Troilite, FeS
	Metal, (Fe-Ni alloy)
Enstatite chondrites	Pyroxene, $MgSiO_3$
	Complex mixed sulfides e.g. (Ca, Mg, Mn, Fe)S
	Metal, (Fe, Ni, Si alloy)
	Nickel silicide, Ni_2Si

The hydrous C1 carbonaceous chondrites have a state of oxidation characteristic of low-pressure condensation to low temperatures. The highly reduced enstatite chondrites are similar to the matter of the endo-Earth, the inner 82% of the Earth.

I showed (i) that there is at most only a single temperature, if any at all, where the ordinary chondrite mineral assemblage can exist in equilibrium with solar matter, and (ii) that condensation of that mineral assemblage would necessitate an atmosphere depleted in oxygen, as well as hydrogen, relative to solar matter (Herndon, 1978). The ordinary chondrite mineral assemblage is not the condensate from an atmosphere of solar composition at hydrogen pressures on the order of 10^{-5} bar. So, what then is the mineral assemblage expected?

From thermodynamic considerations it is possible to make some generalizations related to the condensation process in an atmosphere of solar composition. In that medium, the oxygen fugacity is dominated by the gas-phase reaction $H_2 + 1/2O_2 = H_2O$ which is a function of temperature, but is essentially independent of pressure over a wide range of pressures where ideal gas behavior is approached. Oxygen fugacity controls the condensate state of oxidation at a particular temperature. At high temperatures the state of oxidation is extremely reducing, while at low temperatures it is quite oxidizing. The state of oxidation of the condensate ultimately becomes fixed at the temperature at which reaction with the gas phase ceases and/or equilibrium is frozen-in by the separation of gases from the condensate.

Condensation of an element or compound is expected to occur when its partial pressure in the gas becomes greater than its vapor pressure. Generally, at high pressures in solar matter, condensation is expected to commence at high temperatures. At low pressures, such as a hydrogen pressure of 10^{-5} bar, condensation is expected to progress at relatively low temperatures at a fairly oxidizing range of oxygen fugacity. At low temperatures, all of the major elements in the condensate may be expected to be oxidized because of the great abundance of oxygen in solar matter, relative to the other major condensable elements. Beyond these generalizations, in this low-pressure regime, precise theoretical predictions of specific condensate compounds may be limited by kinetic nucleation dynamics and by gas-grain temperature differences arising because of the different mechanisms by which gases and condensate lose heat.

Among the thousands of known chondrites, only a few, like the Orgueil carbonaceous chondrite, have a state of oxidation and mineral components with characteristics similar to those which might be expected as a condensate from solar matter at low pressures. Essentially all of the major elements in these few chondrites are oxidized, as shown in Table I. The major silicate is not a well-defined crystalline phase like olivine, but is, instead, poorly characterized phyllosilicate, a layer-lattice, claylike, hydrous material. The presence of sharp, angular shards of crystalline olivine and pyroxene in Orgueil (Reid et al., 1970) appear to be an admixed xenolithic component and shows no indication of alteration, suggesting the phyllosilicate is primary, rather than a secondary aqueous alteration product of olivine. Iron occurs, not as metal, but as magnetite, Fe_3O_4, (Hyman et al., 1978) which presents in a variety of unique morphologies including plaquettes and framboids (Jedwab, 1971; Hua and Buseck, 1998). In Orgueil sulfur occurs mainly as epsomite, $MgSO_4 \cdot 7H_2O$, (Endress and Bischoff, 1993) instead of as troilite, FeS.

There is debate as to how much alteration might or might not have occurred on the Orgueil meteorite's parent body (Tomeoka and Buseck, 1988). Nevertheless, that meteorite is the closest chondrite representative to what may be expected as a low-temperature, low-pressure condensate from the oxygen-rich gas of solar composition. Re-melting and/or re-evaporating and re-condensing Orgueil-like matter, after loss of primordial gases, may be expected to yield crystalline minerals, such as olivine and pyroxene, similar in composition to some other, more evolved, carbonaceous chondrites, such as the Allende meteorite which contains so much oxidized iron in its crystalline silicates, that there is very little remaining as the metal. Significantly, reflectance spectroscopy results appear to identify carbonaceous chondrite-like matter on the surfaces of bodies in the Kuiper Belt in the outer regions of the Solar System (Lederer and Vilas, 2003).

The idea of planetary formation from a diffuse solar nebula, with hydrogen pressures on the order of 10^{-5} bar, envisioned that dust would condense at fairly low temperatures, and then would gather into progressively larger grains, and become rocks, then planetesimals, and ultimately planets. In the main, that idea leads to the contradiction of the terrestrial planets having insufficiently massive cores, because the condensate would be far too oxidized for a high proportion of iron metal to exist. But as evidenced by Orgueil and similar meteorites, such low-temperature, low-pressure condensation did in fact occur, perhaps only in the evolution of matter of the outer regions of the Solar System, and thus may contribute to terrestrial planet formation only as a component of late addition veneer.

4. High-Temperature, High-Pressure Condensation

In 1944, on the basis of thermodynamic considerations, Eucken (1944) suggested core-formation in the Earth as a consequence of successive condensation from solar matter, on the basis of volatility, from the central region of a hot, gaseous protoplanet with molten iron metal first raining out at the center. Except for a few investigations initiated in the early 1950s (Kuiper 1951a, b; Urey, 1952; Bainbridge, 1962), that idea languished when interest was diverted to Cameron's low-pressure solar nebula models (Cameron, 1963).

The enstatite chondrites consist of the most highly reduced natural mineral assemblage known (Table 1). The principal silicate mineral, enstatite, $MgSiO_3$, contains very little oxidized iron. The metal phase contains elemental silicon; magnesium and calcium, strongly lithophile (oxyphile) elements, occur in part as sulfides. And, unique nitrogen-containing minerals occur. The Abee enstatite chondrite has virtually the same relative abundance of volatile elements, such as lead and thallium, as the Orgueil carbonaceous chondrite, which consists of hydrous low-temperature minerals. But, in striking contrast, the Abee meteorite shows evidence of having been at melt or near-melt temperatures as evidenced by sub-euhedral crystals of enstatite embayed by iron metal. Interestingly, as Rudee and I have shown by metallurgical experiments, the Abee enstatite chondrite cooled from 700 °C to 200 °C in a matter of about 2 h (Herndon and Rudee, 1978; Rudee and Herndon, 1981).

The formation of enstatite chondrites has posed something of an enigma for those who make models because, for low-temperature condensation at hydrogen pressures of about 10^{-5} bar, solar matter is much too oxidizing for that mineral assemblage. This has led to the suggestion that loss of H_2O or $C/O \geq 0.9$ in solar matter might account for the state of reduction observed (Larimer, 1968).

On the basis of thermodynamic considerations, Suess and I showed at the high-temperatures for condensation at high-pressures, solar matter is sufficiently reducing, i.e., it has a sufficiently low oxygen fugacity, for the stability of some enstatite chondrite minerals. However, formation of enstatite-chondrite-like condensate would necessitate thermodynamic equilibria being frozen-in at near-formation temperatures (Herndon and Suess, 1976). There is much to verify and learn about the process of condensation from near the triple point of solar matter, but the glimpses Suess and I have seen are remarkably similar to the vision of Eucken (1944), i.e., molten iron raining out in the center of a hot, gaseous protoplanet.

At present, there is no adequate published theoretical treatment of solar-matter condensation from near the triple-point. But from thermodynamic and metallurgical considerations, some generalizations can be made. At the high temperatures at which condensation is possible at high pressures, nearly everything reacts with everything else and nearly everything dissolves in everything else. At such pressures, molten iron, together with the elements that dissolve in it, is the most refractory condensate.

There are reasons to associate the highly reduced matter of enstatite chondrites with the inner regions of the Solar System: (i) The regolith of Mercury appears from reflectance spectrophotometric investigations (Vilas, 1985) to be virtually devoid of FeO, like the silicates of the enstatite chondrites (and unlike the silicates of other types of chondrites); (ii) E-type asteroids (on the basis of reflectance spectra, polarization, and albedo), the presumed source of enstatite meteorites, are, radially from the Sun, the inner most of the asteroids (Zellner et al., 1977); (iii) Only the enstatite chondrites and related enstatite achondrites have oxygen isotopic compositions indistinguishable from those of the Earth and the Moon (Clayton, 1993); and, (iv) Fundamental mass ratios of major parts of the Earth (geophysically determined) are virtually identical to corresponding (mineralogically determined) parts of certain enstatite chondrites, especially the Abee enstatite chondrite (Herndon, 1980, 1993, 1996).

In the absence of evidence to the contrary, the observed enstatite-chondritic composition of the terrestrial planets permits the deduction that these planets formed by raining out from the central regions of hot, gaseous protoplanets (Herndon, 2004d). With the possible exception of Mercury, the outer veneer of the terrestrial planets may contain other components derived from carbonaceous-chondrite-like matter and from ordinary-chondrite-like matter.

5. Evidence of Earth Being Like an Enstatite Chondrite

Imagine melting a chondrite in a gravitational field. At elevated temperatures, the iron metal and iron sulfide components will alloy together, forming a dense liquid that will settle beneath the silicates like steel on a steel-hearth. The Earth is like a spherical steel-hearth with a fluid iron-alloy core surrounded by a silicate mantle.

The Earth's core comprises about 32.5% by mass of the Earth as a whole. Only the enstatite chondrites, not the ordinary chondrites, have the sufficiently high proportion of iron-alloy that is observed for the core of the Earth, as shown in Figure 2. Moreover, other components of the interior of the Earth can be identified with corresponding components of an enstatite chondrite meteorite.

Oldham (1906) discovered the Earth's core by determining that beneath the crust the velocities of earthquake-waves increase with increasing depth, but only to a particular depth, below which their velocities abruptly and significantly become slower as they enter the core. When earthquake waves enter and leave the core, they change speed and direction. Consequently, there is a region at the surface, called the shadow zone, where earthquake-waves should be undetectable. But in the early 1930s, earthquake-waves were in fact detected in the shadow zone. Lehmann (1936) discovered the inner core by showing that a small solid object, within the fluid core, could cause earthquake waves to be reflected into the shadow zone.

Four years after its discovery by Inge Lehmann, Birch (1940) pronounced the composition of the inner core to be partially crystallized nickel–iron metal. Birch envisioned the Earth to be like an ordinary chondrite meteorite, the most common type of meteorite observed to fall to Earth. In arriving at that vision, Birch considered neither the rare, oxygen-rich carbonaceous chondrites, which contain little or no iron metal, nor the rare oxygen-poor

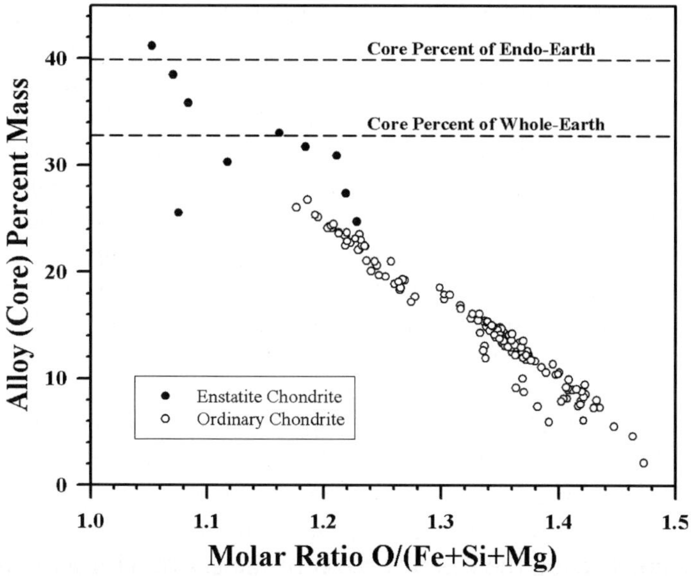

Figure 2. The percent mass of the alloy component of each of nine enstatite chondrites and 157 ordinary chondrites. This figure clearly shows that, if the Earth is chondritic in composition, the Earth as a whole, and especially the endo-Earth, is like an enstatite chondrite and *not* like an ordinary chondrite. The reason is clear from the abscissa which shows the molar ratio of oxygen to the three major elements with which it combines in enstatite chondrites and in ordinary chondrites. This figure also clearly shows that, if the Earth is chondritic in composition, the Earth as a whole, and especially the endo-Earth, has a state of oxidation like an enstatite chondrite and *not* like an ordinary chondrite. Data from Baedecker and Wasson (1975), Jarosewich (1990), Kallemeyn et al. (1989), Kallemeyn and Wasson (1981).

enstatite chondrites, which contain iron metal and also some strange minerals, such as oldhamite, CaS, that are not found in the surface regions of the Earth.

Birch thought that nickel and iron were always alloyed in meteorites and he knew that the total mass of all elements heavier than nickel was too little to comprise a mass as large as the inner core. Birch therefore suggested that the inner core was nickel–iron metal that had begun to crystallize from the melt.

Nearly four decades later, I realized that elemental silicon, discovered in the 1960s in the metal of enstatite chondrites (Ringwood, 1961) under appropriate conditions could cause nickel to precipitate as nickel silicide, an intermetallic compound of nickel and silicon, like the mineral perryite, which had been discovered in the 1960s in enstatite chondrites (Ramdohr, 1964). The abstract of my 1979 paper (Herndon, 1979) states in its totality: "From observations of nature the suggestion is made that the inner core of the Earth consists not of nickel–iron metal but of nickel silicide."

After an inspiring conversation with Inge Lehmann in 1979, I progressed through the following logical exercise: If the inner core is in fact nickel silicide, then the Earth's core must be like the alloy portion of an enstatite chondrite. If the Earth's core is in fact like the alloy portion of an enstatite chondrite, then the Earth's core should be surrounded by a silicate shell like the silicate portion of an enstatite chondrite. This silicate shell, if it exists, should be bounded by a seismic discontinuity, because the silicates of enstatite chondrites have a different and more highly reduced composition than rocks that appear to come from within the Earth's upper mantle (Jagoutz et al., 1979). Using the alloy to silicate ratio of the Abee enstatite chondrite and the mass of the Earth's core, by simple ratio proportion I calculated the mass of that silicate shell. From tabulated mass distributions (Dziewonski and Gilbert, 1972), I then found that the radius of that predicted seismic boundary lies within about 1.2% of the radius at the seismic discontinuity that separates the lower mantle from the upper mantle. This logical exercise led me to discover the fundamental quantitative mass ratio relationships connecting the interior parts of the Earth with parts of the Abee enstatite chondrite that are shown in Table II (Herndon, 1980).

Discovery of the Mohorovičić discontinuity separating the crust from the mantle as well as discovery of the Earth's core and inner core in the first half of the 20th Century resulted from pronounced differences in seismic observables, whereas initially the mantle appeared to be uniform. In the 1960s, improvements in seismic resolution began to indicate difficult-to-observe discontinuities within the mantle (Stacey, 1969). These were initially assumed to result from pressure-induced crystal structure changes, rather than compositional boundaries.

From terrestrial seismic data (Dziewonski and Gilbert, 1972; Dziewonski and Anderson, 1981) the gross features of the inner 82% of the Earth, the lower mantle and core, collectively called the endo-Earth, appear to be relatively simple, consistent with the identification of that part being like an enstatite chondrite (Herndon, 1980, 1982). The upper mantle, on the other

TABLE II

Fundamental mass ratio comparison between the endo-Earth (core plus lower mantle) and the Abee enstatite chondrite (Herndon, 1980)

Fundamental Earth ratio	Earth ratio value	Abee ratio value
Lower mantle mass to total core mass	1.49	1.43
		Theoretical
Inner core mass to total core mass	0.052	0.052 if Ni_3Si
		0.057 if Ni_2Si
Inner core mass to (lower mantle + core) mass	0.021	0.021

hand, displays several seismic discontinuities suggestive of different layers. The oxidized iron content (FeO) of primitive, ultramafic, upper-mantle-derived nodules (Jagoutz et al., 1979) would be out of equilibrium if in contact with the virtually FeO-free $MgSiO_3$ lower mantle, implying one or more layers of yet unknown but chemically different composition within the upper mantle. Such layering is consistent with the addition of carbonaceous-chondrite-like matter and/or ordinary-chondrite-matter during the latter stages of Earth formation (Wetherill, 1980). Indeed, just such a chondritic component is discernable in primitive ultramafic, upper-mantle-derived nodules (Jagoutz et al., 1979).

6. Overview of Solar System Formation

To understand more clearly the implications arising from protoplanetary Earth formation, it is helpful to envision the overall environment as indicated by chondrite chemical evidence and observational data. Although there is an evolutionary pre-history to the origin of the Solar System, involving among other things element nucleosynthesis, that pre-history is not considered here.

There seems to be little doubt that the oxidized, hydrous carbonaceous chondrites, like Orgueil, originate in the outer reaches of our Solar System, regions sufficiently cold to permit the retention of water in the vacuum of space for billions of years. The oxidation state of Orgueil-like carbonaceous chondrites is just what one would expect for solar-matter low pressure condensation at low temperatures.

The highly reduced matter of the inner regions of the Solar System, on the other hand, appears to have originated quite differently. In the main, the terrestrial planets are like the highly reduced enstatite chondrite meteorites. Thermodynamic considerations are consistent with the concept of Eucken (1944) that the terrestrial planets, like the Earth, rained out from the central regions of hot, gaseous protoplanets.

From solar abundances (Anders and Grevesse, 1989), the mass of protoplanetary-Earth was 275–305m_E, not very different from the mass of Jupiter, 318m_E. The formation of early-phase close-in gas giants in our own planetary system, is certainly consistent with observations and implications of near-to-star giant gaseous planets in other planetary systems (Fischer and Valenti, 2005; Santos et al., 2003; Udry et al., 2003), so it is no longer necessary to assume planet migration to explain those observations.

Solar primordial gases and volatile elements were separated from the terrestrial planets soon after planet formation, presumably early during some solar super-luminous event, such as the T-Tauri phase mass-ejections, presumably associated with the thermonuclear ignition of the Sun (Joy, 1945; Herbig, 1962; Lada, 1985; Lehmann et al., 1995). Indeed, there is some reason to think that Mercury was only partially formed at the time of superluminosity.

As I discussed (Herndon, 2004b), the relationship shown in Figure 1 admits the possibility of ordinary chondrites having been derived from mixtures of two components, representative of the other two types of matter, mixtures of a relatively undifferentiated carbonaceous-chondrite-like primitive component and a partially differentiated enstatite-chondrite-like component. All ordinary chondrites are depleted relative to solar matter in siderophile refractory elements, such as iridium and osmium. Siderophile refractory element depletion in individual ordinary chondrites, as I have shown, is proportional to their relative respective proportion of the partially differentiated enstatite-chondrite-like component, indicating a single reservoir source of partially differentiated enstatite chondrite-like matter (Herndon, 2004b).

The high bulk density of planet Mercury indicates that much of the silicate matter for the upper portion of Mercury's mantle was lost at some previous time (Urey, 1951, 1952; Bullen, 1952). I have suggested that some matter from the protoplanet of Mercury, Mercury's complement of lost elements, became that partially differentiated enstatite-chondrite-like planetary component of the ordinary chondrites, presumably separated during the time of Mercury's core formation through dynamic instability and/or expulsion during the Sun's initially violent ignition and approach toward thermonuclear equilibrium. I have suggested that the Mercurian component was then re-evaporated together with a more oxidized component of primitive matter and ended up mainly in the asteroid belt, the presumed source-region for the ordinary chondrites (Chapman, 1996). Such a picture would seem to explain for the ordinary chondrites, their major element compositions, their intermediate states of oxidation, and their ubiquitous deficiencies of refractory siderophile elements, and would explain as well a major, primary source of matter in the asteroid belt.

The approximately seven-fold greater depletion of refractory siderophile elements, within the ordinary chondrites' partially differentiated enstatite chondrite-like planetary component, than other, more volatile, siderophile elements such as nickel, cobalt, and gold, indicates that planetary-scale differentiation, at least in this one instance, progressed in a heterogeneous manner (Herndon, 2004a, b, e).

Although the terrestrial planets appear to have rained out from the central regions of hot, gaseous protoplanets, evidence suggests some outer, minor, secondary accretion of oxidized matter in the grain-growth accumulation way envisioned by, for example, Wetherill (1980). Such secondary accumulation may consist in the main of carbonaceous chondrite-like matter, ordinary chondrite-like matter, and their derivatives, for example, iron meteorites and achondrites. I have estimated that the total mass of ordinary chondrite matter originally present in the Solar System amounts to only 1.83×10^{24} kg (Herndon, 2004e). That amount of mass is insufficient to form a planet as massive as the Earth, but may have contributed significantly to the formation of Mars, as well as adding to the veneer of other planets, including the Earth. Presently, only about 0.1% of that mass remains in the asteroid belt.

7. Implications of Protoplanetary Earth Formation

The principal consequences of Earth's origin from within a giant gaseous protoplanet are profound and affect virtually all areas of geophysics in major, fundamental ways. Principal implications result (i) from Earth having been compressed by about 300 Earth-masses of primordial gases, and (ii) from the deep-interior having a highly reduced state of oxidation. The former provides Earth's main geodynamic driving-energy and leads to a new vision of global dynamics, which I call whole-Earth decompression dynamics (Herndon, 2005b, c) and which, among other things, leads to a new geophysical concept related to heat emplacement at the base of the crust. The latter results in great quantities of uranium and thorium existing within the Earth's core, and leads to the feasibility of the georeactor, a hypothesized natural, nuclear fission reactor at the center of the Earth as the energy source for the geomagnetic field.

8. Evidence of Earth as a Jupiter-Like-Gas-Giant

Planets generally consist of more-or-less uniform, closed, concentric shells of matter, layered according to density. The crust of the Earth, however, is an exception. Approximately 29% of the surface area of the Earth is composed of the portions of continents that presently lie above mean sea level; an additional 12% of the surface area of the Earth is composed of the continental margins, which are submerged to depths of no more than 2 km (Mc Lennan, 1991). The continental crust is less dense and different in composition than the remaining surface area, which is composed of ocean-floor basalt.

To date there has been no adequate geophysical explanation to account for the formation of the non-contiguous, crustal continental rock layer, except the idea put forth by Hilgenberg (1933) that in the distant past for an unknown reason the Earth had a smaller diameter and, consequently, had a smaller surface area. From modern surface area measurements, I calculated that the smaller radius required would be about 64% of its current radius, which would yield a mean density for the Earth of 21 g/cm^3. The reason for Earth's smaller radius, I submit, is that the Earth rained out from within a giant-gaseous protoplanet and originally formed as the rock-plus-alloy kernel of a giant gaseous planet like Jupiter (Herndon, 2004c, 2005c).

The mass of protoplanetary-Earth, calculated from solar abundance data (Anders and Grevesse, 1989) by adding to the condensable-planetary elements their proportionate amount of solar elements that are typically gases (e.g., H, He) or that form volatile compounds (e.g., O, C, N), lies in the range of about 275–305 times the mass of the present-day Earth. That mass is quite similar to Jupiter's mass, $318m_E$.

Pressures at the gas-rock boundary within the interior of Jupiter are estimated to be in the range from 43 Mbar to 60 Mbar (Podolak and Cameron, 1974; Stevenson and Salpeter, 1976). Using a theoretical Thomas–Fermi–Dirac approach (Salpeter and Zapolsky, 1967), I calculated density at Jupiter-model, gas-rock-boundary pressures for matter having the approximate composition of the Earth as a whole. The calculations are based upon eight chemical elements that account about 98% of the Earth's mass, assume volume additivity, and ignore phase separations and transitions. The results of the calculations, presented in Table III, show that a Jovian-like gas envelope is sufficient to compress the protoplanetary alloy-plus-rock core that became the Earth to a mean density of 21 g/cm^3.

The density value of 21 g/cm^3, estimated to result from compression by the great mass of giant-planet gases, is identical to that expected for a smaller Earth with a contiguous, closed, crustal continental shell. That identity, I submit, stands as evidence of the Earth having been a giant, gaseous planet like Jupiter (Herndon, 2004c, d).

9. Whole-Earth Decompression Dynamics

Early in the 20th Century, Wegener (1912) proposed that the continents at one time had been united, but subsequently had separated and drifted through the ocean floor to their present positions. After being ignored for half a century, Wegener's idea of continental drift re-emerged, cast into a new

TABLE III

Published model pressure and density estimates (Podolak and Cameron, 1974; Stevenson and Salpeter, 1976) at the gas-rock boundary of Jupiter, shown for comparison with theoretical calculation of compressed Earth density at the same pressures

Jupiter Model pressure (Mbar)	Jupiter Model density (g/cm^3)	Compressed Earth density (g/cm^3)
43	18	20
46	18	21
60	20	23

form called plate tectonics theory, with more detail and with new supporting observational data.

In plate tectonics, oceanic basalt, observed erupting from the mid-oceanic ridges, is thought to creep slowly across the ocean basin and to subduct, to plunge into the Earth, typically into submarine trenches. This theory appears to explain many geologic features observed at the Earth's surface, such as magnetic striations on the ocean floor, but necessitates solid-state mantle convection (Runcorn, 1965; Davies, 1977; Peltier, 1989), for which there is no unambiguous evidence despite decades of investigations.

Hilgenberg (1933) published a fundamentally different idea about the continents. He imagined that the Earth, for an unknown reason, was initially smaller in diameter, without oceans, and that the continents formed a uniform shell of matter covering the entire surface of the planet. Hilgenberg's idea, that the Earth subsequently expanded, fragmenting the uniform shell of matter into the continents and creating ocean basins in between, is the basis for Earth expansion theory (Carey, 1976, 1988; Scalera, 1990; Scalera and Jacob, 2003).

The principal impediments to the idea of Earth expansion have been (i) the lack of knowledge of a mechanism that could provide the necessary energy (Cook and Eardley, 1961; Beck, 1969) without departing from the known physical laws of nature (Jordan, 1971) and (ii) the ocean floors are less than 200 million years old which would seem to imply very recent expansion. In 1982, Scheidegger stated concisely the prevailing view, "Thus, if expansion on the postulated scale occurred at all, a completely unknown energy source must be found" (Scheidegger, 1982). Recently, I disclosed just such an energy source that follows from fundamental considerations (Herndon, 2004c, 2005b, c), the energy of protoplanetary compression, and set forth a different geodynamic theory, called whole-Earth decompression dynamics, which unifies seemingly disparate elements of plate tectonics theory and Earth expansion theory into one self-consistent description of the dynamics of the Earth as a whole.

After being stripped of its great, Jupiter-like overburden of volatile protoplanetary constituents, presumably by the high temperatures and/or by the violent activity, such as T - Tauri phase solar wind (Joy, 1945; Herbig, 1962; Lada, 1985; Lehmann et al., 1995), associated with the thermonuclear ignition of the Sun, the Earth would inevitably begin to decompress, to rebound toward a new hydrostatic equilibrium. The initial whole-Earth decompression is expected to result in a global system of major *primary* cracks appearing in the rigid crust which persist and are identified as the global, mid-oceanic ridge system, just as explained by Earth expansion theory. But here the similarity with that theory ends. Whole-Earth decompression dynamics sets forth a different mechanism for whole-Earth dynamics which involves the formation of *secondary* decompression cracks and the in-filling of those cracks, a process which is not limited to the last 200 million years.

As the Earth subsequently decompresses and swells from within, the deep interior shells may be expected to adjust to changes in radius and curvature by plastic deformation. As the Earth decompresses, the area of the Earth's rigid surface increases by the formation of secondary decompression cracks often located near the continental margins and presently identified as submarine trenches. These secondary decompression cracks are subsequently in-filled with basalt, extruded from the mid-oceanic ridges, which traverses the ocean floor by gravitational creep, ultimately plunging into secondary decompression cracks, thus emulating subduction.

As viewed today from the Earth's surface, the consequences of whole-Earth decompression dynamics appear very similar to those of plate tectonics, but with some profound differences. In fact, most of the evidence usually presented in support of plate tectonics also supports whole-Earth decompression dynamics. Just as in plate tectonics, one sees seafloor being produced at the mid-oceanic ridge, slowly moving across the ocean basin and disappearing into the Earth. But unlike plate tectonics, the basalt rock is not being re-cycled continuously by convection; instead, it is simply in-filling secondary decompression cracks. From the surface it may be very difficult indeed to discriminate between plate tectonics and whole-Earth decompression dynamics.

Usually arrayed as supporting plate tectonics theory, observations of ocean-floor magnetic striations, transform faults, island arc formation, and the generation and distribution of earthquakes are, I submit, consequences of whole-Earth decompression dynamics. These have the same basis and understanding in whole-Earth decompression dynamics as in plate tectonics.

Moreover, mantle seismic tomography results can be interpreted as imaging in-filled decompression cracks (Bunge et al., 2003). Seismic differences that are used to arrive at such images are not necessarily a reflection of temperatures, as often assumed, but can arise from differences in densities and/or differences in compositions. Moreover, the images are static; motion is only inferred on the basis of anticipations.

But there are global, fundamental differences between whole-Earth decompression dynamics and plate tectonics, especially as pertains to the growth of ocean-floor, to the origin of oceanic trenches, to the fate of down-plunging slabs, to the displacement of continents, and to the emplacement of heat at the base of the crust.

10. Mantle Decompression Thermal-Tsunami

Previously in geophysics, only three heat transport processes have been considered: conduction, radiation, and convection or, more generally, buoyancy-driven mass transport. As a consequence of whole-Earth decompression dynamics, I add a fourth, called mantle decompression thermal-tsunami (Herndon, 2006).

As the Earth decompresses, heat must be supplied to replace the lost heat of protoplanetary compression. Otherwise, decompression would lower the temperature, which would impede the decompression process.

Heat generated within the core from actinide decay or fission or from radioactive decay within the mantle may enhance mantle decompression by replacing the lost heat of protoplanetary compression. The resulting decompression, beginning as low as at the bottom of the mantle, will tend to propagate throughout the mantle, like a tsunami, until it reaches the impediment posed by the base of the crust. There, crustal rigidity opposes continued decompression, pressure builds and compresses matter at the mantle-crust-interface, resulting in compression heating. Ultimately, pressure is released at the surface through volcanism and through secondary decompression crack formation and/or enlargement.

It has been long known through experience in deep mines and with boreholes that temperature increases with depth within the crust. For more than half a century geophysicists have made measurements of continental and oceanic heat flow with the aim of determining the Earth's heat loss (Table IV). Pollack et al. (1993) estimate a global heat loss of 44.2 TW (1 TW = 10^{12} W) based upon 24,774 observations at 20,201 sites.

Previously, numerous attempts have been made to match measured global heat loss with radionuclide heat production from various geophysical models involved with plate tectonics. Usually, models are made to yield the very result they model, but in this case there is a problem. Current models rely upon radiogenic heat for geodynamic processes, geomagnetic field generation, and for the Earth's heat loss. The problem is that radionuclides cannot even satisfy just the global heat loss requirements.

Previous estimates of global heat production invariably come from the more-or-less general assumption that the Earth's current heat loss consists of the steady heat production from long-lived radionuclides (^{235}U, ^{238}U, and ^{40}K).

TABLE IV

Continental and oceanic mean heat flow and global heat loss

Reference	Continental Heat Flow mWm^{-2}	Oceanic Heat Flow mWm^{-2}	Global Heat Flow mWm^{-2}	Global Heat Loss 10^{12}W
Williams et al. (1974)	61	93	84	42.7
Davies (1980)	55	95	80	41.0
Sclater et al. (1980)	57	99	82	42.0
Pollack et al. (1993)	65	101	87	44.2

From Pollack et al. (1993).

Estimates of present-day global radiogenic heat production, based upon chondritic abundances, typically range from 19 TW to 31 TW. These represent an upper limit through the tacit assumption of rapid heat transport irrespective of assumed radionuclide locations. The short-fall in heat production, relative to Earth's measured heat loss (Pollack et al., 1993), has led to speculation that the difference might be accounted for by residual heat from Earth's formation, ancient radiogenic heat from a time of greater heat production, or, perhaps, from a yet unidentified heat source (Kellogg et al., 1999).

One of the consequences of Earth formation as a giant, gaseous, Jupiter-like planet (Herndon, 2004d), as described by whole-Earth decompression dynamics (Herndon, 2004c, 2005b, c), is the existence of a vast reservoir of energy, the stored energy of protoplanetary compression, available for driving geodynamic processes related to whole-Earth decompression. Some of that energy, I submit, is emplaced as heat at the mantle-crust-interface at the base of the crust through the process of mantle decompression thermal-tsunami. Moreover, some radionuclide heat may not necessarily contribute directly to crustal heating, but rather to replacing the lost heat of protoplanetary compression, which helps to facilitate mantle decompression.

11. Precipitation of the Structures of the Endo-Earth

One of the consequences of Earth formation by raining out from the central regions of a hot, gaseous protoplanet is the highly reduced state of oxidation of its interior (Eucken, 1944; Herndon and Suess, 1976). The Earth consists in the main of two distinct reservoirs of matter separated by the seismic discontinuity that occurs at a depth of about 680 km and which separates the mantle into upper and lower parts (Herndon, 1980). The endo-Earth, the inner 82% of the Earth's mass consists of the highly reduced lower mantle and core; the more oxidized exo-Earth is comprised of the components of the upper mantle and crust.

The matter comprising the endo-Earth precipitated from primordial gases under conditions that severely limited its oxygen content, relative to its other elements (Eucken, 1944; Herndon and Suess, 1976; Herndon, 2004d). The oxidation state of the condensate determines not only the relative mass of the core, but the elements the core contains, and the compounds which precipitate from the core and that give it its structure and its energy production capability. The oxidation state of the core cannot be subsequently changed, even by the pressures that prevail in that region.

The seismically deduced structure, divisions, and components of the endo-Earth are essentially identical to corresponding parts of the Abee enstatite chondrite meteorite, as shown by the mass ratio relationships presented in

Table II. The identity of the components of the Abee enstatite chondrite with corresponding components of the Earth (Herndon, 1980, 1993, 1998) means that with reasonable confidence one can understand the composition of the Earth's core by understanding the components of Abee meteorite or of one like it.

Envision highly reduced condensate, like that of the Abee enstatite chondrite and the endo-Earth, raining out from near the triple point of solar matter in the center of a hot giant-gaseous protoplanet (Eucken, 1944; Herndon and Suess, 1976; Herndon, 2004d). The magnesium, silicon, oxygen, and sulfur of enstatite-chondritic-like protoplanetary matter may have all begun their condensate origin dissolved in iron metal, along with minor and trace elements. Because of the extremely low oxygen fugacity in that medium at the high temperatures at which condensation is possible at high pressures, the amount of oxygen in the multi-element condensate would have been severely limited, even though oxygen is more abundant than the sum of all of the readily condensable elements of solar matter.

After raining out in the center of a hot gaseous protoplanet, elements of the condensate would be expected to compete on the basis of chemical activity and, during cooling, would begin to precipitate from the liquid condensate forming the interior parts of the planet. The dominant factors governing subsequent precipitation are oxyphilicity (affinity for oxygen) and incompatibility.

Elements have different chemical affinities for oxygen, which are related to their different oxidation potentials. Generally, oxyphile elements of the initial multi-element protoplanetary condensate will compete for available oxygen and will separate from the iron-alloy like slag separates from steel on a steel-hearth.

In ordinary-chondrite matter, there is more than enough oxygen available for all oxyphile elements (including uranium and thorium) with some left over to combine with iron. Consequently, if the Earth as a whole really were like an ordinary chondrite meteorite, there would be no uranium and thorium in the core and the core would be too small (Figure 2). But that is not the case.

Highly reduced matter, like that of the Abee enstatite chondrite and the endo-Earth, was separated from primordial solar gases under conditions that severely limited the oxygen content (Eucken, 1944; Herndon and Suess, 1976; Herndon, 2004d). For the protoplanetary Earth, elements of the condensate with a high affinity for oxygen (oxyphile elements) would be expected to combine with the limited available oxygen to form, atop the iron-alloy core, a low-density silicate mantle of $MgSiO_3$ which, at lower mantle pressures, is stable in a perovskite crystal structure (Ito and Matsui, 1978; Chaplot et al., 1998; Chaplot and Choudhury, 2001).

As a consequence of its highly reduced state of oxidation, the protoplanetary condensate that became the endo-Earth had insufficient oxygen to accommodate all of its oxyphile elements. As a consequence, certain oxyphile elements, including Si, Mg, Ca, U, and Th, occur in part in the iron-based alloy portion of the Abee enstatite chondrite and in the Earth's core. Oxyphile elements are generally incompatible in an iron-alloy and upon cooling these ultimately tend to precipitate as non-oxides, mainly as sulfides, at the earliest thermodynamically feasible opportunity.

Based upon well-known metallurgical principles (Ribound and Olette, 1978; Inoue and Suito, 1994), the portion of calcium and magnesium, occurring in the core and being incompatible in an iron-based alloy, would be expected to combine with sulfur to form oldhamite, CaS, and niningerite, MgS, low-density, high-temperature precipitates, which would float to the outer surface of the fluid core. These CaS and MgS precipitates, as I have suggested (Herndon, 1993, 1996, 2005a), are responsible for the observed seismic "roughness" at the core-mantle boundary, called D".

Upon further cooling, it is expected that dissolved silicon (Si) in the fluid core will combine with nickel (Ni) and precipitate as nickel silicide, which will settle by gravity, forming the Earth's solid inner core (Herndon, 1979, 1980, 2005a). As shown in Table II, a fully crystallized nickel silicide inner core would have precisely the mass observed, thus providing strong supporting evidence.

12. Radionuclides of the Endo-Earth

For decades there has been much discussion as to the possible existence of ^{40}K in the Earth's core. Although there are some indications from enstatite meteorites of alloy-originated potassium, specifically in the mineral djerfisherite, $K_6(Cu, Fe, Ni)_{25}S_{26}Cl$, the relative proportion of non-oxide potassium appears to represent at most only a few percent of the potassium complement (Fuchs, 1966). In the Abee enstatite chondrite, most of the potassium occurs in the mineral plagioclase, $(Na, Ca)(Si, Al)_4O_8$, which would seem to suggest that most of the endo-Earth's ^{40}K occurs in the lower mantle, perhaps in the region near the boundary of the upper mantle. Additional investigations are needed to be any more precise regarding the distribution of ^{40}K.

Although there may be some intrinsic uncertainty as to amount of ^{40}K, if any, in the Earth's core, current data on the uranium distribution in enstatite chondrites clearly indicate the non-lithophile behavior of that element in EH/E4 enstatite chondrites, like the Abee meteorite, and, by inference, in the endo-Earth. Generally, uranium occurs within the mineral oldhamite, CaS, an indication that in the enstatite chondrite matter, uranium is a high-temperature precipitate. Chemical leaching experiments show that Abee-uranium behaves as a sulfide (Matsuda et al., 1972). The tentative assignment of uranium as the mono-sulfide, US, seems reasonable. As currently available instrumental capability for determining this information quite precisely exists, I have recommended the requisite investigations (Herndon, 1998).

Within the Earth's core, one would expect uranium to precipitate at a high temperature. Just as uranium, a trace element, was swept-up or co-precipitated with a more abundant high-temperature precipitate, oldhamite, CaS, in enstatite chondrites, one might expect to some extent the possibility of a similar fate within the Earth's core. Ultimately, uranium, being the densest substance, would be expected to collect at the Earth's center. Unlike other trace elements such as thorium, uranium masses of at least ~1 kg occurring as nodules early in Earth's history would have been able to maintain sustained nuclear fission chain reactions that could generate sufficient heat to melt their way out of any mineral-occlusion impediment on their descent to the center of the Earth.

Russian scientists (Anisichkin et al., 2003; Rusov et al., 2004) have suggested the possibility of precipitated uranium accumulating in a layer atop the inner core and participating in a slow-burning nuclear-fission wave front reaction. To me, it seems that a uniform layer would be too thin, allowing too great a proportion of neutrons to escape for maintenance of criticality. But uniformity is only one possibility. In this remote and strange frontier, it is a good idea to keep an open mind on all of the possible georeactor variations.

Thorium, like uranium, occurs exclusively in the alloy portion of the Abee enstatite chondrite and by implication in the Earth's core. Also, thorium, like uranium, occurs in that meteorite within the mineral oldhamite, CaS (Murrell and Burnett, 1982), an indication of its being a high-temperature precipitate. Chemical leaching experiments indicate that Abee-thorium behaves in part as a sulfide, and in part as an unknown non-sulfide (Matsuda et al., 1972). Unlike uranium, accumulations of thorium would not have been able to sustain nuclear fission chain reactions.

Thus, it would appear that uranium and thorium may occur at the core-mantle boundary occluded in the core floaters, the low-density, high-temperature precipitate, oldhamite, CaS, atop the fluid core or, alternatively, they may be concentrated at the center of the Earth, depending upon respective precipitation and accumulation dynamics. Presently, there is no methodology by which to predict the relative proportion of these at the two boundaries of the core, its center and its surface. Because of the ability of ~1 kg nodules of uranium to undergo self-sustaining nuclear fission chain reactions, which can melt free of occlusion, one might expect uranium to occur primarily at the center of the Earth and thorium to occur at the core-mantle boundary within oldhamite.

13. Radionuclides of the Exo-Earth

It would be desirable to be able to specify the radionuclide distribution within the exo-Earth, the upper mantle and crust. But at present there is uncertainty in the compositions of the layers of the upper mantle and uncertainty as to the composition of the parent materials for that region of the Earth. Moreover, because of mantle decompression thermal-tsunami, measured heat loss from the crust can no longer be considered a justification for high-radionuclide content of the exo-Earth. As a "ball park" estimate, one might guess that the radionuclide complement of the exo-Earth represents an additional 18% of the endo-Earth complement, with much of the exo-Earth uranium and thorium residing in the crust. Ultimately, it should be possible to refine these estimates by tedious efforts to discover fundamental quantitative relationships that lead logically to that information.

14. Georeactor Nuclear Fission

Nuclear fission produces energy, consumes uranium, and produces neutron-rich fission products which subsequently β decay, yielding antineutrinos. Detection of georeactor-produced antineutrinos is one way to validate the existence of the georeactor (Raghavan, 2002; de Meijer et al., 2004; Domogatski et al., 2004; Fiorentini et al., 2004).

Using Fermi's nuclear reactor theory (Fermi, 1947), in 1993, I demonstrated the feasibility of a planetocentric nuclear fission reactor as the energy source for the geomagnetic field (Herndon, 1993). Initially, I could only postulate that the georeactor would operate as a fast neutron breeder reactor over the lifetime of the Earth (Herndon 1994, 1996). Subsequent state-of-the-art numerical simulations, made at Oak Ridge National Laboratory, verified that the georeactor could indeed function over the lifetime of the Earth as a fast neutron breeder reactor and, significantly, would produce helium in the same range of isotopic compositions observed in oceanic basalts (Hollenbach and Herndon, 2001; Herndon, 2003).

Georeactor numerical simulation calculations are made using the SAS2 analysis sequence contained in the SCALE Code Package from Oak Ridge National Laboratory (SCALE, 1995) that has been developed over a period of three decades and has been extensively validated against isotopic analyses of commercial reactor fuels (England et al., 1984; Hermann and DeHart, 1998). The SAS2 sequence invokes the ORIGEN-S isotopic generation and depletion code to calculate concentrations of actinides, fission products, and activation products simultaneously generated through fission, neutron absorption, and radioactive decay. The SAS2 sequence performs the 1-D transport analyses at selected time intervals, calculating an energy flux spectrum, updating the time-dependent weighted cross-sections for the depletion analysis, and calculating the neutron multiplication of the system.

From nuclear reactor theory (Fermi, 1947), the defining condition for self-sustaining nuclear fission chain reactions is that $k_{\mathrm{eff}} = 1.0$. The value of k_{eff} represents the number of fission neutrons in the current population divided by the number of fission neutrons in the previous population. If $k_{\mathrm{eff}} > 1.0$, the neutron population and the energy output are increasing and will continue until changes in the fuel, moderators, and neutron absorbers cause k_{eff} to decrease to 1.0. If $k_{\mathrm{eff}} < 1.0$, the neutron population and energy output are decreasing and will eventually decrease to zero. If $k_{\mathrm{eff}} = 1.0$, the neutron population and energy output are constant.

Natural uranium consists mainly of the readily fissionable ^{235}U and the essentially non-fissionable ^{238}U. In a natural reactor, the value of k_{eff} is strongly dependent upon the ratio $^{235}\mathrm{U}/^{238}\mathrm{U}$. The reason that thick seams of natural uranium ore are presently unable to undergo self-sustaining nuclear fission chain reactions, i.e., $k_{\mathrm{eff}} < 1.0$, is because the $^{235}\mathrm{U}/^{238}\mathrm{U}$ ratio is too small. The ^{238}U absorbs too high a proportion of neutrons. Because the half-life of ^{235}U is shorter than that of ^{238}U, the ratio of $^{235}\mathrm{U}/^{238}\mathrm{U}$ was higher in the geological past, making possible the condition for natural fission, $k_{\mathrm{eff}} \geq 1.0$.

Main georeactor characteristic operational parameters and uncertainties are illustrated in Figure 3, showing k_{eff} as a function of time for several numerical simulations made at constant fission powers. These show the importance of breeding, fission-product removal, and intrinsic self-regulation.

In Figure 3, the curve labeled "VLP-FPR" shows the necessity for breeding. In this example, the very-low-power fission produced only insignificant amounts of fissionable actinides. Consequently, the k_{eff} was determined almost entirely by the natural decay of uranium, and, by the end of about 2 gigayears of operation, self-sustained nuclear fission chain reactions become impossible.

The "VLP-FPR" and the curve labeled "FPR" were calculated with instantaneous removal of fission products. But the "FPR" curve was calculated at a much higher power level where breeding kept $k_{\mathrm{eff}} > 1.0$. As noted by Herndon (1994) and Seifritz (2003), the principal fuel-breeding takes place by the reaction $^{238}\mathrm{U}(\mathrm{n},\gamma)^{239}\mathrm{U}(\beta^-)^{239}\mathrm{Np}(\beta^-)^{239}\mathrm{Pu}(\alpha)^{235}\mathrm{U}$. Too low an operating power will lead to insufficient breeding, whereas at power levels too high, the uranium fuel would be entirely consumed too early in the lifetime of the Earth.

For the georeactor to be able to operate into the present, fission products must be removed naturally. That necessity is shown quite clearly in Figure 3 by the curve labeled "FPNR," calculated with fission products not removed.

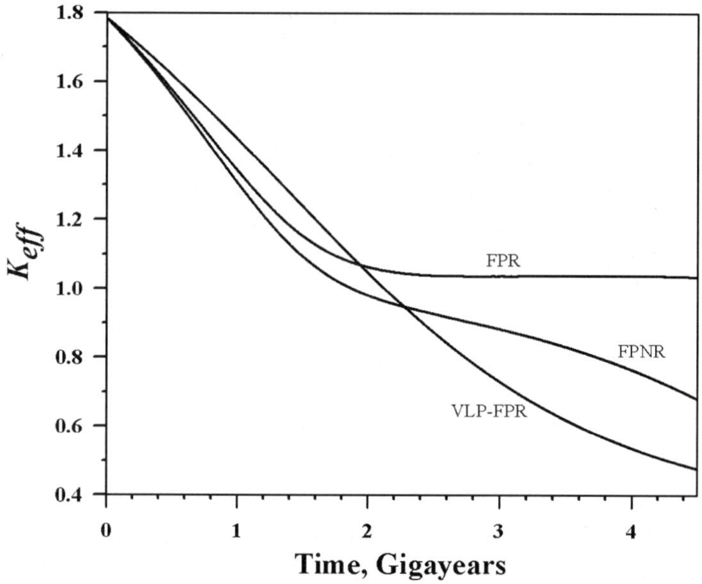

Figure 3. Numerical simulation results, chosen to illustrate main georeactor operational parameters and uncertainties, are presented in terms of k_{eff} over the lifetime of the Earth. The curve labeled "VLP-FPR" is very low power for the case of fission products instantaneously removed. "FPR" is a 3 TW run also for the case of fission products instantaneously removed. "FPNR" is a 3 TW run with fission products not removed.

After operation of about 1.5 gigayears, $k_{\text{eff}} < 1.0$ and self-sustained nuclear fission chain reactions become impossible. As I have discussed (Herndon, 1993, 1994), there is a natural mechanism for georeactor fission product removal: At the center of the Earth, density is a function almost entirely of atomic number and atomic mass. The fission process splits the actinide nucleus into two pieces, each being considerably less dense than its parent. At the high sub-core temperatures, even in the microgravity environment, these would tend to separate on the basis of density. This process may operate as one self-regulation mechanism.

Another, yet unknown, self-regulation mechanism appears evident from the curve labeled "FPNR" in Figure 3. Note that, at the time of Earth formation, the value of k_{eff} is quite high; the uranium mix is "hot." In the numerical simulation, fission power generation was specified and controlled. In nature, without a self-regulation mechanism operating, at this high a value of k_{eff}, the georeactor would have run wild and might have burned out its uranium fuel long before life had existed on Earth. Early on, before about 1.5 gigayears of operation, fission product accumulation alone would not have been an effective self-control mechanism. Some other mechanism must have operated.

15. Radionuclide Abundance and Distribution

Much is yet unknown concerning the distribution of radionuclides within the Earth. Because of the identity between the parts of the endo-Earth and corresponding parts of the Abee enstatite chondrite, it is possible to make direct inferences as to radionuclide states of oxidation and locations within the endo-Earth, although not to the degree of precision that might ultimately be possible given adequate petrologic data with modern instrumentation and appropriate laboratory experiments (Herndon, 1998, 2005a). It is likewise possible to make some rough estimates of current georeactor energy production and uranium consumption, but past georeactor operation is for the most part unknown.

Within those limitations, the following generalizations concerning the endo-Earth radionuclides can be made: (i) Most of the ^{40}K may be expected to exist in combination with oxygen in the silicates of the lower mantle, perhaps being confined to transition-region between the upper and the lower mantle; (ii) Uranium may be expected to exist at the center of the Earth where it may undergo self-sustaining nuclear fission chain reactions, but there is a possibility some non-fissioning uranium may be found scattered diffusely within the CaS core floaters; and, (iii) Thorium may be expected to occur within the core floaters at the core-mantle boundary, although its presence as well at the center of the Earth cannot be ruled out. Thorium is unable to be georeactor fuel or to be converted into fuel for the georeactor (Herndon and Edgerley, 2005).

Radionuclide abundance estimates for the endo-Earth and guesses for the exo-Earth are shown in Table V. Their respective locations are represented schematically in Figure 4. In demonstrating the feasibility of the georeactor, I used very conservative uranium estimates, amounting to approximately 20% of the estimated total possible initial endo-Earth uranium content (Herndon, 1993; Hollenbach and Herndon, 2001). The results shown in Table V, are based upon results of numerical simulations assuming that the entire amount of uranium is available for nuclear fission (Herndon and Edgerley, 2005). These, therefore, provide some boundary conditions on the maximum present-time radionuclide abundances.

In a series of numerical simulations run at successively higher power levels, Edgerley and I found that, with the same maximum initial endo-Earth uranium content, the georeactor could operate at a constant power level of as much as 30 TW and still be operating (Herndon and Edgerley, 2005). The question of power level, especially in times past, is the greatest unknown. Measurements of geo-antineutrinos pose the possibility of revealing the current distribution of radioactive nuclides and fission products.

TABLE V

Estimates of the maximum present-day radionuclide content within the endo-Earth and guessed amounts in the exo-Earth

Nuclide	Kilograms	
Endo-Earth (estimate)		
^{40}K	5.001×10^{17}	
^{232}Th	1.322×10^{17}	
^{235}U	3.065×10^{14}	(2.504×10^{14})
^{238}U	3.373×10^{15}	(3.456×10^{16})
Exo-Earth (guess)		
^{40}K	1.100×10^{17}	
^{232}Th	2.908×10^{16}	
^{235}U	4.629×10^{15}	
^{238}U	1.528×10^{16}	

Endo-Earth values of uranium in parentheses, given for reference only, assume no fission activity. Data from Baedecker and Wasson (1975), Murrell and Burnett (1982).

16. Georeactor Variability

Seated deep within the Earth, the geomagnetic field varies in intensity and reverses polarity frequently, but quite irregularly, with an average time between reversals of about 200,000 years. Previously envisioned deep-Earth energy sources, including natural radioactivity, change only gradually and in only one direction over time. Variations, in the geomagnetic field, therefore, have previously only been ascribed to some mechanical instability in its production mechanism. I have suggested that the variable and intermittent changes in the intensity and direction of the geomagnetic field may have their origin in nuclear reactor variability (Herndon, 1993). Generally, variability in nuclear fission reactors arises from changes in composition and/or position of fuel, moderators, and neutron absorbers. Although as yet there is no irrefutable evidence of planetocentric nuclear reactor variability, circumstantial evidence certainly invites inquiry.

Upon considering observations of Jupiter's internally generated energy, I demonstrated the feasibility of planetocentric nuclear fission reactors as energy sources for the giant planets (Herndon, 1992) in part using the same type of calculations employed by Kuroda (1956) to predict conditions for the natural reactors that were later discovered at Oklo, Republic of Gabon (Bodu et al., 1972; Fréjacques et al., 1975; Hagemann et al., 1975). The near-surface natural reactors at Oklo, which were critical about 1.8 gigayears ago, operated intermittently (Maurette, 1976). Recent investigations suggest quite rapid cycling periods with 0.5 h of operation followed by 2.5 h of dormancy

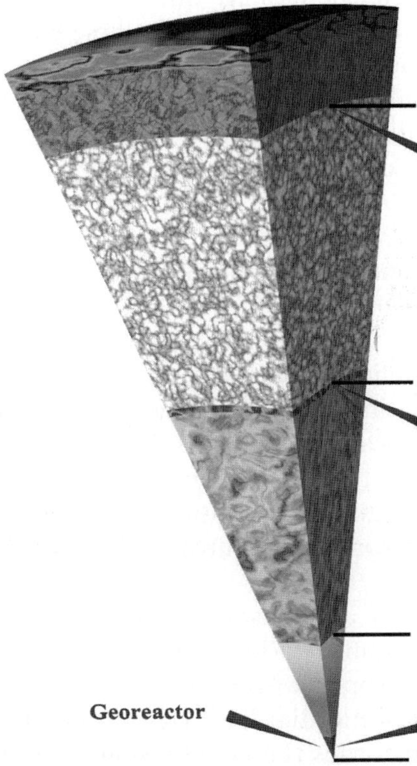

Figure 4. Schematic representation of the interior of the Earth showing regions in the endo-Earth where radionuclides may be expect to be concentrated.

(Meshik et al., 2004). While the specific control mechanism, presumably involving water, may not be directly applicable to the planetocentric reactors, the observations nevertheless demonstrate the potential variability of natural nuclear reactors.

Atmospheric turbulence in the giant planets appears to be driven by their internal energy sources. Jupiter, Saturn, and Neptune produce prodigious amounts of energy and display prominent turbulent atmospheric features. Uranus, on the other hand, radiates little, if any, internally generated energy and appears featureless. In the summer of 1878, Jupiter's Great Red Spot increased to a prominence never before recorded and, late in 1882, its prominence, darkness, and general visibility began declining so steadily that by 1890 astronomers thought that the Great Red Spot was doomed to extinction. Changes have been observed in other Jovian features, including the formation of a new lateral belt of atmospheric turbulence (Peek, 1958).

Jupiter, 98% of which consists of a mixture of H and He, an excellent heat transfer medium, is capable of rapid thermal transport. It is important to establish whether these atmospheric changes are due to changes in planetocentric nuclear reactor output as it seems, especially as these would represent short-period variability (Herndon, 1994). Ultimately, one may hope to understand the nature and possible variability of georeactor energy production by making fundamental discoveries and by discovering fundamental quantitative relationships in nature.

17. Deep-Earth Helium Evidence of the Georeactor

Clarke et al. (1969) discovered that ^{3}He and ^{4}He are venting from the Earth's interior. The ^{3}He/^{4}He ratio of helium released to the oceans at the mid-oceanic ridges is about eight times greater than in the atmosphere ($R/R_A = 8 \pm 1$, where R is the measured value of ^{3}He/^{4}He and R_A is the same ratio measured in air $= 1.4 \times 10^{-6}$), and, therefore, cannot be ascribed to atmospheric contamination. High helium ratios, e.g., ~37 R_A (Hilton et al., 1999), have been observed from deep-source plumes, such as Iceland and Hawaii.

Previously, lacking knowledge of a deep-source ^{3}He production mechanism, deep-Earth ^{3}He has been assumed to be of primordial origin (Clarke et al., 1969; Hilton et al., 1999), trapped within the mantle at the time that the Earth formed. But the ratio of primordial ^{3}He/^{4}He is thought to be ~10^{-4}, a value inferred from gas-rich meteorites (Pepin and Singer, 1965), which is about one order of magnitude greater than helium released from the mantle. In ascribing a primordial origin to the observed deep-Earth ^{3}He/^{4}He, the assumption implicitly made is that the primordial component is diluted by a factor of about 10 with ^{4}He produced by the natural radioactive decay of uranium and thorium in the mantle and/or in the crust. The alternative suggestion (Anderson, 1993), that the ^{3}He/^{4}He arises instead from cosmic dust, subducted into the mantle, necessitates assuming that the influx of interplanetary dust particles was considerably greater in ancient times than at present and also assuming a ten-fold dilution by ^{4}He.

Helium isotope fission products from georeactor numerical simulations made at the Oak Ridge National Laboratory are shown in Figure 5. The data shown are values of the ^{3}He/^{4}He ratio, relative to the same ratio in air (R_A), at each 2×10^6 year time step for each power level of the numerical simulations. For comparison, the range of values of the same ratio, measured in oceanic basalts, is shown in Table VI at a 2σ confidence level. The entire range of ^{3}He/^{4}He values from oceanic basalts, shown in Table VI, is produced by self-sustaining nuclear fission chain reactions, as demonstrated by the georeactor numerical simulations results presented in Figure 5.

I have suggested that the observed deep-source helium is georeactor-produced and is in fact strong evidence for the georeactor's existence

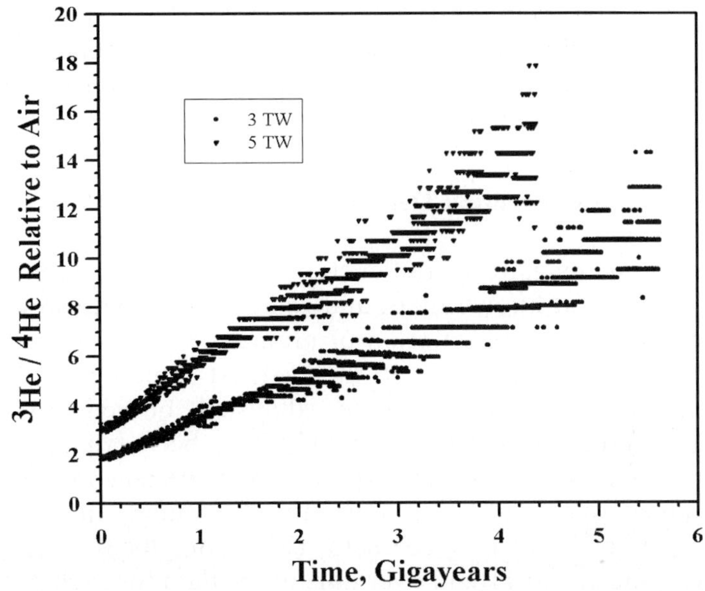

Figure 5. Nuclear georeactor numerical simulation results for 3 TW and 5 TW power levels showing the $^3He/^4He$ ratios relative to air (R_A) produced during 2×10^6 year increments over the lifetime of the georeactor. Each data point represents the ratio of the ^3He and ^4He fission yields for a single time step. The pronounced upward trend of the data results from the continuing reduction of ^{238}U, the principle source of ^4He, by fission and by breeding. From Herndon (2003).

TABLE VI

Statistics of $^3He/^4He$ relative to air (R_A) of basalts from along the global spreading ridge system at a two standard deviation (2σ) confidence level

Propagating Lithospheric Tears	11.75 ± 5.13 R_A
Manus Basin	10.67 ± 3.36 R_A
New Rifts	10.01 ± 4.67 R_A
Continental Rifts or Narrow Oceans	9.93 ± 5.18 R_A
South Atlantic Seamounts	9.77 ± 1.40 R_A
MORB	8.58 ± 1.81 R_A
EM Islands	7.89 ± 3.63 R_A
North Chile Rise	7.78 ± 0.24 R_A
Ridge Abandoned Islands	7.10 ± 2.44 R_A
South Chile Rise	6.88 ± 1.72 R_A
Central Atlantic Islands	6.65 ± 1.28 R_A
HIMU Islands	6.38 ± 0.94 R_A
Abandoned Ridges	6.08 ± 1.80 R_A

Anderson (2000).

(Hollenbach and Herndon, 2001; Herndon, 2003). Rao (2002) has provided background information and described the georeactor as being the solution to the riddles of relative abundances of helium isotopes and geomagnetic field variability.

18. Eventual Demise of the Georeactor

Energy production by natural radioactive decay is predictable over time, decreasing gradually at known rates, and will continue to do so well into the future. By contrast, the consumption of uranium by georeactor-nuclear-fission may not have been constant in the past. At some point, the uranium supply of the georeactor may become exhausted, burned out by nuclear fission, possibly much sooner than it would have been exhausted by radioactive decay alone. The high $^3He/^4He$ values in certain measurements of so-called plumes, specifically Icelandic and Hawaiian, may indicate the approach of the demise of the georeactor (Herndon, 2003).

In Figure 5, the upward trend over time of the data for each power level is principally the consequence of the diminishment by natural decay and by fuel breeding of ^{238}U, the principle source of ^4He. For a particular power level, the highest values represent the most recent production, especially near the end of the nuclear fission lifetime of the georeactor.

The limitation on the upper limits for $^3He/^4He$ depends upon the georeactor being critical, *i.e.*, $k_{eff} \geq 1.0$, as its actinide fuel approaches depletion. The main factors affecting that circumstance are the amount and nature of the initial actinide sub-core and the operating history of the georeactor. One may reasonably expect, therefore, that the high values for $^3He/^4He$, shown in Figure 5, may not be true upper limits. It seems reasonable, though, that the high helium isotope ratios, measured in Hawaiian and in Icelandic basalt (Hilton et al., 1999), may signal the approach of the end of georeactor lifetime, although one may presently only speculate as to the time-frame involved.

One shortcoming of oceanic basalt helium isotopic measurements is that the time of formation of the helium is unknown. But from Figure 5, one can see that helium time-of-formation is important for assessing the time of demise of the georeactor. Future precision measurements of geo-antineutrinos may help to address that shortcoming.

19. Grand Overview and Generalizations

Only three processes, operant during the formation of the Solar System, are responsible for the diversity of compositions observed in planets, asteroids, and comets and are directly responsible for planetary internal-structures and dynamical processes, including and especially, geodynamics. These processes are: (i) Low-pressure, low-temperature condensation from solar matter in the remote reaches of the Solar System or in the interstellar medium; (ii) High-pressure, high-temperature condensation from solar matter associated with planetary-formation by raining out from the interiors of giant-gaseous protoplanets, and; (iii) Stripping of the primordial volatile components from the inner portion of the Solar System by super-intense solar wind associated with T-Tauri phase mass-ejections, presumably during the thermonuclear ignition of the Sun.

Low-pressure, low-temperature condensation from solar matter in the remote reaches of the Solar System or in the interstellar medium is the process responsible for cometary matter, and is responsible for one of the two components from which ordinary chondrite meteorites are composed. It is responsible for the primitive Orgueil-like carbonaceous chondrite meteorites and, after separation from primordial volatile components and being melted and/or re-evaporated and re-condensed, it is responsible for the more crystallized and evolved carbonaceous chondrites, such as the Allende meteorite. This type of matter contributes to the terrestrial planets only as a late-addition veneer component.

High-pressure, high-temperature condensation from solar matter, associated with planetary-formation by raining out from the interiors of giant-gaseous protoplanets, is the process responsible for the bulk of planetary formation and for establishing the highly reduced state of oxidation of planetary interiors. Internal planetary structures are produced as a consequence of the highly reduced state of planetary interiors, including the occurrence of major quantities of uranium in planetary cores, leading to planetocentric nuclear fission reactors. That same condensation process is responsible for Earth formation as a giant gaseous Jupiter-like planet and for storing vast amounts of the energy of protoplanetary compression in the rock-plus-alloy kernel that became Earth as we know it.

Stripping of the primordial volatile components from the inner portion of the Solar System by super-intense solar wind associated with T-Tauri phase mass-ejections, presumably during the thermonuclear ignition of the Sun, is the process responsible for removal of gaseous components associated with the formation of terrestrial planets, including removal of part of the protoplanet of Mercury, which became the other of the two components from which ordinary-chondrite matter formed in the region of the asteroid belt. It is the process responsible removing approximately 300 Earth-masses of primordial volatile gases from the Earth, which began Earth's decompression process, making available vast amounts of energy for driving geodynamic processes which I have described by the new whole-Earth decompression dynamics, and which is responsible for emplacing heat at the mantle-crust-interface at the base of the crust through the process I have described, called mantle decompression thermal-tsunami.

The three processes, operant during the formation of the Solar System, lead logically, in a causally related manner, to a coherent vision of planetary formation with profound implications. Consequently, there is reason to suppose that each planet and, perhaps, each of the larger moons, has at its center, a region of highly reduced enstatite-chondrite-like matter and a uranium sub-core at one time capable of self-sustained nuclear fission reactions. The vision of planetary formation presented here is consistent with observations of near-to-star gas-giants in other planetary systems. The geodynamic processes for the terrestrial planets may differ from one another to some extent, not so much due to their interiors, but as a consequence of the circumstances of their accumulation and removal of primordial volatile components.

These are exciting times in the natural physical sciences. Along with the new understanding of Solar System formation and whole-Earth geodynamics described above, new experimental advances are being made that, I submit, will inevitably confirm and perhaps extend these concepts. Already, astronomers are beginning to image remote planetary systems and finding close-to-star gas giants like Earth at a very early stage. Neutrino physicists, with decades of experience measuring neutrinos from the Sun and from outer space, are beginning to detect anti-neutrinos from within our own planet. To image the interior of the Earth using anti-neutrinos, physicists face great challenges in attempting to attain the high resolution and directionality needed. But facing great challenges and making important discoveries is what science is all about.

References

Anders, E. and Grevesse, N.: 1989, *Geochim. Cosmochim. Acta* **53**, 197–214.

Anderson, D. L.: 1993, *Science* **261**, 170–176.

Anderson, D. L.: 2000, *Geophys. Res. Lett.* **27**, 2401–2404.

Anisichkin, V. F., Ershov, A. P. and Bezborodov, A. A., et al.: 2003, ed. *Proc. Int. Conf. VII Zababakkin's Scientific' Lectures* Snezhinsk, Russia, p. 117.

Aumann, H. H., Gillespie, C. M. J. and Low, F. J.: 1969, *Astrophys. J.* **157**, L69.

Baedecker, P. A. and Wasson, J. T.: 1975, *Geochim. Cosmochim. Acta* **39**, 735–765.

Bainbridge, J.: 1962, *Astrophys. J.* **136**, 202–210.

Beck, A. E.: 1969, in S. K. Runcorn (ed.), *Energy changes in an expanding Earth*, Wiley, London, pp. 77–83.

Birch, F.: 1940, *Am. J. Sci.* **238**, 192–211.

Bodu, R., Bouzigues, H., Morin, N. and Pfiffelmann, J. P.: 1972, *C. R. Acad. Sci., Paris* **D275**, 1731–1736.

Bullen, K. E.: 1952, *Nature* **170**, 363–364.

Bunge, H.-P., Hagelberg, C. R. and Travis, B. J.: 2003, *Geophys. J. Int.* **152**, 280–301.

Cameron, A. G. W.: 1963, *Icarus* **1**, 339–342.

Carey, S. W.: 1976, *The Expanding Earth*, Elsevier, Amsterdam.

Carey, S. W.: 1988, *Theories of the Earth and Universe – A History of Dogma in the Earth Sciences*, Stanford University Press, Stanford.

Chaplot, S. L. and Choudhury, N.: 2001, *Am. Min.* **86**, 752–761.

Chaplot, S. L., Choudhury, N. and Rao, K. R.: 1998, *Am. Min.* **83**, 937–941.

Chapman, C. R.: 1996, *Meteor. Planet. Sci.* **31**, 699–726.

Clarke, W. B., Beg, M. A. and Craig, H.: 1969, *Earth Planet. Sci. Lett.* **6**, 213–220.

Clayton, R. N.: 1993, *Annu. Rev. Earth Planet. Sci.* **21**, 115–149.

Conrath, B. J., Pearl, J. C., Appleby, J. F., Lindal, J. F., Orton, G. S. and Bezard, B.: 1991, in J. T. Bergstralh, E. D. Miner and M. S. Mathews (eds.), *Uranus*, University of Arizona Press, Tucson.

Cook, M. A. and Eardley, A. J.: 1961, *J. Geophys. Res.* **66**, 3907–3912.

Davies, G.: 1980, *Rev. Geophys.* **18**, 718–722.

Davies, G. F.: 1977, *Geophys. J. R. Astron. Soc.* **49**, 459–486.

de Meijer, R. J., van der Graaf, E. R. and Jungmann, K. P.: 2004, *Nucl. Phys. News Int.* **14**, 20–25.

Domogatski, G., Kopeikin, L., Mikaelyan, L. and Sinev, V.: 2004, arXiv:hep-ph/0401221 v1 28 Jan 2004.

Dziewonski, A. M. and Anderson, D. A.: 1981, *Phys. Earth Planet. Inter.* **25**, 297–356.

Dziewonski, A. M. and Gilbert, F.: 1972, *Geophys. J. R. Astr. Soc.* **72**, 393–446.

Endress, M. and Bischoff, A.: 1993, *Meteoritics* **28**, 345–345.

England, T. R., Wilson, R. E., Schenter, R. E. and Mann, F. M.: 1984, *Summary of ENDF/B-V Data for Fission Products and Actinides, EPRI NP-3787 (LA-UR 83–1285) (ENDF-322), Electric Power Research Institute.*

Eucken, A.: 1944, *Nachr. Akad. Wiss. Goettingen, Math.-Kl.* **1**, 1–25.

Fermi, E.: 1947, *Science* **105**, 27–32.

Fiorentini, G., Lissia, M., Mantovani, F. and Vannucci, R.: 2004, arXiv.org/hep-ph/0409152 14 Sept 2004.

Fischer, D. A. and Valenti, J.: 2005, *Ap. J.* **622**, 1102.

Flügge, F.: 1939, *Die Naturwissenschaften* **27**, 402.

Fréjacques, C., Blain, C., Devilers, C., Hagemann, R. and Ruffenbach, J.-C. 1975, in *The Oklo Phenomenon*, I.A.E.A., Vienna, p. 509.

Fuchs, L. H.: 1966, *Science* **153**, 1711–1719.

Hagemann, R., Devillers, C., Lucas, M., Lecomte, T. and Rufferbach, J.-C. 1975, in *The Oklo Phenomenon*, I.A.E.A., Vienna, p. 415.

Hahn, O. and Strassmann, F.: 1939, *Die Naturwissenschaften* **27**, 11–15.

Herbig, G. H.: 1962, *Adv. Astron. Astrophys.* **1**, 47.

Hermann, O. W. and DeHart, M. D.: 1998, Validation of SCALE (SAS2H) Isotopic Predictions for BWR Spent Fuel, ORNL/TM-13315, Lockheed Martin Energy Research Corp., Oak Ridge National Laboratory.

Herndon, J. M.: 1978, *Proc. R. Soc. Lond.* **A363**, 283–288.

Herndon, J. M.: 1979, *Proc. R. Soc. Lond.* **A368**, 495–500.

Herndon, J. M.: 1980, *Proc. R. Soc. Lond.* **A372**, 149–154.

Herndon, J. M.: 1982, *Naturwissenschaften* **69**, 34–37.

Herndon, J. M.: 1992, *Naturwissenschaften* **79**, 7–14.

Herndon, J. M.: 1993, *J. Geomag. Geoelectr.* **45**, 423–437.

Herndon, J. M.: 1994, *Proc. R. Soc. Lond.* **A455**, 453–461.

Herndon, J. M.: 1996, *Proc. Natl. Acad. Sci. USA* **93**, 646–648.

Herndon, J. M.: 1998, *Phys. Earth Plan. Inter.* **105**, 1–4.

Herndon, J. M.: 2003, *Proc. Natl. Acad. Sci. USA* **100**, 3047–3050.

Herndon, J. M.: 2004a, arXiv:astro-ph/0410009 1 Oct 2004.

Herndon, J. M.: 2004b, arXiv:astro-ph/0405298 15 May 2004.
Herndon, J. M.: 2004c, arXiv:astro-ph/0408539 30 Aug 2004.
Herndon, J. M.: 2004d, arXiv:astro-ph/0408151 9 Aug 2004.
Herndon, J. M.: 2004e, arXiv: astro-ph 0410242 8 Oct 2004.
Herndon, J. M.: 2005a, *Curr. Sci.* **88**, 1034–1037.
Herndon, J. M.: 2005b, arXiv:physics/0510090 30 Sept. 2005.
Herndon, J. M.: 2005c, *Curr. Sci.* **89**, 1937–1941.
Herndon, J. M.: 2006, *Curr. Sci.* **90**, 1605–1606.
Herndon, J. M. and Edgerley, D. A.: 2005, arXiv:hep-ph/0501216 24 Jan 2005.
Herndon, J. M. and Rudee, M. L.: 1978, *Earth. Planet. Sci. Lett.* **41**, 101–106.
Herndon, J. M. and Suess, H. E.: 1976, *Geochim. Cosmochim. Acta* **40**, 395–399.
Herndon, J. M. and Suess, H. E.: 1977, *Geochim. Cosmochim. Acta* **41**, 233–236.
Hilgenberg, O. C.: 1933, *Vom wachsenden Erdball*, Giessmann and Bartsch, Berlin.
Hilton, D. R., Grönvold, K., Macpherson, C. G. and Castillo, P. R.: 1999, *Earth Planet. Sci. Lett.* **173**, 53–60.
Hollenbach, D. F. and Herndon, J. M.: 2001, *Proc. Natl. Acad. Sci. USA* **98**, 11085–11090.
Hua, X. and Buseck, P. R.: 1998, *Meteor. Planet. Sci.* **33**, A215–A220.
Hubbard, W. B. 1990, in J. K. B. A. A. Chaikin (ed.), *Interiors of the Giant Planets*, Sky Publishing Corp., Cambridge, MA, pp. 134–135.
Hyman, M., Rowe, M. W. and Herndon, J. M.: 1978, *Geochem. J.* **13**, 37–39.
Inoue, R. and Suito, H.: 1994, *Steel Res.* **65**, 403–409.
Ito, E. and Matsui, Y.: 1978, *Earth Planet. Sci. Lett.* **38**, 443–450.
Jagoutz, E., Palme, H., Baddenhausen, H., Blum, K., Cendales, M., Dreibus, G., Spettel, B., Lorenz, V. and Wanke, H.: 1979, *Proc. Lunar Planet. Sci. Conf.* **10**, 2031–2050.
Jarosewich, E.: 1990, *Meteoritics* **25**, 323–337.
Jedwab, J.: 1971, *Icarus* **15**, 319.
Jordan, P.: 1971, *The Expanding Earth: Some Consequences of Dirac*'s Gravitation Hypothesis, in International Series of Monographs in Natural Philosophy, Pergamon, Oxford, p. 202.
Joy, A. H.: 1945, Astrophys. J. **102**, 168–195 (plus 164 unnumbered pages of plates).
Kallemeyn, G. W., Rubin, A. E., Wang, D. and Wasson, J. T.: 1989, *Geochim. Cosmochim. Acta* **53**, 2747–2767.
Kallemeyn, G. W. and Wasson, J. T.: 1981, *Geochim. Cosmochim. Acta* **45**, 1217–1230.
Kellogg, L. H., Hager, B. H. and van der Hilst, R. D.: 1999, *Science* **283**, 1881–1884.
Kuiper, G. P.: 1951a, *Proc. Natl. Acad. Sci. USA* **37**, 383–393.
Kuiper, G. P.: 1951b, *Proc. Natl. Acad. Sci. USA* **37**, 1–14.
Kuroda, P. K.: 1956, *J. Chem. Phys.* **25**, 781–782.
Lada, C. T.: 1985, *Ann. Rev. Astron. Astrophys.* **23**, 267–317.
Larimer, J. W.: 1968, *Geochim. Cosmochim. Acta* **32**, 965–982.
Larimer, J. W. and Anders, E.: 1970, *Geochim. Cosmochim. Acta* **34**, 367–387.
Lederer, S. M. and Vilas, F.: 2003, *Earth, Moon Planets* **92**, 193–199.
Lehmann, I.: 1936, *Publ. Int. Geod. Geophys. Union, Assoc. Seismol., Ser. A, Trav. Sci.* **14**, 87–115.
Lehmann, T., Reipurth, B. and Brander, W.: 1995, *Astron. Astrophys.* **300**, L9–L12.
Mason, B.: 1962, *Am. Museum Novitates* **2085**, 1–20.
Matsuda, H., Shima, M. and Honda, M.: 1972, *Geochem. J.* **6**, 37–42.
Maurette, M.: 1976, *A. Rev. Nucl. Sci.* **26**, 319–350.
Mc Lennan, S. M.: 1991, in *Encyclopedia of Earth System Science*, Academic Press, Inc., San Diego, p. 581.
Meshik, A. P., Hohenberg, C. M. and Pravdivtseva, O. V.: 2004, *Phys. Rev. Lett.* **93**, 182301–182304.

Murrell, M. T. and Burnett, D. S.: 1982, *Geochim. Cosmochim. Acta* **46**, 2453–2460.
Oldham, R. D.: 1906, *Q. T. Geol. Soc. Lond.* **62**, 456–476.
Peek, B. M.: 1958, *The Planet Jupiter*, Faber and Faber, London.
Peltier, W. B.: 1989, *Mantle Convection*, Gordon and Breach, New York.
Pepin, R. O. and Singer, P.: 1965, *Science* **149**, 253–265.
Podolak, M. and Cameron, A. G. W.: 1974, *Icarus* **22**, 123–148.
Pollack, H. N., Hurter, S. J. and Johnson, J. R.: 1993, *Rev. Geophys.* **31**, 267–280.
Raghavan, R. S.: 2002, arXiv:hep-ex/0208038 Aug 2002.
Ramdohr, P.: 1964, *Abh. D. Akad. Wiss. Ber., Kl. Chem., Geol., Biol.* **5**, 1–20.
Rao, K. R.: 2002, *Curr. Sci.* **82**, 126–127.
Reid, A. M., Bass, M. N., Fyjita, H., Kerridge, J. F. and Fredriksson, K.: 1970, *Geochim. Cosmochim. Acta* **34**, 1253–1254.
Ribound, P. and Olette, M.: 1978, in *Physical Chemistry and Steelmaking*, Versailles, France.
Ringwood, A. E.: 1961, *Geochim. Cosmochim. Acta* **25**, 1–13.
Rudee, M. L. and Herndon, J. M.: 1981, *Meteoritics* **16**, 139–140.
Runcorn, S. K.: 1965, *Phil. Trans. Roy. Soc. Lond.* **A258**, 228–251.
Rusov, V. D., Pavlovich, V. N., Vaschenko, V. N., et al.: 2004, arXiv.org/hep-ph/0402039 4 Feb 2004.
Salpeter, E. E. and Zapolsky, H. S.: 1967, *Phys. Rev.* **158**, 876–886.
Santos, N. C., Israelian, G., Mayor, M., Rebolo, R. and Udry, S.: 2003, *Astron. Astrophys.* **426**, L19.
SCALE.: 1995, SCALE: A Modular Code System for Performing Standardized Analyses for Licensing Evaluations, NUREG/CR-0200, Rev. 4, (ORNL/NUREG/CSD-2/R4), Vols. I, II, and III, April 1995. Available from Radiation Safety Information Computational Center at Oak Ridge National Laboratory as CCC-545.
Scalera, G.: 1990, *Phys. Earth Plan. Inter.* **62**, 126–140.
Scalera, G. and Jacob, K.-H.: 2003, Why Expanding Earth, INGV Publisher, Rome, p 465.
Scheidegger, A. E.: 1982, *Principles of Geodynamics*, Springer-Verlag, Heidelberg.
Sclater, J. C., Jaupart, C. and Galson, D.: 1980, *Rev. Geophys.* **18**, 269–311.
Seifritz, W.: 2003, *Kerntechnik* **68**, 193–196.
Stacey, F. D.: 1969, *Physics of the Earth*, John Wiley & Sons, New York.
Stevenson, D. J.: 1982, *Planet. Space Sci.* **30**, 755–764.
Stevenson, D. J. and Salpeter, E. E.: 1976, in T. Gehrels (ed.), *Jupiter*, University of Arizona Press, Tucson, p. 85.
Stevenson, D. J.: 1978, in *The Origin of the Solar System*, S. F. Dermott (ed.), Wiley, New York, pp. 395–431.
Tomeoka, K. and Buseck, P. R.: 1988, *Geochim. Cosmochim. Acta* **52**, 1627–1640.
Udry, S., Mayor, M. and Santos, N. C.: 2003, *Astron. Astrophys.* **407**, 369.
Urey, H. C.: 1951, *Geochim. Cosmochim. Acta* **1**, 36–82.
Urey, H. C.: 1952, *The Planets*, Yale University Press, New Haven.
Urey, H. C. and Craig, H.: 1953, *Geochim. Cosmochim. Acta* **4**, 36–82.
Vilas, F.: 1985, *Icarus* **64**, 133–138.
Wegener, A. L.: 1912, *Geol. Rundschau* **3**, 276–292.
Wetherill, G. W.: 1980, *Ann. Rev. Astron. Astrophys.* **18**, 77–113.
Wiik, H. B.: 1969, *Commentationes Physico-Mathematicae* **34**, 135–145.
Williams, D. L., von Herzen, R. P., Sclater, J. C. and Anderson, R. N.: 1974, *Geophys. J. R. Astron. Soc.* **38**, 587–608.
Zellner, B., Leake, M., Morrison, D. and Williams, J. G.: 1977, *Geochim. Cosmochim. Acta* **41**, 1759–1767.

Earth, Moon, and Planets (2006) 99:91–110
DOI 10.1007/s11038-006-9115-5

Geo-Neutrinos: from Theory to the KamLAND Results

GIOVANNI FIORENTINI
Dipartimento di Fisica, Università di Ferrara, I-44100, Ferrara, Italy
Istituto Nazionale di Fisica Nucleare, Sezione di Ferrara, I-44100, Ferrara, Italy

MARCELLO LISSIA
Istituto Nazionale di Fisica Nucleare, Sezione di Cagliari, I-09042, Monserrato (CA), Italy
Dipartimento di Fisica, Università di Cagliari, I-09042, Monserrato (CA), Italy

FABIO MANTOVANI
Dipartimento di Scienze della Terra, Università di Siena, I-53100, Siena, Italy
Centro di GeoTecnologie CGT, I-52027, San Giovanni Valdarno, Italy
Istituto Nazionale di Fisica Nucleare, Sezione di Ferrara, I-44100, Ferrara, Italy
(E-mail: fabio.mantovani@unisi.it)

BARBARA RICCI
Dipartimento di Fisica, Università di Ferrara, I-44100, Ferrara, Italy
Istituto Nazionale di Fisica Nucleare, Sezione di Ferrara, I-44100, Ferrara, Italy

(Received 1 March 2006; Accepted 19 July 2006)

Abstract. Earth shines in antineutrinos produced from long-lived radioactive elements: detection of this signal can provide a direct test of the Bulk Silicate Earth (BSE) model and fix the radiogenic contribution to the terrestrial heat flow. In this paper we present a systematic approach to geo-neutrino production based on global mass balance, supplemented by a detailed geochemical and geophysical study of the region near the detector, in order to build theoretical constraints on the expected signal. We show that the prediction is weakly dependent on mantle modeling while it requires a good description of the crust composition in the region of the detector site. In 2005 the KamLAND experiment proved that the technique for exploiting geo-neutrinos in the investigation of the Earth's interior is now available. After performing an analysis of KamLAND data which includes recent high precision measurements of the $^{13}C(\alpha, n)^{16}O$ cross section, we discuss the potential of future experiments for assessing the amount of uranium and thorium in different reservoirs (crust, mantle and core) of the Earth.

1. Introduction

The KamLAND collaboration has recently published (Araki et al., 2005) new experimental results, claiming some 28 geo-neutrino events from uranium and thorium decay chains in a two-year exposure. This important step shows that the technique for exploiting geo-neutrinos in the investigation of the Earth's interior is now available. In order to understand where to go with geo-neutrinos, we have to know where we stand in the light of the available data.

The most interesting feature of geo-neutrinos is that they bring to Earth's surface information coming from the entire planet concerning the amount of long-lived radioactive elements. Their detection, allowing for a quantitative determination of global elemental abundances, can provide a direct observational test of a classical geochemical paradigm, the Bulk Silicate Earth (BSE) model. Furthermore, geo-neutrinos can reveal the radiogenic contribution to terrestrial heat flow, providing thus an important contribution to the understanding of Earth's energetics.

In this review, mainly based on the results of our group (Fiorentini et al., 2005b; Mantovani et al., 2004; Fiorentini et al., 2005a; Fiorentini et al., 2003b; Fiorentini et al., 2003d), we shall address the following questions:

- What do we know about U, Th and ^{40}K in the Earth?
- What are the predictions and their uncertainties of a *reference* model for geo-neutrino production, i.e. of a model based on the current geochemical and geo-physical information?
- What is the contribution of the region near the detector? A close look at the nearby region is relevant in order to subtract from the geo-neutrino signal the local contribution, with the aim of determining the global component.
- How do we relate the geo-neutrino signal with the total mass of long-lived radioactive elements in the Earth?
- What are the implications of the KamLAND result?

Finally, we discuss the potential of future experiments for assessing the amounts of U and Th in different reservoirs (crust, mantle and core) of the Earth.

2. U, Th and K in the Earth: How much and where?

Earth's global composition is generally estimated from that of chondritic meteorites by using geochemical arguments which account for losses and fractionation during planet formation. Along these lines the Bulk Silicate Earth (BSE) model is built, which describes the "primitive mantle", i.e., the outer portion of the Earth after core separation and before the differentiation between crust and mantle. The model is believed to describe the present crust plus mantle system. Since lithophile elements should be absent in the core[1],

[1] One needs to be careful, since the definition of an element's behaviour, i.e., lithophile or not, depends on the surrounding system; there exist models of the Earth's core suggesting it is a repository for radioactive elements.

TABLE I
U, Th and K according to BSE, from (Fiorentini et al., 2003b)

	m [10^{17} kg]	H_R [10^{12} W]	L_ν [10^{24} s^{-1}]
U	0.8	7.6	5.9
Th	3.1	8.5	5.0
^{40}K	0.8	3.3	21.6

the BSE provides the total amounts of U, Th and K in the Earth, estimates from different authors being concordant within 10–15% (McDonough, 2003). From the estimated masses, the present radiogenic heat production rate H_R and anti-neutrino luminosity L_ν can be immediately calculated, see Table I and, e.g., (Fiorentini et al., 2005a).

The BSE is a fundamental geochemical paradigm. It is consistent with most observations, which regard mostly the crust and an undetermined portion of the mantle. The measurement of quantities – such as the geo-neutrino signals – which are directly related to the global amounts of radioactive elements in the Earth will provide a direct test of this model of the composition and origin of our planet.

Indeed, heat released from radiogenic elements is a major source of the terrestrial heat flow, however its role is not understood at a quantitative level. The masses estimated within the BSE account for a present radiogenic production of 19 TW, which is about one half of the estimated heat flow from Earth (McDonough, 2003, Hofmeister and Criss, 2005). Anderson refers (Anderson, 2005) to this difference as the missing heat source mystery and summarizes the situation with the following words: "Global heat flow estimates range from 30 to 44 TW ... Estimates of the radiogenic contribution ... based on cosmochemical considerations, vary from 19 to 31 TW. Thus, there is either a good balance between current input and output ... or there is a serious missing heat source problem, up to a deficit of 25 TW ..." If one can determine the amounts of radioactive elements by means of geo-neutrinos, an important ingredient of Earth's energetics will be fixed.

Concerning the distribution of radiogenic elements, estimates for uranium in the continental crust based on observational data are in the range:

$$m_C(U) = (0.3 - 0.4) \times 10^{17}\,\text{kg} \tag{1}$$

The extreme values have been obtained in (Fiorentini et al., 2003b) by taking the lowest (highest) concentration reported in the literature for each layer of the Earth's crust, see Table II of (Mantovani et al., 2004), and integrating over a 2° × 2° crust map. The main uncertainty comes from the uranium mass abundance a_{LC} in the lower crust, with estimates in the range (0.2–1.1) ppm. Estimates for the abundance in the upper crust, a_{UC}, are more concordant,

TABLE II
Average uranium abundance in the continental crust, from (Fiorentini et al., 2003b)

Reference	$\langle a_{CC} \rangle$ [ppm]
Taylor and McLennan, 1985	0.91
Weaver and Tarney, 1984	1.3
Rudnick and Fountain (1995)	1.42
Wedepohl (1995)	1.7
Shaw et al. (1986)	1.8
This work, minimal	1.3
This work, reference	1.54
This work, maximal	1.8

ranging from 2.2 to 2.8 ppm. The crust – really a tiny envelope – should thus contain about one half of the BSE prediction of uranium in the Earth.

About the mantle, observational data are scarce and restricted to the uppermost part, so the best estimate for its uranium content m_M is obtained by subtracting the crust contribution from the BSE estimate:

$$m_M = m_{BSE} - m_C \tag{2}$$

A commonly held view is that there is a vertical gradient in the abundances of incompatible elements in the mantle, with the top being most depleted. A minimum gradient model has a fully mixed and globally homogeneous mantle; the other extreme is a model where all the uranium is at the bottom of the mantle.

Geochemical arguments are against the presence of radioactive elements in the completely unexplored core, as discussed by McDonough in a recent review of compositional models of the Earth (McDonough, 2003).

Similar considerations hold for thorium and potassium, the relative mass abundance with respect to uranium being globally estimated as:

$$a(\mathrm{Th}) : a(\mathrm{U}) : a(\mathrm{K}) \approx 4 : 1 : 10{,}000 \tag{3}$$

We remark that the well-fixed ratios[2] in Eq. (3) imply that detection of geo-neutrinos from uranium will also bring important information on the amount of thorium and potassium in the whole Earth.

[2] We shall always refer to element abundances in mass and we remind the reader that the natural isotopic composition is $^{238}\mathrm{U/U} = 0.993$, $^{232}\mathrm{Th/Th} = 1$ and $^{40}\mathrm{K/K} = 1.2 \times 10^{-4}$.

3. A reference model and its uncertainties

3.1. Uranium, thorium and potassium distribution

Our aim is to build a reference model (labeled as "ref."), which incorporates the best available knowledge of U, Th and K distributions inside Earth. Concerning Earth's crust, we distinguish oceans and seawater, the continental crust, subdivided into three sublayers (upper, middle, and lower), sediments and oceanic crust. These seven layers have been mapped in (Bassin et al., 2000), which provides values of density and depth over the globe on a grid with 2° steps. We distinguish then the upper mantle (extending down to about 600 km), the lower mantle (down to about 2900 km), and the core: we use the preliminary reference earth model (PREM) (Dziewonski and Anderson, 1981) for the values of the densities of the mantle, assuming spherical symmetry.

For each component, one has to adopt a value for the abundances a(U), a(Th), and a(K). In the literature of the last twenty years one can find many estimates of abundances for the various components of the crust (OC, upper CC, lower CC,...), generally without an error value (see Tables II–IV of Mantovani et al., 2004), two classical reviews being in Refs. (Taylor and McLennan, 1985; Wedepohl, 1995) and a most useful source being provided by the GERM Reservoir Database (GERM, 2003).

For the upper mantle we are aware of several estimates by Jochum et al. (1986), White (1993), O'Nions and McKenzie (1993), Hofmann (1988), and Zartman and Haines (1988). In this respect data obtained from material emerged from unknown depths are assumed to be representative of the average composition down to about 600 km.

For each (sub)layer of the crust and for the upper mantle, we adopt as reference value for the uranium abundance $a^{\mathrm{ref}}(\mathrm{U})$ the average of the values reported in Tables II, III, and IV of (Mantovani et al., 2004). Concerning Th and K, we observe that the abundance ratios with respect to uranium are much more consistent among different authors than the corresponding absolute abundances. We shall thus take the average of ratios and from these construct the reference abundances for thorium and potassium:

$$a^{\mathrm{ref}}(\mathrm{Th}) = \langle \mathrm{Th}/\mathrm{U} \rangle a^{\mathrm{ref}}(\mathrm{U}) \quad \text{and} \quad a^{\mathrm{ref}}(\mathrm{K}) = \langle \mathrm{K}/\mathrm{U} \rangle a^{\mathrm{ref}}(\mathrm{U}) \tag{4}$$

For the lower mantle, where no observational data are available, we resort to the BSE model, which – we recall – describes the present crust-plus-mantle system based on geochemical arguments.

The mass of each element (X = U, Th, K) in the lower mantle $m_{\mathrm{LM}}(\mathrm{X})$ is thus obtained by subtracting from the BSE estimate the mass calculated for the crust and upper mantle:

TABLE III

Total yields. N_{no} is the total number of geoevents (U + Th) in the absence of oscillations predicted from the reference model for 10^{32} proton yr (or in TNU) and $\Delta N_{\rm no}$ is the "1σ" error

Location	$N_{\rm no}$	$\Delta N_{\rm no}$	$N_{\rm no}^{\rm low}$	$N_{\rm no}^{\rm high}$
Baksan	91	13	51	131
Hawaii	22	6	10	49
Homestake	91	13	51	130
Kamioka	61	10	33	96
Gran Sasso	71	11	39	106
Pyhasalmi	92	13	51	131
Sudbury	87	13	48	125
Curacao	57	10	30	92

$N_{\rm no}^{\rm low}$ ($N_{\rm no}^{\rm high}$) is the minimal (maximal) prediction. For $\delta m^2 > 4 \times 10^{-5}{\rm eV}^2$ the geoevent yield is $N = N_{\rm no}\left[1 - 0.5\sin^2(2\theta)\right]$, from (Mantovani et al., 2004).

TABLE IV

Errors from the regional geophysical and geochemical uncertainties at Kamioka, from (Fiorentini et al., 2005b)

Source	ΔS[TNU]	Remarks
Composition of upper-crust samples	0.96	3σ error
Upper-crust discretization	1.68	
Lower-crust composition	0.82	Full range
Crustal depths	0.72	3σ error
Subducting slab	2.10	Full range
Japan Sea	0.31	Full range
Total	3.07	Full range

$$m_{\rm LM}({\rm X}) = m_{\rm BSE}({\rm X}) - m_{\rm CC}({\rm X}) - m_{\rm OC}({\rm X}) - m_{\rm UM}({\rm X}) \tag{5}$$

Reference abundances for the lower mantle are then obtained by dividing these values by its mass $m_{\rm LM} = 2.9 \times 10^{24}$ kg. According to geochemical arguments, negligible amounts of U, Th and K should be present in the core. The resulting choice of input values for the reference model is collected in Tables II–IV of (Mantovani et al., 2004).

3.2. The uncertainties of the reference model

Since the abundance ratios look relatively well determined, we concentrate on the uncertainties of the uranium abundances in the different layers and propagate them to the other elements. For the reference model, we have $m_{CC}(U) = 0.345 \times 10^{17}$ kg, $m_{OC}(U) = 0.005 \times 10^{17}$ kg, the total mass of CC being $m_{CC} = 2.234 \times 10^{22}$ kg. According to our model, the average uranium abundance in the CC is thus $a_{CC}(U) = 1.54 \times 10^{-6}$.

We determine a range of acceptable values of $a_{CC}(U)$ by taking the lowest (highest) concentration reported in the literature for each layer, see Table II of (Mantovani et al., 2004). The main source of uncertainty is from the abundance in the lower crust, estimated at 0.20 ppm in (Rudnick and Fountain, 1995) and at 1.1 ppm in (Shaw et al., 1986). Estimates for the abundance in the upper crust are more concordant, ranging from 2.2 ppm (Condie, 1993) to 2.8 ppm (Taylor and McLennan, 1985). We remark that, within this approach, the resulting average crustal U abundance $\langle a_{CC} \rangle$ is in the range 1.3–1.8 ppm, which encompasses all estimates reported in the literature (Rudnick and Fountain, 1995; Shaw et al., 1986; Wedepohl, 1995; Weaver and Tarney, 1984) except for that of (Taylor and McLennan, 1985), $\langle a_{CC} \rangle = 0.91$ ppm, see Table II.[3]

$$\text{low}: a_{CC}(U) = 1.3 \times 10^{-6}; a_{CC}(Th) = 5.2 \times 10^{-6}; a_{CC}(K) = 1.3 \times 10^{-2}$$

$$\text{high}: a_{CC}(U) = 1.8 \times 10^{-6}; a_{CC}(Th) = 7.6 \times 10^{-6}; a_{CC}(K) = 1.97 \times 10^{-2}$$

For the upper mantle, we take as extrema the two values known to us (Jochum et al., 1986; Zartman and Haines, 1988) for uranium and we deduce thorium and potassium by rescaling

$$\text{low}: a_{UM}(U) = 5 \times 10^{-9}; a_{UM}(Th) = 13 \times 10^{-9}; a_{UM}(K) = 6 \times 10^{-5}$$

$$\text{high}: a_{UM}(U) = 8 \times 10^{-9}; a_{UM}(Th) = 21 \times 10^{-9}; a_{UM}(K) = 9.6 \times 10^{-5}$$

Concerning the lower mantle, we fix the mass of radiogenic elements by requiring that the BSE constraint (3.2) is satisfied and we assume uniform abundance.

3.3. Predicted yields

The no oscillation yields, calculated with the fluxes of the reference model, are shown in Table XII of (Mantovani et al., 2004). In the same table we also present the estimated 1σ errors. The geo-neutrino signal is expressed in

[3] Note that this paper quotes ranges of mass and fluxes tighter than in (Mantovani et al., 2004), which used the value $\langle a_{CC} \rangle = 0.91$ ppm from (Taylor and McLennan, 1985) as lower limit.

Terrestrial Neutrino Units, one TNU corresponding to 10^{-32} geo-neutrino captures per target proton per year.

For the Kamioka site the prediction of the reference model is $N_{no} = 61$ TNU in good agreement with the "best model" of (Fiorentini et al., 2003a; Fiorentini et al., 2003c), $N_{no} = 67$ TNU, in between the values of (Rothschild et al., 1998), $N_{no} = 43$ TNU, and of model 1b of (Raghavan et al., 1998), $N_{no} = 75$ TNU.

4. A closer look for Kamioka

The geo-neutrino signal depends on the total uranium mass of radioactive elements in the Earth and on the geochemical and geophysical properties of the region around the detector (Fiorentini et al., 2003a). For KamLAND, we estimated (Mantovani et al., 2004) that about one half of the signal originates within 200 km from the detector. This region, although containing a globally negligible amount of uranium and thorium, produces a large contribution to the signal as a consequence of its proximity to the detector. This contribution has to be determined on the grounds of a detailed geochemical and geophysical study of the region, if one wants to extract from the total signal the remaining part which carries the relevant information. The study of the region around Kamioka, including the possible effects of the subducting plates below the Japan Arc and a discussion of the contribution from of the Japan Sea, is in (Fiorentini et al., 2005b).

Starting from the $2° \times 2°$ world crustal map, we isolated six "tiles", around Kamioka and we performed a detailed study of their uranium content, see Figure 1. The seismic velocity structure of the crust beneath the Japan Islands has been determined in (Zhao et al., 1992) from the study of some 13,000 arrival times of 562 local shallow earthquakes. By applying an inversion method, the depth distribution of the Conrad and Moho discontinuities beneath the whole of the Japan Islands are derived, with an estimated standard error of $\pm$ 1 km over most of Japan territory. This allows distinguishing two layers in the crust: an upper crust extending down to the Conrad and a lower part down to the Moho discontinuity.

The upper-crust chemical composition of Japan Islands has been studied in (Togashi et al., 2000), based on 166 representative specimens, which can be associated with 37 geological groups based on age, lithology and province. By combining the base geological map of Figure 2 of (Togashi et al., 2000) – which distinguishes 10 geological classes – with the abundances reported in Table I of the same paper, one can build a map of uranium abundance in the upper crust, under the important assumption that the composition of the whole upper crust is the same as that inferred in (Togashi et al., 2000) from the study of the exposed portion.

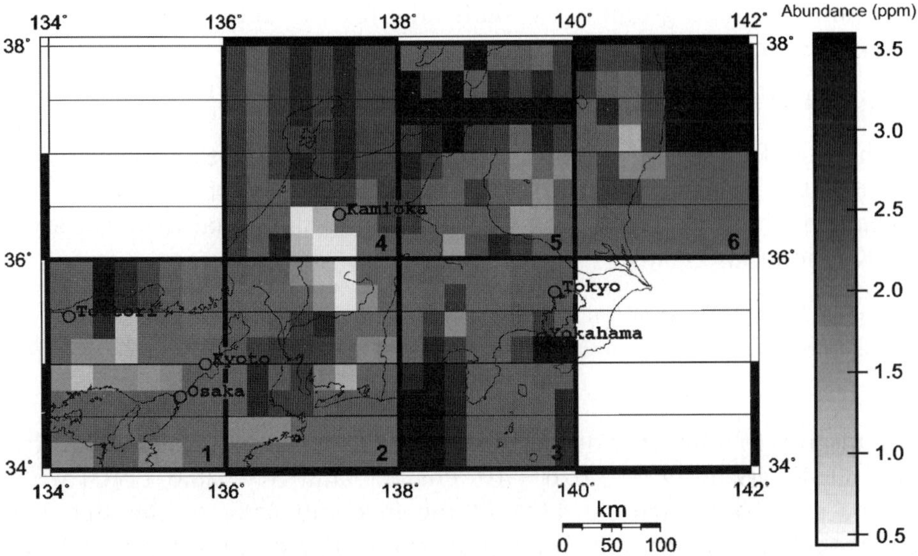

Figure 1. Uranium abundance in the upper crust of Japan (Fiorentini et al., 2005b).

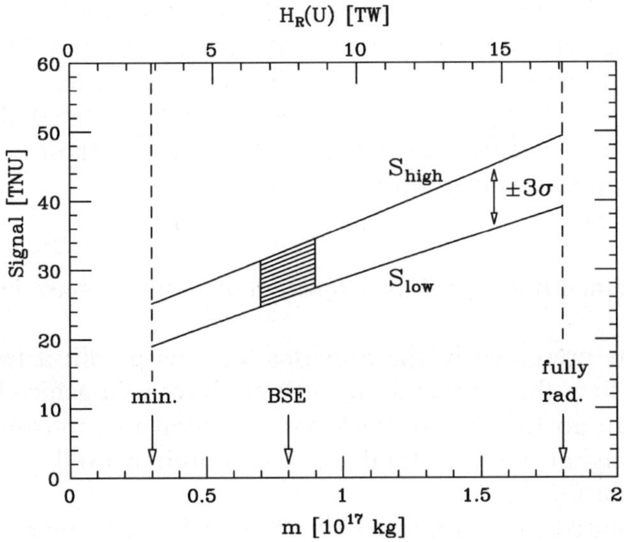

Figure 2. The predicted signal from uranium geo-neutrinos at KamLAND (Fiorentini et al., 2005b).

We are not aware of a specific study of the lower part of the Japan crust, however, it is well known that there are similarities between the composition of the Japanese crust and that of the Sino-Korean block. In an extensive compositional study of East China crust (Gao et al., 1998), the estimated uranium abundance in the lower part is between 0.63 and

1.08 ppm. On these grounds we shall take for the abundance in the lower crust of Japan:

$$a_{\mathrm{LC}} = 0.85 \pm 0.23 \quad \mathrm{ppm}$$

For the this discussion, we use the asymptotic value of the survival probability and the best fit value of the mixing angle, i.e. $\langle P_{ee} \rangle = 0.59$.[4]

The contributions to the produced flux and to the signal from the six tiles near Kamioka are:

$$\Phi_6 = 1.59 \times 10^6 \quad \mathrm{cm}^{-2}\ \mathrm{s}^{-1}$$

$$S_6 = 12.74 \quad \mathrm{TNU}$$

With respect to our previous estimate from the whole globe (Mantovani et al., 2004), giving $\Phi = 3.676 \times 10^6\ \mathrm{cm}^{-2}\mathrm{s}^{-1}$ and $S = 28.6$ TNU, we find that the six tiles contribute 43% of the flux and 45% of the signal: this justifies the close scrutiny of the region within the six tiles. Some 3/4 of the contribution arises from the upper crust.

In more detail, the tile hosting Kamioka generates 29% and 30% of the total produced flux and signal, respectively. The host cell, i.e., the cell where Kamioka is located, contributes 9% to the total produced flux.

The uranium mass contained in the six tiles is about $m_6 = 3.3 \times 10^{13}$ kg, really negligible (less then 0.05%) with respect to that estimated for the whole Earth. We have considered several sources of the uncertainties affecting this estimate of the local contribution, see Table IV.

5. The geo-neutrino signal as a function of uranium mass in the Earth

The arguments presented in the previous sections permit a test of the BSE model, which fixes the total amount of long-lived radiogenic elements in the Earth. One can go further, and ask for a general relationship between the geo-neutrino signal and the total mass of uranium (and other radiogenic elements) in the Earth.

The main ingredient is what we call "the proximity argument", i.e. the fact that for a fixed mass the maximal (minimal) signal is obtained by placing the sources as close to (as far from) the detector as possible. We already isolated the contribution from the region near the detector and thus we concentrate on the contribution from different reservoirs in the rest of the world (RW), by

[4] For more details of the dependence of the survival probability on the distance with δm^2 (see Mantovani et al., 2004); the discussion of the errors on the oscillation parameters can be found at the end of section 5.4.

TABLE V
Minimal and maximal estimated uranium abundances for the continental crust in ppm, from (Fiorentini et al., 2005b)

	Min	Max
Upper crust	2.2	2.8
Lower crust	0.2	1.1

supplementing the proximity argument with the constraint that the distribution of radiogenic elements are consistent with geochemical and geophysical information on the globe.

5.1. The crusts contribution

For the Earth's crust, we use again the $2° \times 2°$ map of (Bassin et al., 2000) distinguishing several crustal layers which are known to contain different amounts of radioactive elements. For each layer minimal and maximal estimates of uranium abundances found in the literature are adopted, so as to obtain a range of acceptable fluxes, see Table V.

Depending on the adopted values, the uranium mass[5] in the crust $m_C(\mathrm{U})$ is in the range (0.3–0.4) in units – here and in the following – of 10^{17} kg. Clearly a larger mass means a bigger signal, the extreme values being:

$$S_C^{(\min)}(\mathrm{U}) = 6.448 \quad \mathrm{TNU} \quad \text{for} \quad m_C(\mathrm{U}) = 0.3 \quad \text{and}$$
$$S_C^{(\max)}(\mathrm{U}) = 8.652 \quad \mathrm{TNU} \quad \text{for} \quad m_C(\mathrm{U}) = 0.4$$

5.2. The contribution from the mantle

Concerning uranium in the mantle, we assume spherical symmetry and that the uranium mass abundance is a non-decreasing function of depth. It follows that, for a fixed uranium mass in the mantle $m_M(\mathrm{U})$, the extreme predictions for the signal are obtained by:

(i) placing uranium in a thin layer at the bottom and
(ii) distributing it with uniform abundance in the mantle.

[5] We are discussing uranium, however similar considerations hold for thorium.

These two cases give, respectively:

$$S_{\mathrm{M}}^{(\mathrm{min})}(\mathrm{U}) = 12.15 \times m_{\mathrm{M}}(\mathrm{U}) \quad \mathrm{TNU} \quad \text{and}$$
$$S_{\mathrm{M}}^{(\mathrm{max})}(\mathrm{U}) = 17.37 \times m_{\mathrm{M}}(\mathrm{U}) \quad \mathrm{TNU}$$

5.3. Crust and mantle

By using again the proximity argument, we can combine the contributions from crust and mantle so as to obtain extreme predictions: for a fixed total $m(\mathrm{U}) = m_{\mathrm{C}}(\mathrm{U}) + m_{\mathrm{M}}(\mathrm{U})$, the highest signal is obtained by assigning to the crust as much material as consistent with observational data ($m_{\mathrm{C}}(\mathrm{U}) = 0.4$) and putting the rest, $m(\mathrm{U}) - m_{\mathrm{C}}(\mathrm{U})$, in the mantle with a uniform distribution. Similarly, the minimal flux/signal is obtained for the minimal mass in the crust ($m_{\mathrm{C}}(\mathrm{U}) = 0.3$) and the rest in a thin layer at the bottom of the mantle. In conclusion, the contribution from the rest of the world is within the range:

$$S_{\mathrm{RW}}^{(\mathrm{min})} = [6.448 + 12.15(m - 0.3)] \quad \mathrm{TNU} \text{ and}$$
$$S_{\mathrm{RW}}^{(\mathrm{max})} = [8.652 + 17.37(m - 0.4)] \quad \mathrm{TNU}$$

5.4. Geo-neutrino signal and uranium mass

By combining the regional contribution, we get the uranium geo-neutrino signal as a function of uranium mass in the Earth:

$$S(\mathrm{U}) = S_0(\mathrm{U}) \pm \Delta(\mathrm{U})$$

where

$$S_0(\mathrm{U}) = 17.66 + 14.76 \times m(\mathrm{U})$$

and

$$\Delta^2(\mathrm{U}) = (3.07)^2 + [2.61 \times m(\mathrm{U}) - 0.55]^2 \tag{6}$$

This error is obtained by combining in quadrature all geochemical and geophysical uncertainties discussed in the preceding paragraphs. All of them have been estimated so as to cover $\pm 3\sigma$ intervals of experimental measurements and total ranges of theoretical predictions.

However, this error does not account for present uncertainties on neutrino oscillation parameters and on the cross section of the scattering

TABLE VI
Effect of the oscillation parameters on the signal. The relative/absolute variation is computed with respect to the prediction for the best fit values ($\delta m^2 = 7.9 \times 10^{-5}$ eV^2 and $\tan^2\theta = 0.40$), from (Fiorentini et al., 2005b)

Parameter	Signal variation
$\tan^2\theta = 0.26$	+13.5%
$\tan^2\theta = 0.67$	−12.2%
$\delta m^2 = 6.9 \times 10^{-5}$ eV^2	+0.11 TNU
$\delta m^2 = 9.3 \times 10^{-5}$ eV^2	−0.10 TNU

antineutrino-proton. For the sake of discussing the potential of geo-neutrinos, we shall ignore for the moment these error sources.

The expected signal from uranium geo-neutrinos at KamLAND is presented as a function of the total uranium mass m(U) in Figure 2. The upper horizontal scale indicates the corresponding radiogenic heat production rate from uranium ($H_R = 9.5 \times m$).

The predicted signal as a function of m(U) is between the two lines denoted as S_{high} and S_{low}, which correspond, respectively, to $S_0 \pm \Delta$.

Since the minimal amount of uranium in the Earth is 0.3×10^{17} kg (corresponding to the minimal estimate for the crust and the assumption of negligible amount in the mantle), we expect a signal of at least 19 TNU. On the other hand, the maximal amount of uranium tolerated by Earth's energetics[6], 1.8×10^{17} kg, implies a signal not exceeding 49 TNU.

For the central value of the BSE model, $m(\text{U}) = 0.8 \times 10^{17}$ kg, we predict $S(\text{U}) = 29.5 \pm 3.4$ TNU, i.e., with an accuracy of 12% at "3σ". We remark that estimates by different authors for the uranium mass within the BSE are all between (0.7–0.9) $\times 10^{17}$ kg. This implies that the uranium signal has to be in the interval (24.7–34.5) TNU. The measurement of geo-neutrinos can thus provide a direct test of an important geochemical paradigm.

The effect of uncertainties about the oscillation parameters is presented in Table VI. In this respect the mixing angle is most important. Figure 4 (b) of (Araki et al., 2005) shows a 3σ range $0.26 < \tan^2\theta < 0.67$ (central value 0.40): the corresponding range for the average survival probability is $0.52 < P_{ee} < 0.67$ (central value 0.59), with a 3σ relative error on the signal $\Delta S/S \approx 13\%$, which is comparable to the geological uncertainty in Eq. (5.1). The mixing angle should be determined more precisely for fully exploiting the geo-neutrino signal.

[6] For an uranium mass $m = 1.8 \times 10^{17}$ kg and relative abundances as in Eq. (3), the present radiogenic heat production rate from U, Th and K decays equals the maximal estimate for the present heat flow from Earth, $H_E^{max} = 44$ TW (Pollack et al., 1993).

5.5. Extension to thorium

The same analysis was extended to thorium in (Mantovani et al., 2004) assuming global chondritic uranium-to-thorium mass ratio, $m(\mathrm{Th})/m(\mathrm{U}) = 3.9 \pm 0.1$, so that we can now connect the combined signal at Kamioka due to geo-neutrinos from uranium and thorium progenies, S(U + Th), with the radiogenic heat production rate from these elements, H(U + Th), see Figure 3.

The allowed band in Figure 3 is estimated by considering rather extreme models for the distributions of radioactive elements, chosen so as to maximize or minimize the signal for a given heat production rate, see (Fiorentini et al., 2005b).

We also remark that, in comparison with the present experimental error, the width of the band is so narrow that we can limit the discussion to the median line of the allowed band in Figure 3, which represents our best estimate for the relationship between signal and radiogenic power.

By using the Bulk Silicate Earth (BSE) model, giving $H(\mathrm{U} + \mathrm{Th}) = 16$ TW, our prediction for Kamioka is centered at 37 TNU.

By assuming that uranium and potassium in the Earth are in the ratio 1/10,000 and that there is no potassium in the core, the total radiogenic power is $H(\mathrm{U} + \mathrm{Th} + \mathrm{K}) = 1.18\, H(\mathrm{U} + \mathrm{Th})$. With these assumptions, a maximal and fully radiogenic heat production rate, $H(\mathrm{U} + \mathrm{Th} + \mathrm{K}) = 44$ TW, corresponds to $H(\mathrm{U} + \mathrm{Th}) = 37$ TW, which gives signal $S(\mathrm{U} + \mathrm{Th}) \approx 56$ TNU.

6. Discussion of the KamLAND results

The KamLAND collaboration has reported (Araki et al., 2005) data from an exposure of $N_\mathrm{p} = (0.346 \pm 0.017) \times 10^{32}$ free protons over time $T = 749$ days with detection efficiency $\varepsilon = 69\%$: the effective exposure is thus $E_\mathrm{eff} = N_\mathrm{p} \times T \times \varepsilon = (0.487 \pm 0.025)\, 10^{32}$ protons· year. In the energy region where geo-neutrinos are expected, there are $C = 152$ counts, implying statistical fluctuation of ± 12.5. Of these counts, number $R = 80.4 \pm 7.2$ are attributed to reactor events, based on an independent analysis of higher energy data. Fake geo-neutrino events, originating from $^{13}\mathrm{C}(\alpha, \mathrm{n})^{16}\mathrm{O}$ reactions following the alpha decay of contaminant $^{210}\mathrm{Po}$, are estimated to be $F = 42 \pm 11$, where the error is due to 20% uncertainty on the $^{13}\mathrm{C}(\alpha, \mathrm{n})^{16}\mathrm{O}$ cross section and 14% uncertainty on the number of $^{210}\mathrm{Po}$ decays in the detector. Other minor backgrounds account for $B = 4.6 \pm 0.2$ events. The number of geo-neutrino events is estimated by subtraction, $N(\mathrm{U}+\mathrm{Th}) = C - R - F - B$, with an uncertainty obtained by combining the independent errors: $N(\mathrm{U}+\mathrm{Th}) = 25^{+19}_{-18}$. The geo-neutrino signal is thus

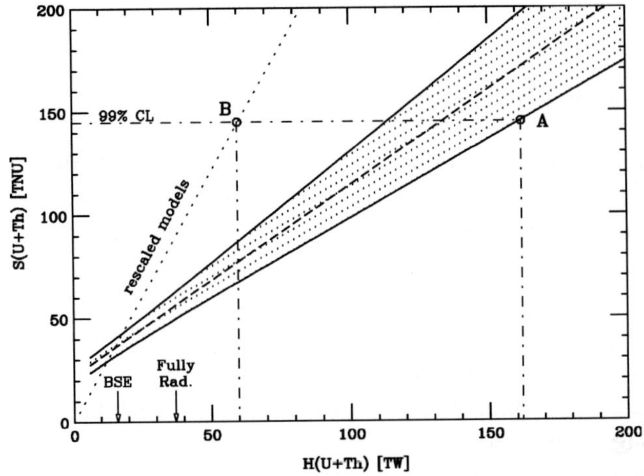

Figure 3. Predictions on the combined signal S(U + Th) from uranium and thorium geo-neutrinos at Kamioka as a function of the radiogenic heat production rate H(U + Th). The shaded area denotes the region allowed by geochemical and geophysical constraints. The dashed median line represents our best estimate for the relationship between signal and radiogenic power (Fiorentini et al., 2005a).

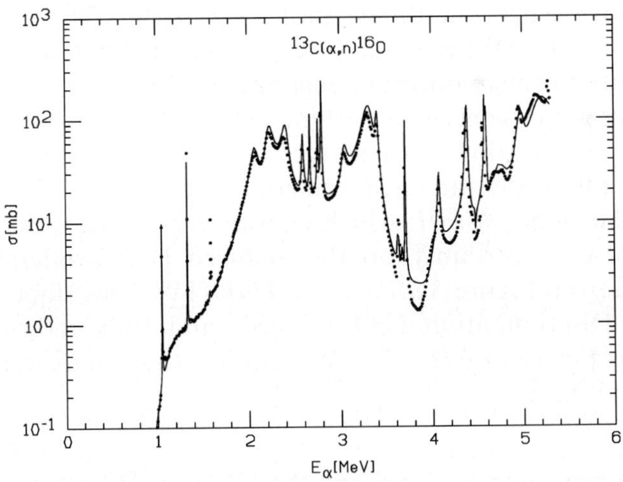

Figure 4. Cross section of $^{13}C(\alpha, n)^{16}O$. The solid line corresponds to the JENDL compilation, dots are the experimental points from (Harissopulos et al., 2005).

$S(\mathrm{U+Th}) = N(\mathrm{U+Th})/E_{\mathrm{eff}} = 51^{+39}_{-36}$ TNU. From the median line in Figure 3 one finds

$$H(\mathrm{U+Th}) = 31^{+43}_{-31}\mathrm{TW} \quad \text{(rate only)}$$

This "rate only" study has been improved in (Araki et al., 2005) by exploiting the shape of the spectrum. A likelihood analysis of the unbinned spectrum yields $N(\mathrm{U+Th}) = 28^{+16}_{-15}$, see Figure 4b of (Araki et al., 2005). This implies $S(\mathrm{U+Th}) = 57^{+33}_{-31}$ TNU and

$$H(\mathrm{U+Th}) = 38^{+35}_{-33}\mathrm{TW} \quad (\text{rate + spectrum})$$

The best fit value is close to the maximal and fully radiogenic model, however the BSE is within 1σ.

By using the median line in Figure 3, the 99% confidence limit on the signal (145 TNU) corresponds to 133 TW. If we include the uncertainty band of the theoretical models, we find an upper bound of 162 TW, see point in Figure 3. This point corresponds to a model with total uranium mass $m(\mathrm{U}) = 80 \times 10^{16}$ kg, an uranium poor crust, $m_c(\mathrm{U}) = 3 \times 10^{16}$ kg, the rest of the uranium being placed at the bottom of the mantle, and global chondritic thorium-to-uranium ratio.

This 162 TW upper bound is much higher than the 60 TW upper bound claimed in (Araki et al., 2005), which was obtained by using a family of geological models which are too narrow and are also incompatible with well-known geochemical and geophysical data, see (Fiorentini et al., 2005a).

We remark that the bound $H(\mathrm{U+Th}) < 162$ TW which we have extracted from KamLAND data does not add any significant information on Earth's interior, since anything exceeding $H(\mathrm{U+Th}) = 37$ TW [i.e. $H(\mathrm{U+Th+K}) = 44$ TW] is unrealistic. The upper limit simply reflects the large uncertainty in this pioneering experiment.

On the other hand, what is important for deciding the potential of future experiments is the relationship between geo-neutrino signal and heat production in the physically interesting region, $H(\mathrm{U+Th}) \leq 37$ TW. The basic parameter is the slope, dS/dH, which expresses how the experimental error translates into an uncertainty on the deduced heat production. For our models we find from Figure 3 $dS/dH \sim 1$ TNU/TW. This slope is the same at any location. Discrimination between BSE and fully radiogenic models, which demands precision $\Delta H \sim 7$ TW, requires thus an experiment with an accuracy $\Delta S \sim 7$ TNU.

7. The geo-neutrino signal and the ^{13}C(α, n)^{16}O cross section

As already remarked, a major uncertainty for extracting the geo-neutrino signal originates from the ^{13}C(α, n)^{16}O cross section. The values used in (Araki et al., 2005) are taken from the JENDL (2005) compilation, which provides an R-matrix fit of relatively old data. A 20% overall uncertainty has been adopted in (Araki et al., 2005), corresponding to the accuracy claimed in the original experimental papers, see e.g. (Bair and Haas, 1973).

Recently, a series of high precision measurements for this cross section has been performed (Harissopulos et al., 2005). In the relevant energy range

$(1.0 \div 5.3)$ MeV, the absolute normalization has been determined within 4% accuracy. The measured values are generally in very good agreement with those recommended in JENDL, see Figure 4; however, we find that the neutron yield per alpha particle is 5% smaller. It follows that the number of fake neutrinos is lower, $F = 40 \pm 5.8$, and geo-neutrino events obviously increase.

The "rate only" analysis gives now 27^{+16}_{-15} geo-neutrino events, corresponding to $S(\mathrm{U+Th}) = 55^{+33}_{-31}$ TNU. From the median line of Figure 3, the radiogenic power is now:

$$H(\mathrm{U+Th}) = 36^{+35}_{-33}\mathrm{TW} \quad (\text{rate spectrum} + \text{new}\,^{13}\mathrm{C}(\alpha,\mathrm{n})^{16}\mathrm{O})$$

We also performed an analysis of the binned spectrum reported in Figure 3 of (Araki et al., 2005). This analysis gives $N(\mathrm{U+Th}) = 31^{+14}_{-13}$ counts, corresponding to $S(\mathrm{U+Th}) = 63^{+28}_{-25}$ TNU and thus:

$$H(\mathrm{U+Th}) = 44^{+31}_{-27}\mathrm{TW} \quad (\text{rate spectrum} + \text{new}\,^{13}\mathrm{C}(\alpha,\mathrm{n})^{16}\mathrm{O})$$

8. Future prospects

The present situation can be summarized in the following points:

- KamLAND has shown that the technique for exploiting geo-neutrinos in the investigation of the Earth's interior is now available.
- New data on $^{13}\mathrm{C}(\alpha, \mathrm{n})^{16}\mathrm{O}$ corroborate the evidence for geo-neutrinos in KamLAND data, which becomes close to 2.5σ.
- On the other hand, the determination of radiogenic heat power from geo-neutrino measurements is still affected by a 70% uncertainty. The best fit of $H(\mathrm{U+Th})$ is close to the prediction of maximal and fully radiogenic model, however the BSE prediction is within 1σ.
- The universal slope $dS/dH \sim 1$ TNU/TW means that for determining the radiogenic heat within ± 7 TW the experimental error has to be ± 7 TNU, i.e. factor four improvement with respect to present.

It looks to us that the following questions are relevant for the future:

- How shall we have definite (at least 3σ) evidence of geo-neutrinos?
- How much uranium and thorium are in the Earth's crust?
- How much in the mantle?
- What about the core?

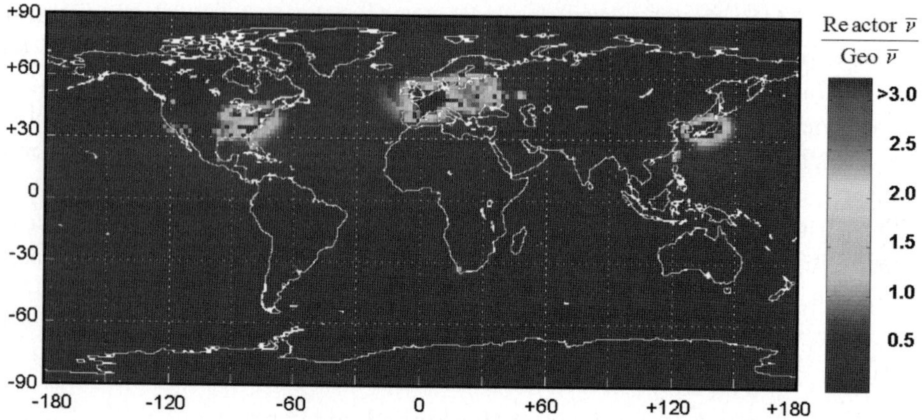

Figure 5. The ratio of reactor anti-neutrino events (in the geo-neutrino energy region) to the expected geo-neutrino events all over the globe.

A preliminary point for establishing suitable detector locations is the reactor background. Figure 5 shows the ratio of reactor events (in the geo-neutrino energy region) to the expected geo-neutrino events all over the globe. The location of KamLAND is obviously one of the worst for the study of geo-neutrinos.

The potential of different locations is summarized in Table VII, where we present the separate contributions to the geo-neutrino signal from crust and mantle according to our reference model, as well as the merit figure r = geo-neutrino events/reactor events.

With more statistics KamLAND should be capable of providing three sigma evidence of geo-neutrinos, but discrimination between BSE and fully radiogenic models definitely requires new detectors, with class and size

TABLE VII

The signal (U + Th) expected from the crust S_C, from the mantle S_M and the total signal S_{TOT} in Terrestrial Neutrino Units [TNU]

Location	S_C(U + Th)	S_M(U + Th)	S_{TOT}(U + Th)	r
Baksan	43.3	9.3	52.6	5.0
Hawaii	3.6	9.3	12.9	10.0
Homestake	43.8	9.3	53.1	5.0
Kamioka	26.4	9.3	35.7	0.1
Gran Sasso	32.8	9.3	42.1	1.1
Pyhasalmi	44.0	9.3	53.3	2.0
Sudbury	43.3	9.3	52.6	0.9
Curacao	24.3	9.3	33.6	10.0

The r factor is the ratio between the geo-neutrino events and reactor events. For this discussion, we use the asymptotic value of the survival probability $\langle P_{ee} \rangle = 0.59$.

similar to that of KamLAND, far away from nuclear power plants. Borexino should reach the 3σ evidence, but cannot go much further due to its relatively small size.

SNO^+ with liquid scintillator will have excellent opportunities to determine the uranium mass in the crust, which accounts for about 80% of the geo-neutrino signal at Sudbury. This will provide an important test of models for the Earth's crust.

A detector at Hawaii, very far from the continental crust and reactors, will be mainly sensitive to the mantle composition. We note that the amount of radioactive materials in this reservoir is the main uncertainty of geological models of the Earth. The expected signal, however, is rather small and this demands a several kilotons size.

For the very long term future, one can speculate about completely new detectors, capable of providing (moderately) directional information. These should allow identification of different geo-neutrino sources (crust, mantle and possibly core) in the Earth; in summary, "se son rose fioriranno".

Acknowledgments

We are grateful to C. Rolfs and his group for useful discussions and for allowing us to use their results. We thank for their useful comments A. Bottino, L. Carmignani, M. Contorti, E. Lisi, W. F. McDonough, K. Inoue, G. Ottonello, R. Raghavan, E. Sanshiro and R. Vannucci.

References

Anderson, D.: 2005, Energetics of the Earth and the Missing Heat Source Mystery, available at www.mantleplumes.org/Energetics.html.

Araki, T. *et al.*: 2005, (KamLAND coll.), *Nature* **436**, 499.

Araki, T. *et al.*: 2005, (KamLAND Collaboration), *Phys. Rev. Lett.* **94,** 081801 hep-ex/0406035.

Bair, J. K. and Haas, F. X.: 1973, *Phys. Rev.* **7**, 1356.

Bassin, C., Laske, G. and Masters, G.: 2000, *EOS Trans. Am. Geophys.* Union 81, F897 [http://mahi.ucsd.edu/Gabi/rem.html].

Condie, K. C.: 1993, *Chem. Geol.* **104**, 1.

Dziewonski, A. M. and Anderson, D. L.: 1981, *Earth Planet. Interact.* **25**, 297.

Fiorentini, G., Mantovani, F. and Ricci, B.: 2003a, *Phys. Lett. B* **557**, 139 [arXiv:nucl-ex/0212008].

Fiorentini, G., Lissia, M., Mantovani, F. and Vannucci, R.: 2003b, Astroparticle and high energy physics AHEP2003/035 [arXiv:hep-ph/0401085].

Fiorentini, G., Lissia, M., Mantovani, F. and Ricci, B.: 2003c, arXiv:physics/0305075.

Fiorentini, G., Lasserre, T., Lissia, M., Ricci, B. and Schönert, S.: 2003d, *Phys. Lett. B* **558**, 15 [arXiv:hep-ph/0301042].

Fiorentini, G., Lissia, M., Mantovani, F. and Ricci, B.: 2005a, *Phys. Lett. B* **629**, 77–82 hep-ph/0508048.

Fiorentini, G., Lissia, M., Mantovani, F. and Vannucci, R.: 2005b, *Phys. Rev. D* **72**, 03317 [arXiv:hep-ph/0501111].

Gao, S. et al.: 1998, *Geochimica et Cosmochimica Acta* **62**, 1959–1975.

GERM: 2003, The Geochemical Earth Reference Model, is available on the web at http://earthref.org.

Harissopulos, S. et al.: 2005, *Phys. Rev. C* **72**, 06281 [arXiv:nucl-ex/0509014].

Hofmann, A. W.: 1988, *Earth Planet Sci. Lett.* **90**, 297.

Hofmeister, A. M. and Criss, R. E.: 2005, *Tectonophysics* **395**, 159–177.

JENDL: 2005, Japanese Evaluated Nuclear Data Library, http://wwwndc.tokai.jaeri.go.jp/jendl/.

Jochum, K. P. et al.: 1986, *Nature* (London) **322**, 221.

Mantovani, F., Carmignani, L., Fiorentini, G. and Lissia, M.: 2004, *Phys. Rev. D* **69**, 013001 [arXiv:hep-ph/0309013].

McDonough, W. F.: 2003, in R. W. Carlson (ed.), *The Mantle and Core* Vol. 2, H. D. Holland and K. K. Turekian (eds.), *Treatise on Geochemistry*, Elsevier-Pergamon, Oxford, pp. 547–568.

O'Nions, R. K. and McKenzie, D.: 1993, *Philos. Trans. R. Soc. London A* **342**, 65.

Pollack, H. N., Hunter, S. J. and Johnson, J. R.: 1993, *Rev. Geophys.* **31**, 267–280.

Raghavan, R. S., Schönert, S., Enomoto, S., Shirai, J., Suekane, F., and Suzuki, A.: 1998, *Phys. Rev. Lett.* **80**, 635.

Rothschild, C. G., Chen, M. C. and Calaprice, F. P.: 1998, *Geophys. Res. Lett.* **25**, 1083 [arXiv:nucl-ex/9710001].

Rudnick, R. L. and Fountain, D. M.: 1995, *Rev. Geophys.* **33**, 267.

Shaw, D. M., Cramer, J. J., Higgins, M. D. and Truscott, M. G.: 1986, in J. Dawson *et al* (eds.), *The Nature of the Lower Continental Crust*, Geological Society of London, London, p. 275.

Taylor, S. R. and McLennan, S. M.: 1985. *The Continental Crust: its Composition and Evolution*, Blackwell Scientic, Oxford.

Togashi, S. *et al.*: 2000, *Geochem. Geophys. Geosyst. Electronic J. Earth Sci.* **1**, 2000GC00083.

Weaver, B. L. and Tarney, J.: 1984, *Nature (London)* **310**, 575.

Wedepohl, K. H.: 1995, *Geochim. Cosmochim. Acta* **59**, 1217.

White, W. M.: 1993, *Earth Planet Sci. Lett.* **115**, 211 [arXiv:hep-ph/0212202].

Zartman, R. E. and Haines, S.: 1988, *Geochim. Cosmochim. Acta* **52**, 1327.

Zhao, D., Horiuchi, S. and Hasegawa, A.: 1992, *Tectonophysics* **212**, 289–301.

Earth, Moon, and Planets (2006) 99:111–130
DOI 10.1007/s11038-006-9107-5

Geo-Neutrinos: A Systematic Approach to Uncertainties and Correlations

G. L. FOGLI, E. LISI, A. PALAZZO and A. M. ROTUNNO
Dip. di Fisica and Sezione INFN di Bari, Via Amendola 173, 70126, Bari, Italy
(E-mail: eligio.lisi@ba.infn.it)

A. PALAZZO
Astrophysics, Denys Wilkinson Building, Keble Road, OX13RH, Oxford, UK

(Received 28 March 2006; Accepted 28 June 2006)

Abstract. Geo-neutrinos emitted by heat-producing elements (U, Th and K) represent a unique probe of the Earth interior. The characterization of their fluxes is subject, however, to rather large and highly correlated uncertainties. The geochemical covariance of the U, Th and K abundances in various Earth reservoirs induces positive correlations among the associated geo-neutrino fluxes, and between these and the radiogenic heat. Mass-balance constraints in the Bulk Silicate Earth (BSE) tend instead to anti-correlate the radiogenic element abundances in complementary reservoirs. Experimental geo-neutrino observables may be further (anti)correlated by instrumental effects. In this context, we propose a systematic approach to covariance matrices, based on the fact that all the relevant geo-neutrino observables and constraints can be expressed as linear functions of the U, Th and K abundances in the Earth's reservoirs (with relatively well-known coefficients). We briefly discuss here the construction of a tentative "geo-neutrino source model" (GNSM) for the U, Th, and K abundances in the main Earth reservoirs, based on selected geophysical and geochemical data and models (when available), on plausible hypotheses (when possible), and admittedly on arbitrary assumptions (when unavoidable). We use then the GNSM to make predictions about several experiments ("forward approach"), and to show how future data can constrain *a posteriori* the error matrix of the model itself ("backward approach"). The method may provide a useful statistical framework for evaluating the impact and the global consistency of prospective geo-neutrino measurements and Earth models.

Keywords: Bulk Silicate Earth, covariance, earth interior, error matrix, heat-producing elements, neutrinos, statistical analysis

1. Introduction

Electron antineutrinos emitted in the decay chains of the heat-producing elements (HPE) U, Th, and K in terrestrial rocks – the so-called geo-neutrinos – represent a truly unique probe of the Earth interior; see Fiorentini et al. (2005a) for a recent review and Krauss et al. (1984) for earlier discussions and references. The first indications for a (U + Th) geo-neutrino signal at $>2\sigma$ confidence level in the KamLAND experiment by Araki et al. (2005) have boosted the interest in this field, and have started to bridge the two communities of particle physicists and Earth scientists – as exemplarily testified by this Workshop (Neutrino Geophysics, Honolulu, Hawaii, 2005).

The hope is that future measurement of geo-neutrino fluxes can put statistically significant constraints to the global abundances of HPEs and to their associated heat production rates, which are currently subject to highly debated Earth model assumptions (McDonough, 2003; Sleep, 2005). This goal, despite being experimentally very challenging, is extremely important and deserves dedicated (possibly joint) studies from both scientific communities.

One methodological difficulty is represented by the different "feeling" for uncertainties by particle physicists versus Earth scientists. First important attempts to systematize (U, Th, K) abundance uncertainties in a format convenient for geo-neutrino analyses have been performed in Enomoto (2005) and particularly in Mantovani et al. (2004, 2005), Fiorentini et al. (2005b), where errors have been basically assessed from the spread in published estimates (consistently with mass balance constraints).

We propose to make a further step, by systematically taking into account the ubiquitous error *covariances*, i.e., the fact that several quantities happen to vary in the same direction (positive correlations) or in opposite directions (negative correlations) in the geo-neutrino context. For instance, in a given Earth reservoir (say, the mantle), the U, Th and K abundances are typically positively correlated. However, they may be anticorrelated in two complementary reservoirs constrained by mass balance arguments, such as the mantle and the crust in Bulk Silicate Earth (BSE) models. Experimental geo-neutrino observables may be further (anti)correlated by instrumental effects.

An extensive discussion of our approach to these problems is beyond the scope of this contribution and will be presented elsewhere (Fogli et al., 2006a). Here we briefly report about some selected issues and results, according to the following scheme. In Sec. 2, we discuss the general aspects and the statistical tools related to covariance analyses, with emphasis on geo-neutrino observables. In Sec. 3, we construct a tentative model for the source distribution of (U, Th, K) in global Earth reservoirs (Geo Neutrino Source Model, GNSM). In Sec. 4, we discuss some issues related to the characterization of local sources around geo-neutrino detector sites. In Secs. 5 and 6, we show examples of "forward" error estimates (i.e., propagation of GNSM errors to predicted geo-neutrino rates) and shortly discuss "backward" error updates (i.e., GNSM error reduction through prospective geo-neutrino data). We draw our conclusions in Sec. 7.

2. Covariance and Correlations: General Aspects

In this section, we discuss some general aspects of covariance analyses in geochemistry and in neutrino physics, and then present the basic tools relevant for geo-neutrino physics. We remind that, for any two quantities P and Q, estimated as

$$P = \overline{P} \pm \sigma_P, \tag{1}$$

$$Q = \overline{Q} \pm \sigma_Q, \tag{2}$$

the correlation index $\rho_{PQ} \in [-1, +1]$ between the $1-\sigma$ errors of P and Q parameterizes the degree of "covariation" of the two quantities: $\rho > 0$ (< 0) if they change in the same (opposite) direction, while $\rho = 0$ if they change independently; see, e.g., Eadie et al. (1971). The covariance (or squared error) matrix of P and Q contains σ_P^2 and σ_Q^2 as diagonal elements, and $\rho_{PQ}\sigma_P\sigma_Q$ as off-diagonal ones. For more than two variables with errors σ_i and correlations σ_{ij}, thc covariance matrix is $\sigma_{ij}^2 = \rho_{ij}\sigma_i\sigma_j$ (symmetric, with $\rho_{ii} = 1$ on the diagonal).

2.1. Covariance analyses in geochemistry

In 1998, the Geochemical Earth Reference Model (GERM) initiative was officially launched (Staudigel, 1998), in order to provide a "consensus model" for the elemental abundances, together with their errors and correlations, in all relevant Earth reservoirs. Although a lot of work has been done in this direction, e.g., through rich compilations of data and estimates (http://www.earthref.org), the correlation matrices have not yet been estimated – not even for subsets of elements such as (U, Th, K). To our knowledge, only a few regional studies discuss HPE covariances. These difficulties can be in part overcome by using the (more frequently reported) elemental ratio information. For instance, if the ratio of two abundances P and Q is reported together with its error $\sigma_{P/Q}$, the correlation between P and Q can be inferred through the following statistical relation, valid at first order in error propagation:

$$\left(\frac{\sigma_{P/Q}}{P/Q}\right)^2 = \left(\frac{\sigma_P}{P}\right)^2 + \left(\frac{\sigma_Q}{Q}\right)^2 - 2\rho_{PQ}\left(\frac{\sigma_P}{P}\right)\left(\frac{\sigma_Q}{Q}\right). \tag{3}$$

Although the "ratio" and "correlation" information appear to be interchangeable through the above formula, from a methodological viewpoint it is better to use the latter rather than the first, since the ratio of two Gaussian variables is a Cauchy distribution with formally infinite variance (Eadie et al., 1971) – a rather tricky object in statistical manipulations.

2.2. Covariance analyses in neutrino physics

Neutrino physics has undergone a revolution in recent years, after the discovery of neutrino flavor oscillations. Phenomenological fits to neutrino oscillation data have become increasingly refined, and now routinely include covariance analyses (see, e.g., Fogli et al., 2006b). For our purposes, a relevant example is also given by the Standard Solar Model (Bahcall et al., 2005), which provides, among other things, errors and correlations for solar neutrino sources. We shall try to apply a similar "format" to a Geo-Neutrino Source Model (GNSM) in Sec. 3.

Correlations arise not only at the level of neutrino sources, but also at the detection level, as a consequence of instrumental effects. For instance, the KamLAND experiment is currently more sensitive to the sum (U + Th) of geo-neutrino fluxes rather than to the separate U and Th components. As a consequence, the measured U and Th geo-neutrino event rates are anticorrelated: if one rate increases the other one tends to decrease, in order to keep the total rate constant (within errors).

Figure 1 shows explicitly the anticorrelation between the U and Th experimental rates through their 1, 2, and 3-σ contours (solid lines) taken from our KamLAND data analysis (Fogli et al., 2006a). The contours in

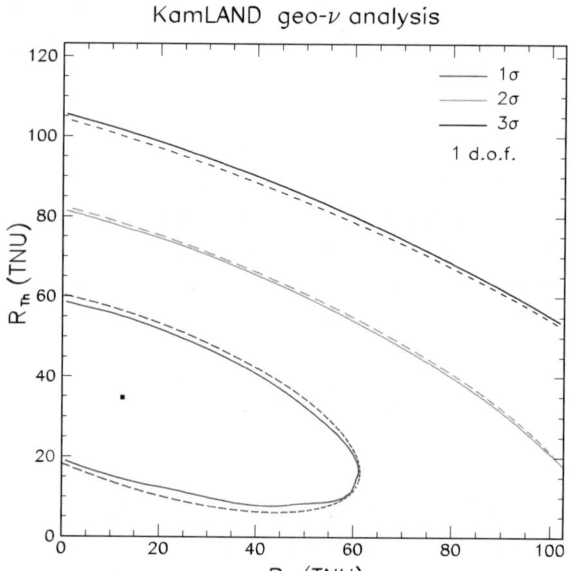

Figure 1. Best-fit U and Th event rates and error contours (solid lines) from our analysis of KamLAND geo-neutrino data (Fogli et al., 2006a). The contours are very close to two-dimensional Gaussian confidence levels (dashed ellipses). Units: 1 TNU = 1 event/year/10^{32} target protons.

Fig. 1 are well approximated by a bivariate gaussian (dashed lines) with parameters:

$$R_{\mathrm{U}} = 12.5 \pm 48.9\,\mathrm{TNU}, R_{\mathrm{Th}} = 34.7 \pm 28.5\,\mathrm{TNU}, \rho = -0.645. \tag{4}$$

We have also verified that our KamLAND data analysis reproduces the confidence level contours in the alternative plane spanned by $R_{\mathrm{U}} + R_{\mathrm{Th}}$ and $(R_{\mathrm{U}} - R_{\mathrm{Th}})/(R_{\mathrm{U}} + R_{\mathrm{TH}})$ (Araki et al., 2005) (not shown); however, as explained in the previous subsection, we prefer to avoid any "ratio" and to use just (R_{U}, R_{Th}) and their correlation.

2.3. General tools for geo-neutrinos

In the context of geo-neutrinos, statistical analyses are greatly simplified by the fact that all relevant observables are linear combinations of the HPE abundances a_i^{S} in different reservoirs (S = U, Th, K; i = reservoir index), with coefficient determined by known physics and by the geometry of the Earth mass distribution.

In particular we consider: (i) the total radiogenic heat H of the Earth (decay energy absorbed per unit of time); (ii) the geo-neutrino flux Φ_D at a given detector site D (number of $\overline{\nu}_e$ per unit of area and time); and (iii) the corresponding event rate R_D at a given detector site D (number of events from $\overline{\nu}_e + p \rightarrow n + e^+$, per unit of time and of target protons). Such quantities can be written as:

$$H_R = \sum_{\mathrm{S}} h_{\mathrm{S}} \sum_i M_i a_i^{\mathrm{S}}, \tag{5}$$

$$\Phi_D = \langle P_{ee} \rangle \sum_{\mathrm{S}} \phi_{\mathrm{S}} \sum_i f_i^D a_i^{\mathrm{S}}, \tag{6}$$

$$R_D = \langle P_{ee} \rangle \sum_{\mathrm{S}} r_{\mathrm{S}} \sum_i f_i^D a_i^{\mathrm{S}}, \tag{7}$$

where the universal coefficients h_{S}, ϕ_{S}, and r_{S}, according to our calculations (Fogli et al., 2006a), are given in Table I. In the above equations, M_i is the mass of the i-th reservoir, while $\langle P_{ee} \rangle \simeq 0.57$ is the average survival oscillation probability of geo-$\overline{\nu}_e$. The geometrical coefficients f_i^D represent the mass-weighted average of the inverse square distance of the detector site D from the ith reservoir, necessary to account for the flux decrease with distance; their numerical values will be reported elsewhere (Fogli et al., 2006a).

TABLE I

Universal (reservoir-independent and detector site-independent) coefficients for the calculation of the total heat (H) of the Earth, and of the total $\bar{\nu}_e$ flux (Φ_D) and event rates from inverse beta decay (R_D) at any detector site D

S	h_S (μW/kg)	ϕ_S (10^{12} $\bar{\nu}_e$/cm^2/s)	r_S (10^8 TNU)
U	98.0	123	15.2
Th	26.3	26.1	1.06
K	34.9×10^{-4}	45.4×10^{-3}	0

Conversion factors: 1 TNU = 1 event/year/10^{32} target protons; 1 year = 3.15576×10^7 s. Natural abundances of isotopes are assumed.

The Earth mass distribution necessary to compute the M_i's and f_i^D's is taken from the Preliminary Earth Reference Model (PREM) (Dziewonski and Anderson, 1981), properly matched with a crustal model defined over a grid of $2° \times 2°$ tiles (Bassin et al., 2000). The Earth is assumed to be partitioned into the following homogeneous reservoirs: core, lower mantle (LM), upper mantle (UM), continental crust (CC) (in three layers – upper, middle, lower) and oceanic crust (OC) (lumped into one layer). The distinction (if any) between LM and UM is strongly debated and will be commented later. For a set of possible geo-neutrino detector sites (Kamioka, Gran Sasso, Sudbury, Hawaii, Pyhäsalmi, Baksan), we consider "local" reservoirs, defined as the nine (three-by-three) tiles of the $2° \times 2°$ model which surround each detector site – except for Kamioka, where 13 tiles are considered, corresponding to the "Japanese arc and forearc" as defined in the crustal model (Bassin et al. 2000). Due to the inverse-squared-distance decrease of the neutrino flux, it turns out that local and global reservoirs can provide comparable contributions to the geo-neutrino event rates, at least for detectors sitting on the CC.

The main task is then to build a model for the abundances a_i^S, embedding covariances. In other words, by switching to a single-index vector notation for simplicity,

$$\{a_i^S\}_{i=1,\dots,N}^{S=U,Th,K} \rightarrow \mathbf{a} = \{a_i\}_{i=1,\dots,3N} \quad (8)$$

(where N is the number of reservoirs), an Earth model should provide, for any entry in the HPE abundance vector $\mathbf{a}$, both a central value $\bar{a}_i$ and a standard deviation $\pm\sigma_i$,

$$a_i = \bar{a}_i \pm \sigma_i \quad (9)$$

together with the error correlation matrix $\boldsymbol{\rho}$. The components of the covariance matrix $\boldsymbol{\sigma}^2$ are then

$$[\boldsymbol{\sigma}^2]_{ij} = \rho_{ij}\sigma_i\sigma_j. \quad (10)$$

Given any two quantities P and Q (as in Eqs. (5)–(7)) defined as linear combinations of the a_i's (with T = transpose),

$$P = \sum_i p_i a_i = \boldsymbol{p}^{\mathrm{T}} \boldsymbol{a}, \tag{11}$$

$$Q = \sum_i q_i a_i = \boldsymbol{q}^{\mathrm{T}} \boldsymbol{a}, \tag{12}$$

it turns out that their $1 - \sigma$ errors σ_P and σ_Q are simply given by

$$\sigma_P^2 = \sum_{ij} p_i p_j \rho_{ij} \sigma_i \sigma_j = \boldsymbol{p}^{\mathrm{T}} \boldsymbol{\sigma}^2 \boldsymbol{p}, \tag{13}$$

$$\sigma_Q^2 = \sum_{ij} q_i q_j \rho_{ij} \sigma_i \sigma_j = \boldsymbol{q}^{\mathrm{T}} \boldsymbol{\sigma}^2 \boldsymbol{q}, \tag{14}$$

with (P, Q) correlation given by

$$\rho_{PQ} = \frac{\sum_{ij} p_i q_j \rho_{ij} \sigma_i \sigma_j}{\sigma_P \sigma_Q} = \frac{\boldsymbol{p}^{\mathrm{T}} \boldsymbol{\sigma}^2 \boldsymbol{q}}{\sigma_P \sigma_Q}. \tag{15}$$

The above equations will be used below to compute correlations among experimental event rates, or between an experimental rate and the radiogenic heat.

3. Towards a Geo-Neutrino Source Model

In this Section we briefly discuss our methodology to provide entries for Eqs. (9, 10), i.e., a GNSM for HPE abundances in the Earth. We remind that, concerning the entries for Eq. (9) (errors only, no correlations), our approach overlaps in part with earlier relevant work performed in Fiorentini et al. (2005a, b), Mantovani et al. (2005).

3.1. The bulk silicate earth

Bulk Silicate Earth models (McDonough, 2003) provide global constraints on elemental abundances (especially in the primitive mantle), under a set of hypotheses. In particular, BSE models include the plausible assumption that elements which have both high condensation temperature ("refractory") and that are preferentially embedded in rocks rather in iron ("lithophile") should be found in the primitive mantle (i.e., in the undifferentiated mantle + crust reservoir) in the same ratio as in the parent, pristine meteoritic material. Among the three main HPEs (U, Th, K), the first two are also refractory lithophile elements (RLE), so the Th/U global ratio should be the same in BSE and in the (supposedly) parent and most primitive meteoritic material (carbonaceous chondrites CI). Our summary (Fogli et al., 2006a) of the recent and detailed works on absolute (Palme and O'Neill, 2003; Lodders, 2003) and relative (Rocholl and Jochum, 1993; Goreva and Burnett, 2001) U and Th abundances in CI meteorites (with 1σ errors) is:

$$a_{\mathrm{CI}}^{\mathrm{Th}} = 30.4(1 \pm 0.10) \times 10^{-9}, \tag{16}$$

$$a_{\mathrm{CI}}^{\mathrm{U}} = 8.10(1 \pm 0.10) \times 10^{-9}, \tag{17}$$

$$\left(a_{\mathrm{CI}}^{\mathrm{Th}}/a_{\mathrm{CI}}^{\mathrm{U}}\right) = 3.75(1 \pm 0.05), \tag{18}$$

which implies a (Th, U) error correlation $\rho_{\mathrm{CI}} = 0.875$ through Eq. (15).

The BSE/CI abundance ratio is expected to be the same for all RLEs, if indeed they did not volatilize during the Earth formation history. The benchmark is usually provided by a major RLE element such as Al, which, being much more abundant than the trace elements Th and U, can be more robustly constrained, both by mass-balance arguments and by direct sampling. Our summary (Fogli et al., 2006a) for the BSE/CI abundance ratio of Al from three detailed BSE models (McDonough and Sun, 1995; Allègre et al., 2001; Palme and O'Neill, 2003) is

$$\left(a_{\mathrm{BSE}}^{\mathrm{Al}}/a_{\mathrm{CI}}^{\mathrm{Al}}\right) = 2.7(1 \pm 0.10). \tag{19}$$

The previous arguments and estimates imply that

$$a_{\mathrm{BSE}}^{\mathrm{Th}} = a_{\mathrm{CI}}^{\mathrm{Th}} \cdot \left(a_{\mathrm{BSE}}^{\mathrm{Al}}/a_{\mathrm{CI}}^{\mathrm{Al}}\right) = 82.1(1 \pm 0.14) \times 10^{-9}, \tag{20}$$

$$a_{\mathrm{BSE}}^{\mathrm{U}} = a_{\mathrm{CI}}^{\mathrm{Th}} \cdot \left(a_{\mathrm{BSE}}^{\mathrm{Al}}/a_{\mathrm{CI}}^{\mathrm{Al}}\right) = 21.9(1 \pm 0.14) \times 10^{-9}, \tag{21}$$

with (Th, U) error correlation $\rho_{\mathrm{BSE}} = 0.936$.

The K element, being (moderately) volatile, needs a separate discussion. In Jochum et al. (1983) it was argued that U is a good "global" proxy for K, since: (1) the K/U abundance ratio was found to be nearly constant in 22 samples of Mid-Ocean Ridge Basalt (MORB) from the Atlantic and Pacific ocean floor (thought to be representative of the whole mantle); (2) the MORB K/U ratio was found to be (accidentally) similar to the K/U ratio estimate in the crust from an older model (Wasserburg et al., 1964). The

MORB K/U ratio (1.27×10^4) was then boldly generalized to the whole Earth, with small estimated errors (1.6%) (Jochum et al., 1983). However, it should be noticed there are no geochemical arguments to presume that disparate elements such as U and K should have the same partition coefficients between melt (=crust) and residual mineral (=depleted mantle); indeed, analogous alleged coincidences have later been disproved (Hofmann, 2003). Therefore, we think that the "canonical K/U = 12,700" ratio, so often quoted in the geochemical literature, should be critically revisited in future studies. Provisionally, from a survey of recent literature about the abundances of K and U (and of another possible K-proxy element, La (Palme and O'Neill, 2003)) in MORB databases, CC samples and estimates, and BSE models, we are inclined to Fogli et al. (2006a): (1) increase significantly – although subjectively – the K/U uncertainty; and (2) slightly lower the central value (as compared with Jochum et al., 1983). More precisely, we take

$$\left(a^{\mathrm{K}}_{\mathrm{BSE}}/a^{\mathrm{U}}_{\mathrm{BSE}}\right) = 1.2 \times 10^4(1 \pm 0.15), \tag{22}$$

which, by proper error propagation, gives the absolute K abundance as

$$a^{\mathrm{K}}_{\mathrm{BSE}} = 263 \times 10^{-6}(1 \pm 0.21), \tag{23}$$

with (K, Th) and (K, U) correlations equal to 0.648 and 0.701, respectively. Table II presents a summary of the BSE (U, Th, K) abundances, errors and correlation matrix, together with similar information about the main BSE sub-reservoirs (as discussed below).

Needless to say, all the above BSE estimates may be significantly altered, if possible indications for non-zero HPE abundances in the Earth core (Rama Murthy, 2005) are corroborated by further studies. For the sake of simplicity, we do not consider such possibility in this work.

3.2. The continental crust (cc)

Average elemental abundances in CC, and their vertical distribution in the three main identifiable layers (upper, middle, lower crust = UC, MC, LC), have been presented in a recent comprehensive review (Rudnick and Gao, 2003), together with a wealth of data and with a critical survey of earlier literature on the subject. In particular, it is stressed in Rudnick et al. (2003) that some previous CC models are not consistent with known crustal heat production constraints (Jaupart and Mareschal, 2003). This fact shows that: (1) the spread of published values for elemental abundances is not necessarily indicative of the real uncertainties, since some estimates can be invalidated by new and independent data; (2) heat production estimates in the CC provide a relevant constraint (linear in the U, Th, K crustal abundances) which might help, together with geo-neutrino measurements, to reduce the HPE abundance

TABLE II

Geo-Neutrino Source Model (GNSM): abundances, errors and correlations of radiogenic elements (U, Th, K) in global reservoirs

Geo-Neutrino Source Model (GNSM) for global reservoirs				Correlation matrix														
				BSE			CC			OC			UM			LM		
Reser.	Elem.	Abundance	±1σ	U	Th	K	U	Th	K	U	Th	K	U	Th	K	U	Th	K
BSE	U	21.9×10^{-9}	±14%	1	+0.936	+0.701	0	0	0	0	0	0	0	0	0	+0.908	+0.893	+0.690
	Th	82.1×10^{-9}	±14%		1	+0.648	0	0	0	0	0	0	0	0	0	+0.850	+0.954	+0.638
	K	26.3×10^{-5}	±21%			1	0	0	0	0	0	0	0	0	0	+0.636	+0.618	+0.985
CC	U	1.46×10^{-6}	±17%				1	+0.906	+0.906	0	0	0	0	0	0	−0.409	−0.263	−0.146
	Th	6.29×10^{-6}	±10%					1	+0.595	0	0	0	0	0	0	−0.371	−0.291	−0.096
	K	1.62×10^{-2}	±10%						1	0	0	0	0	0	0	−0.371	−0.173	−0.161
OC	U	1.00×10^{-7}	±30%							1	+0.906	+0.868	0	0	0	−0.012	−0007	−0007
	Th	2.20×10^{-7}	±30%								1	+0764	0	0	0	−0011	−0007	−0006
	K	1.25×10^{-3}	±28%									1	0	0	0	−0.010	−0.006	−0.008
UM	U	3.95×10^{-9}	±30%										1	+0.906	+0.868	−0.093	−0.65	−0.058
	Th	10.8×10^{-9}	±30%											1	+0.764	−0.084	−0.071	−0.051
	K	5.02×10^{-5}	±28%												1	−0.081	−0.054	−0.066
LM	U	17.3×10^{-9}	±27%													1	+0.924	+0.692
	Th	60.4×10^{-9}	±27%														1	+0.640
	K	21.7×10^{-5}	±36%															1

estimates in reference models. The latter point will be further elaborated elsewhere (Fogli et al., 2006a).

We basically adopt the results in Rudnick and Gao (2003) for the UC, MC, LC abundances of (U, Th, K) and their uncertainties, with the following differences: (1) since no error estimates are given for the LC, we conservatively (but arbitrarily) assume fractional 1σ errors of 40% in this layer; (2) our reference crustal model (Bassin et al., 2000) and the one in Rudnick and Gao (2003) provide mass ratios among layers (UC:MC:LC), respectively, equal to 0.359:0.330:0.311 and 0.317:0.296:0.387. This difference is somewhat disappointing, since it induces weighted-average HPE abundance shifts in the CC of order 10%, which are definitely non-negligible. "Consensus values" for the mass distribution in the three CC layers (upper, middle, lower) would thus be desirable in the future. Provisionally, we assume that CC elemental abundance errors cannot be smaller than the "mass distribution" induced error (10%). We also assume, from a survey of the relevant literature, a 9% fractional error for each of the K/U, Th/U, and K/Th ratios in the crust – which in turn provide the (U, Th, K) correlations (Fogli et al., 2006a). Given such inputs, the CC abundances (central values, errors, and correlations) turn out to be as shown in Table II.

3.3. The upper mantle (UM)

We assume a homogeneous composition for UM (defined as the sum of transition zone + low-velocity zone + "lid" in the PREM model, Dziewonski and Anderson, 1981). Global and detailed analyses of all the available UM samples and constraints have been performed in two recent papers (Workman and Hart, 2005; Salters and Stracke, 2006) which, unfortunately, do not really agree in their conclusions, despite being based in part on the same petrological database (http://www.petdb.org). Concerning HPE's, we then take as central values the average of Salters and Stracke ((2006) and Workman and Hart (2005), but we attach the most conservative error estimates of Salters and Stracke (2006), which are large enough to cover the spread between Salters and Stracke (2006) and Workman and Hart (2005). We assume a K/U ratio error in UM of the same size as for the BSE ($\pm 15\%$), and a Th/U ratio error of $\pm 13\%$, as suggested from the scatter of points in Fig. 2 of Salters and Stracke (2006). Given such inputs, the UM abundances turn out to be Fogli et al. (2006a) as shown in Table II.

3.4. The oceanic crust (OC)

The OC is difficult to sample and, not surprisingly, only a few papers (to our knowledge) deal with its average trace-element composition (see, e.g., Taylor

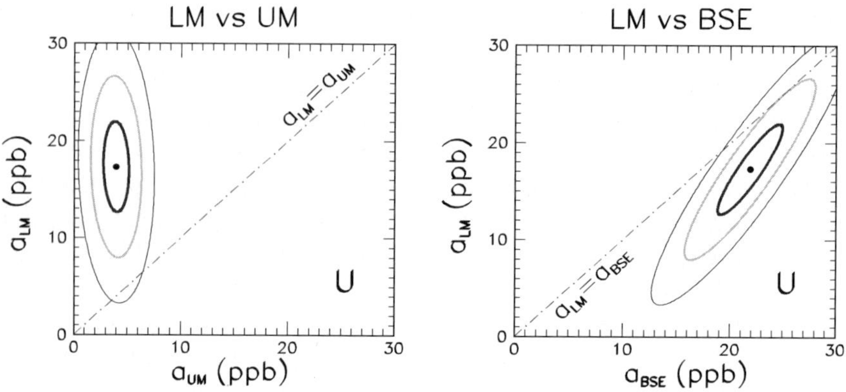

Figure 2. Comparison between LM, UM, and BSE abundances of Uranium, in ppb (10^{-6}) units. Our GNSM estimates are shown as 1, 2, and 3σ error ellipses. The slanted lines represent the cases of whole mantle convection (in the left panel) and of decoupled LM (in the right panel).

and McLennan, 1985; Hofmann, 1988, Wedepohl and Hartman, 1994). We adopt the "intermediate" central values of Taylor and McLennan (1985), which suggest a HPE enrichment of the OC by a factor 20–25 with respect to the (parent) UM. Since the enrichment is approximately uniform for all three HPEs, we think it reasonable to assume that the same relative spread of abundances is transferred from the UM (parent mineral) to the OC (melt). Therefore, in the absence of other information, we attach to the (U, Th, K) abundances in the OC the same fractional errors and correlations as for the UM, see Table II.

3.5. The lower mantle (LM)

The consistent derivation of LM abundances, errors and correlations is a qualifying result of our work. The abundances in the lower mantle (LM) are obtained by subtraction (LM = BSE − UM − CC − OC), namely, by the mass balance constraint:

$$a^{S}_{LM} = (a^{S}_{BSE}M_{BSE} - a^{S}_{UM}M_{UM} - a^{S}_{CC}M_{CC} - a^{S}_{OC}M_{OC})/M_{LM}, \tag{24}$$

for S = U, Th, K. Since the three HPE abundances a^{S}_{LM} are linear combinations of BSE, UM, CC, and OC abundances, it is possible to apply the formalism of Sec. 2.3 and to obtain their errors and correlations, whose numerical values are listed in Table II (last three columns). It turns out that the LM fractional uncertainties are comparable to those of the UM (~30%), and that the LM abundances are strongly correlated with the BSE ones but moderately anticorrelated with the CC ones, due to the subtraction procedure. The LM anticorrelation with the UM and OC is instead very small, since the latter two reservoirs contain relatively small absolute amounts of HPEs, as compared to the CC and BSE.

Once the LM contents of HPEs are obtained, the BSE information becomes redundant, and one can proceed with the information contained in the 12 entries for the (CC, OC, UM, LM) abundances, and the corresponding 12×12 correlation matrix, which are reported in Table II. Notice that, within the quoted uncertainties, the abundances in Table II are consistent with those reported in Mantovani et al. (2004).

3.6. What about mantle convection?

There is currently a strong debate about the nature and extent of mantle convection, with scenarios ranging from two-layer models (with geochemically decoupled UM and LM) to whole mantle convection (with completely mixed UM and LM), and many intermediate possibilities and variants (McDonough, 2003). Two extreme possibilities are: (1) a geochemically homogeneous mantle (i.e., no difference between UM and LM, $a^{\mathrm{S}}_{\mathrm{LM}} = a^{\mathrm{S}}_{\mathrm{UM}}$); and (1) a strict two-layer model (i.e., a lower mantle conserving primitive mantle abundances, $a^{\mathrm{S}}_{\mathrm{LM}} = a^{\mathrm{S}}_{\mathrm{BSE}}$).

Our estimates in Table II are intermediate between such two cases, and thus agree better with models predicting partial mantle mixing. The two extreme cases are anyway recovered by stretching the uncertainties to roughly $\pm 3\sigma$. Figure 2 shows the 1, 2, and 3σ error ellipses in the (LM, UM) and (LM, BSE) Uranium abundance planes; within 3σ, both cases $a^{\mathrm{U}}_{\mathrm{LM}} = a^{\mathrm{U}}_{\mathrm{UM}}$ and $a^{\mathrm{U}}_{\mathrm{LM}} = a^{\mathrm{U}}_{\mathrm{BSE}}$ (slanted lines) are allowed. Similar results are obtained for Th and K (not shown). Therefore, our GNSM estimates are sufficiently conservative to cover, within $\pm 3\sigma$, a wide spectrum of mantle mixing scenarios (two-layer convection, partial UM–LM mixing, whole mantle convection).

4. Issues Related to "Local" Reservoirs

In our work, local reservoirs have been arbitrarily defined as the nine crustal tiles around each detector (except for Kamioka), see Sec. 2.3. Here we discuss some issues related to this or other choices for the "local" contribution to geo-neutrino fluxes.

4.1. What is a "local" reservoir?

It is necessary to define in some way "local" reservoirs, since the crust (and perhaps the mantle) within a few hundred km from each detectors site may well be different from the average crust defined in the previous section. This fact has already been recognized in Enomoto (2005) and Fiorentini et al. (2005b), where "local" HPE abundances for the Kamioka site have been estimated. The boundaries of the local crust are matter of convention – but any convention is not without consequences, however. In particular, in our approach, we observe that the correlation between "local" reservoirs and "global" ones (LM, UM, OC, CC) is expected to vanish: the uncertainty of the U abundance near the Kamioka mine has probably nothing to do with the errors of the whole CC and OC crust estimates. However, only dedicated studies, which should take into account all locally homogenous geochemical micro-reservoirs and their correlation lengths with farther geo-structures, can provide a physically motivated distinction between local and global reservoirs – a task much beyond the scope of this work. For simplicity, we just assume that all local-to-global abundance correlations are exactly zero.

4.2. Horizontal and vertical (U, Th, K) distribution uncertainties

Assuming that a "local" reservoir is defined in some way, its volumetric distributions of HPE's can significantly affect the estimated geo-ν fluxes, due to the inverse square distance dependence. In principle, one would like to have such detailed information around each detector site, both horizontally and vertically. In practice, however, one usually has mainly scattered "surface" samples and only weak constraints about the vertical HPE distribution. Although the HPE abundances are expected to decrease with depth, the decrease may be highly site-dependent and non-monotonic (see Furukawa and Shinjoe (1997) as an example for the Japanese crust). In some cases (e.g., Schneider et al., 1987) the horizontal and vertical distributions of HPE's are correlated by the fact that the crust is locally "tilted" – a situation which may represent both a complication and an opportunity.

In all cases, significant progress in the characterization of the HPE volumetric distribution in local reservoirs can be obtained only by dedicated geophysical and geochemical studies, which should collect all the (currently sparse and partly unpublished) relevant pieces of data, including representative rock samples, local crustal models, and heat flow measurements. Some interesting work in this direction has been done for the Kamioka site (Enomoto, 2005; Fiorentini et al., 2005b), showing that a $O(10\%)$ uncertainty in the local geo-neutrino flux (at 1σ) may perhaps be reachable. We think that 10% should be the "target error" for the characterization of the local geo-neutrino flux at each detector site. Much larger errors would hide information coming from farther reservoirs, and in particular from the mantle.

4.3. Provisional assumptions

In order to make provisional numerical estimates, we make the following assumptions for local HPE abundances in the crust: (1) we assume the same numerical values and errors as for the average upper, middle, and lower crust estimates (Rudnick and Gao, 2003) discussed in Sec. 3.2, except for the Kamioka site where the average upper crust abundances are taken from the thorough geochemical study in Togashi et al. (2000); (2) we assume that the correlations between local and global abundances, as well between local crust layers, are zero; (3) we assume a plausible (but arbitrary) hierarchy of correlations between HPE abundances in each layer: ρ(U, Th) = 0.8, ρ(U, K) = 0.7, ρ(K, Th) = 0.6, implying that Th is a good proxy for U and that K is a somewhat worse proxy for both U and Th (as it generally happens in other reservoirs). Further comments about such choices are given elsewhere (Fogli et al., 2006a). As previously remarked, the admittedly arbitrary assumptions characterizing local contributions to geo-ν fluxes can and should be improved by dedicated inter-disciplinary studies.

5. Forward Propagation of Uncertainties

We have described in Sec. 2 a possible path towards the definition of a GNSM, i.e., of a set of HPE abundances, errors and correlations in a given partition of the Earth into global and local reservoirs. We now show examples of propagation of such uncertainties, according to Eqs. (11)–(15).

5.1. Errors and correlations at a specific site (Kamioka)

Figure 3 shows our estimated geo-neutrino event rates from U and Th decays at the Kamioka site (including neutrino oscillations with $\langle P_{ee}\rangle = 0.57$), superposed to the same experimental (gaussian) contours as in Fig. 1. The numerical values for the GNSM predictions are:

$$R_{\mathrm{U}} = 24.9 \pm 2.0 \text{ TNU}, R_{\mathrm{Th}} = 6.7 \pm 0.5 \text{ TNU}, \rho = +0.902. \tag{25}$$

The correlation between the theoretical rates is positive, since Th and U are good proxies of one another in each reservoir. The total (U + Th) rate in Kamioka is also positively correlated with the total (U + Th + K) radiogenic heat H in the Earth. We estimate:

$$R_{\mathrm{U+Th}} = 31.6 \pm 2.5 \text{ TNU}, H = 21.1 \pm 3.0 \text{ TW}, \rho = +0.858. \tag{26}$$

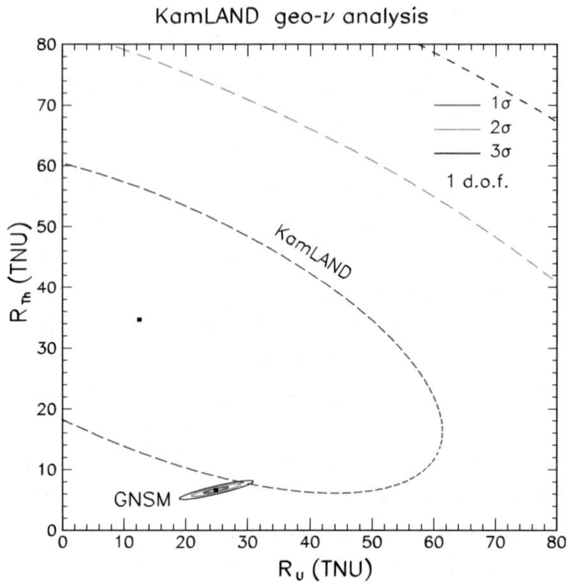

Figure 3. U and Th geo-neutrino event rate predictions from our tentative GNSM at Kamioka (small ellipses with positive correlations), superposed to the same KamLAND experimental constraints as in Fig. 1. In both cases, the 1, 2, and 3σ contours are shown. Current experimental errors appear to be significantly larger than the "theoretical" GNSM ones.

The strong correlation between $R_{\mathrm{U+Th}}$ and H implies that a precise measurement of the former would yield a robust constraint on the latter. Unfortunately, the experimental errors in Fig. 3 are still much larger than the theoretical (GNSM) ones, implying that, at present, the first KamLAND data do not significantly constrain plausible Earth models and the associated radiogenic heat (see also Fiorentini et al., 2005c). Patient accumulation of statistics, significant reduction of background and systematics, and new independent experiments, are required to test and constrain typical Earth model predictions.

5.2. Errors and correlations among different sites

Table III shows our estimates for the total (U + Th) rates (central values and $\pm 1\sigma$ error) at different possible detector sites, together with their correlation matrix. Correlations are always positive (when one rate increases, any other is typically expected to do the same), but can either be strong (such as between Gran Sasso and Pyhäsalmi, both located in somewhat similar CC settings), or relatively weak (such as between any "continental crustal site" and the peculiar "oceanic site" at Hawaii, which sits on the mantle). Such

TABLE III

Expected total neutrino event rates (U + Th), together with their errors and correlations, as calculated for different sites within the GNSM, assuming $\langle P_{ee} \rangle = 0.57$

Site	Rate (U + Th) $\pm 1\sigma$ (TNU)	Correlation matrix					
		Kam.	Gra.	Sud.	Haw.	Pyh.	Bak.
Kamioka (Japan)	31.60 ± 2.46	1.000	0.722	0.649	0.825	0.630	0.624
Gran Sasso (Italy)	40.55 ± 2.86		1.000	0.707	0.641	0.734	0.700
Sudbury (Canada)	47.86 ± 3.23			1.000	0.554	0.688	0.652
Hawaii (USA)	13.39 ± 2.21				1.000	0.484	0.510
Pyhäsalmi (Finland)	49.94 ± 3.45					1.000	0.692
Baksan (Russia)	50.73 ± 3.41						1.000

correlations should be taken into account in the future, when data from two or more detectors will be compared with Earth models.

6. Backward Update of the GNSM Error Matrix

As shown in Fig. 3, the first KamLAND data do not yet constrain our tentative GNSM. However, it is tempting to investigate the impact of future, high-statistics and multi-detector geo-neutrino data on the model. In particular, one might try to estimate what are the HPE abundances which best fit *both* the starting GNSM *and* a set of prospective, hypothetical experimental data ("backward" update of GNSM errors). It can be shown that the covariance formalism allows to reduce this problem to matrix algebra (Fogli et al., 2006a).

Here we give just a relevant example of possible results, in an admittedly optimistic future scenario where all six detectors in Table III are operative and collect separately U and Th events for a total exposure of 20 kilo-ton year, at exactly the predicted rate, with no background and no systematics. In such scenario, the mantle and BSE uranium abundance errors would be reduced as in Fig. 4, which should be compared with the previous estimates in Fig. 2. It can be seen that, in principle, the depicted scenario might allow to reject at $\gg 3\sigma$ the case $a_{\mathrm{LM}} = a_{\mathrm{UM}}$, i.e., of global mantle convection, which would be a really relevant result in geophysics and geochemistry. Needless to say, more realistic (and less optimistic) simulations of prospective data need to be performed in order to check if similar goals can be experimentally reached. In any case, our approach may provide a useful template for such numerical studies.

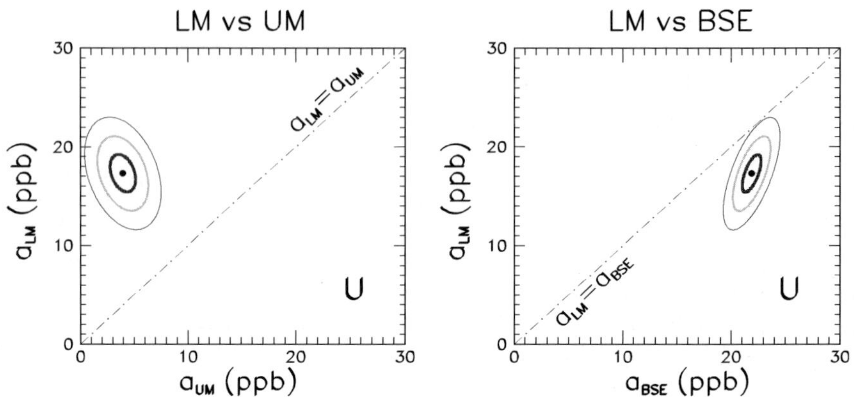

Figure 4. As in Fig. 2, but including constraints from a hypothetical data set from six detectors running for 20 kTy in ideal conditions. See the text for details.

7. Conclusions and Prospects for Further Work

In this contribution to the Hawaii Workshop on Neutrino Geophysics (2005) we have briefly presented a systematic approach to the ubiquitous issue of covariances in geo-neutrino analyses. Correlations among the abundances of (U, Th, K) in each reservoir and among different reservoirs, as well as covariances between any two linear combinations of such abundances (including neutrino fluxes, event rates, heat production rates) have been treated in a statistically consistent way. A tentative GNSM – embedding a full error matrix for the (U, Th, K) abundances in relevant local and global reservoirs – has been built, based on published data (when available) and on supplementary assumptions (when needed). The construction of the GNSM highlights some crucial issues that should be solved by dedicated studies, in order to get the most from future geo-neutrino data. Applications of our approach have been given in terms of predictions for future experiments ("forward" propagation of errors) and of GNSM error reduction through prospective data ("backward" update of uncertainties). Inter-disciplinary studies of more refined geochemical and geophysical Earth models, and of future possible observations of geo-neutrino signals, are needed to quantify more realistically both the assumed uncertainties and the future impact of geo-neutrino data in Earth sciences.

Acknowledgements

E.L. thanks the organizers of the Hawaii Workshop on Neutrino Geophysics (2005), where results of this work were presented, for kind hospitality and support. E.L. also thanks and W.F. McDonough and G. Fiorentini for interesting discussions about the BSE model. This work is supported in part by the Italian Istituto Nazionale di Fisica Nucleare (INFN) and Ministero dell'Istruzione, Università e Ricerca (MIUR) through the "Astroparticle Physics" research project.

References

Allègre, C., Manhés, G. and Lewin, É.: 2001, *Earth Planet. Sci. Lett.* **185**, 49.

Araki, T. et al. (KamLAND Collaboration): 2005, *Nature* **436**, 499.

Bahcall, J. N., Serenelli, A. and Basu, S.: 2005, preprint astro-ph/0511406.

Bassin, C., Laske, G. and Masters, G.: 2000, *Eos, Trans. Am. Geoph. Union*, **81**, F897. See also the website mahi.ucsd.edu/Gabi/rem.dir/crust/crust2.html.

Dziewonski, A. M. and Anderson, D. L.: 1981, *Phys. Earth Planet. Int.* **25**, 297.

Eadie, T., Drijard, D., James, F. E., Roos, M. and Sadoulet, B.: 1971. *Statistical Methods in Experimental Physics*, North Holland, Amsterdam.

Enomoto, S.: 2005, PhD thesis, Tohoku University, Japan. Available at http://www.awa.tohoku.ac.jp/KamLAND/publications/Sanshiro_thesis.pdf.

Fiorentini, G., Lissia, M., Mantovani, F. and Ricci, B.: 2005a, *Earth Planet. Sci. Lett.* **238**, 235.

Fiorentini, G., Lissia, M., Mantovani, F. and Vannucci, R.: 2005b, *Phys. Rev. D* **72**, 033017.

Fiorentini, G., Lissia, M., Mantovani, F. and Vannucci, R.: 2005c, *Phys. Lett. B* **629**, 77.

Fogli, G. L., Lisi E., Palazzo A. and Rotunno A. M.: 2006a, preprint in preparation.

Fogli, G. L., Lisi, E., Marrone, A. and Palazzo, A.: 2006b, *Prog. Part. Nucl. Phys.*, **57**, 742.

Furukawa, Y. and Shinjoe, H.: 1997, *Geophys. Res. Lett.* **24**, 1279.

Goreva, J. S. and Burnett, D. S.: 2001, *Meteorit. Planet. Sci.* **36**, 63.

Hofmann, A. W.: 1988, *Earth Planet. Sci. Lett.* **90**, 297.

Hofmann, A. W.: 2003, in R. W. Carlson (ed.), *Treatise On Geochemistry*, Vol. 2, Elsevier-Pergamon, Oxford, pp. 61–101.

Jaupart, C. and Mareschal, J. C.: 2003, in R. L. Rudnick (ed.), *Treatise on Geochemistry*, Vol. 3, Elsevier-Pergamon, Oxford, pp. 65–84.

Jochum, K. P., Hofmann, A. W., Ito, E., Seufert, H. M. and White, W. M.: 1983, *Nature* **306**, 431.

Krauss, L. M., Glashow, S. L. and Schramm, D. N.: 1984, *Nature* **310**, 191.

Lodders, K.: 2003, *Astrophys. J.* **591**, 1220.

Mantovani, F. et al.: 2005, Neutrino Geophysics Workshop (Honolulu, Hawaii), in these Proceedings.

Mantovani, F., Carmignani, L., Fiorentini, G. and Lissia, M.: 2004, *Phys. Rev. D* **69**, 013001, 12 pp.

McDonough, W. F. and Sun, S.-S.: 1995, *Chem. Geol.* **120**, 223.

McDonough, W. F.: 2003, in R. W. Carlson (ed.) *The Mantle and Core*, pp. 547–568, Holland, H. D. and Turekian, K. K. (eds.), Vol. 2 of "*Treatise on Geochemistry*", Elsevier-Pergamon, Oxford.

Palme, H. and O'Neill, H.St.C.: 2003, in R. W. Carlson (ed.), *Treatise on Geochemistry*, Vol. 2, Elsevier-Pergamon, Oxford, pp. 1–38.

Rama Murthy, V.: 2005, Neutrino Geophysics Workshop (Honolulu, Hawaii), in these Proceedings.

Rocholl, A. and Jochum, K. P.: 1993, *Earth Planet. Sci. Lett.* **117**, 265.

Rudnick, R. L. and Gao, S.: 2003, in R. L. Rudnick (ed.), *Treatise On Geochemistry*, Vol. 3, Elsevier-Pergamon, Oxford, pp. 1–64.

Salters, V. and Stracke, A.: 2006, *Geochem. Geophys. Geosyst.* **5**, Q05004, doi:10.1029/2003GC000597.
Schneider, R. V., Roy, F. R. and Smith, A. R.: 1987, *Geophys. Res. Lett.* **14**, 264.
Sleep, N.: 2005, Neutrino Geophysics Workshop (Honolulu, Hawaii), in these Proceedings.
Staudigel, H. et al.: 1998, *Chem. Geol.* **145**, 153–159.
Taylor, S. R. and McLennan, S. M.: 1985. *The Continental crust: Its Composition and Evolution*, Blackwell, Oxford, 312 pp.
Togashi, S. et al.: 2000, *Geochem. Geophys. Geosyst.* **1** 2000 GC000083.
Wasserburg, G. J., MacDonald, G. J. F., Hoyle, F. and Fowler, A. W.: 1964, *Science* **143**, 465.
Wedepohl, K. H. and Hartmann, G.: 1994, in H. O. A. Meyer and O. H. Leonardos (eds.), *Kimberlites, Related Rocks and Mantle Xenoliths*, Vol. 1, Companhia de Pesquisas de Recursos Minerais, Rio de Janeiro, pp. 486–495.
Workman, R. K. and Hart, S. R.: 2005, *Earth Planet. Sci. Lett.* **231**, 53.

Earth, Moon, and Planets (2006) 99:131–146
DOI 10.1007/s11038-006-9120-8

Experimental Study of Geoneutrinos with KamLAND

SANSHIRO ENOMOTO
Research Center for Neutrino Science, Tohoku University, Aramaki Aoba, Aobaku, Miyagi, Sendai, 980-8578, Japan
(E-mail: sanshiro@awa.tohoku.ac.jp)

(Received 10 June 2006; Accepted 22 August 2006)

Abstract. The Kamioka liquid scintillator antineutrino detector (KamLAND), which consists of 1000 tones of ultra-pure liquid scintillator surrounded by 1879 photo-multiplier tubes (PMT), is the first detector sensitive enough to detect geoneutrinos. Earth models suggest that KamLAND observes geoneutrinos at a rate of 30 events/10^{32}-protons/year from the ^{238}U decay chain, and 8 events/10^{32}-protons/year from the ^{232}Th decay chain. With 7.09×10^{31} proton-years of detector exposure and detection efficiency of 0.687 ± 0.007, the 'rate-only' analysis gives 25^{+19}_{-18} geoneutrino candidates. Assuming a Th/U mass concentration ratio of 3.9, the 'rate + shape' analysis gives the 90% confidence interval for the total number of geoneutrinos detected to be from 4.5 to 54.2. This result is consistent with predictions from the Earth models. The 99% C.L. upper limit is set at 1.45×10^{-31} events per target proton per year, which is 3.8 times higher than the central value of the model prediction that gives 16 TW of radiogenic heat production from ^{238}U and ^{232}Th. Although the present data have limited statistical power, they provide by direct means an upper limit for the Earth's radiogenic heat of U and Th.

Keywords: Geoneutrino, KamLAND, earth energetics, bulk silicate earth

1. Neutrino Geophysics with KamLAND

It is widely accepted that radiogenic heat contributes a large part to the Earth's heat budget. Due to direct relation between the number of radioactive decays in the Earth and the neutrino flux from the decays, neutrinos are expected to provide fruitful information on Earth's energetics. Using neutrinos to study processes inside the Earth was first suggested by Eder (1966) and Marx (1969) in 1960's, and has been reviewed a number of times (Avilez et al., 1981; Krauss et al., 1984; Kobayashi and Fukao, 1991; Raghavan et al., 1998; Rothschild et al., 1998; Mantovani et al., 2004). The Kamioka liquid scintillator antineutrino detector (KamLAND), which detects few MeV of electron antineutrinos with 1000 tones of ultra-pure liquid scintillator, is the first detector sensitive enough to measure geologically produced antineutrinos (geoneutrinos).

^{238}U, ^{235}U and ^{232}Th generate radiogenic heat via a series of alpha and beta decays. ^{40}K generates radiogenic heat via either beta decay or electron capture with branching ratios of 0.893 and 0.107, respectively. In addition to a daughter nucleus, each beta decay produces an electron and an antineutrino. Electron capture produces a neutrino and a daughter nucleus. These decays can be summarized by

$$\begin{aligned} {}^{238}\mathrm{U} &\rightarrow {}^{206}\mathrm{Pb} + 8\,{}^{4}\mathrm{He} + 6e^{-} + 6\bar{\nu}_e + 51.7\,\mathrm{MeV}, \quad \tau_{1/2} = 4.468 \times 10^{9}\ \mathrm{yr} \\ {}^{235}\mathrm{U} &\rightarrow {}^{207}\mathrm{Pb} + 7\,{}^{4}\mathrm{He} + 4e^{-} + 4\bar{\nu}_e + 46.0\ \mathrm{MeV}, \quad \tau_{1/2} = 7.038 \times 10^{8}\ \mathrm{yr} \\ {}^{232}\mathrm{Th} &\rightarrow {}^{208}\mathrm{Pb} + 6\,{}^{4}\mathrm{He} + 4e^{-} + 4\bar{\nu}_e + 42.7\ \mathrm{MeV}, \quad \tau_{1/2} = 1.405 \times 10^{10}\ \mathrm{yr} \\ {}^{40}\mathrm{K} &\rightarrow {}^{40}\mathrm{Ca} + e^{-} + \bar{\nu}_e + 1.31\ \mathrm{MeV}\ (89.3\%), \quad \tau_{1/2} = 1.277 \times 10^{9}\ \mathrm{yr} \\ {}^{40}\mathrm{K} + e^{-} &\rightarrow {}^{40}\mathrm{Ar} + \nu_e + 1.51\ \mathrm{MeV}\ (10.7\%) \end{aligned} \tag{1}$$

Antineutrino luminosity $L_{\bar{\nu}}$ (number of antineutrino emissions per unit time) and heat generation H of ^{238}U, ^{235}U, ^{232}Th and ^{40}K can be directly calculated from their mass M by

$$\begin{aligned} {}^{238}\mathrm{U} : L_{\bar{\nu}}[1/\mathrm{sec}] &= 7.46 \times 10^{7} \cdot M[\mathrm{kg}] = 7.84 \times 10^{11} \cdot H[\mathrm{W}] \\ {}^{235}\mathrm{U} : L_{\bar{\nu}}[1/\mathrm{sec}] &= 3.20 \times 10^{8} \cdot M[\mathrm{kg}] = 5.67 \times 10^{11} \cdot H[\mathrm{W}] \\ {}^{232}\mathrm{Th} : L_{\bar{\nu}}[1/\mathrm{sec}] &= 1.62 \times 10^{7} \cdot M[\mathrm{kg}] = 6.18 \times 10^{11} \cdot H[\mathrm{W}] \\ {}^{40}\mathrm{K} : L_{\bar{\nu}}[1/\mathrm{sec}] &= 2.31 \times 10^{8} \cdot M[\mathrm{kg}] = 8.18 \times 10^{12} \cdot H[\mathrm{W}] \end{aligned} \tag{2}$$

where the neutrino kinetic energy is subtracted from H, since almost all this energy escapes the Earth. In order to calculate the energies taken by neutrinos on beta decays, the allowed beta transition formula with the Fermi correction is used except for ^{40}K and ^{210}Bi in the ^{238}U series; the 3rd unique forbidden transition formula is used for ^{40}K, and the spectrum tabulated in (Aleksankin, 1989) is used for ^{210}Bi. The isotope decay data used in this calculation is taken from (Firestone and Shirley, 1996). Based on chondritic abundances and cosmochemical consideration of the volatility of elements, the current model of the bulk silicate Earth (BSE) (McDonough and Sun, 1995) gives heat generation of 8 TW by ^{238}U and ^{235}U, 8 TW by ^{232}Th, and 3 TW by ^{40}K, resulting in total of 20 TW for the radiogenic heat generated in the Earth's interior. Table I summarizes the neutrino luminosity, heat generation of these elements, among with total mass in the Earth given by the BSE model. Figure 1 shows the calculated antineutrino spectra from these elements. There are even more radioactive elements in the Earth, as also listed in Table I, making negligible contribution to the total radiogenic heat production.

The neutrino flux at a position $\vec{r}$ for each isotope can be calculated from the isotope distribution (isotope mass per unit rock mass) $a(\vec{r})$ by integrating the contribution over the entire Earth,

TABLE I

Radiogenic heat and geoneutrino luminosity. ^{238}U, ^{235}U and ^{232}Th represent the whole decay chains, ^{40}K to ^{187}Re are beta-decaying or electron-capture (EC) isotopes, ^{123}Te is EC only, and ^{152}Gd to ^{186}Os are alpha-decaying isotopes. $M_{\mathrm{Earth}} = 5.97 \times 10^{24}$ kg, $M_{\mathrm{BSE}} = 4.05 \times 10^{24}$ kg, and $M_{\mathrm{Core}} = 1.93 \times 10^{24}$ kg is used

Isotope	Isotope abundance (%)	H/M_{Isotope} (μW/kg)	$L_{\bar{\nu}}/M_{\mathrm{Isotope}} \times 10^7$ (1 s^{-1} kg^{-1})	BSE (Core) Model prediction		
				a_{element} (ppm)	Total H (TW)	Total $L_{\bar{\nu}} \times 10^{24}$ (1 s^{-1})
^{238}U	99.28	95.2	7.46	0.02	7.7	6.0
^{232}Th	100	26.3	1.62	0.08	8.5	5.2
^{235}U	0.72	564	3.20	0.02	0.33	0.19
^{40}K	0.0117	28.2	2.31	240	3.2	2.6
^{176}Lu	2.59	2.5×10^{-1}	2.0×10^{-1}	0.068	1.8×10^{-3}	1.4×10^{-2}
^{115}In	95.7	6.4×10^{-6}	2.6×10^{-5}	0.01	2.5×10^{-7}	1.0×10^{-5}
^{113}Cd	12.22	1.8×10^{-7}	1.3×10^{-6}	0.04 (0.15)	3.6×10^{-9} (6.5×10^{-9})	2.5×10^{-7}(4.5×10^{-7})
^{87}Rb	27.835	4.2×10^{-2}	3.2×10^{-1}	0.60	2.8×10^{-2}	2.2
^{138}La	0.0902	1.8×10^{-1}	3.1×10^{-2}	0.65	4.3×10^{-4}	7.4×10^{-4}
^{187}Re	62.6	1.3×10^{-4}	1.6×10^{-1}	0.0003 (0.23)	1.0×10^{-7}(3.6×10^{-5})	1.2×10^{-3} (4.5×10^{-1})
^{123}Te	0.908	0	0	0.012 (0.85)	0 (0)	0 (0)
^{152}Gd	0.20	2.8×10^{-4}	0	0.54	1.2×10^{-6}	0
^{148}Sm	11.3	4.1×10^{-6}	0	0.41	7.6×10^{-7}	0
^{144}Nd	23.8	1.2×10^{-5}	0	1.25	1.5×10^{-5}	0
^{147}Sm	15.0	3.1×10^{-1}	0	0.41	7.8×10^{-2}	0
^{174}Hf	0.162	1.5×10^{-5}	0	0.28	2.8×10^{-8}	0
^{190}Pt	0.01	5.6×10^{-2}	0	0.007 (5.7)	1.6×10^{-7} (6.1×10^{-5})	0 (0)
^{186}Os	1.58	1.6×10^{-5}	0	0.003 (2.8)	3.1×10^{-9} (1.4×10^{-6})	0 (0)

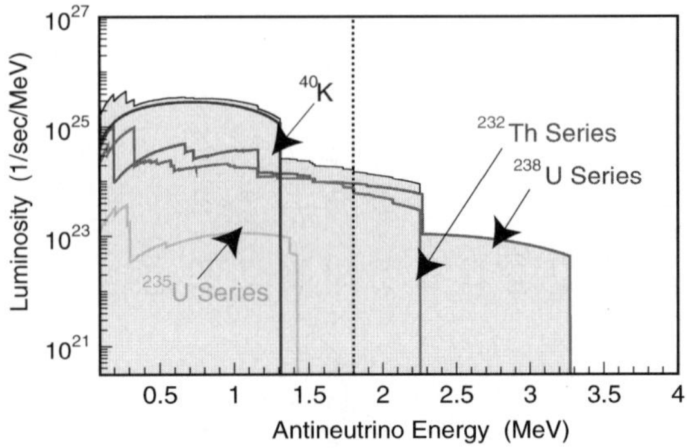

Figure 1. Geoneutrino Spectrum. KamLAND can only detect geoneutrinos with energies greater than 1.8 MeV (right of the vertical dotted line); hence it is insensitive to ^{235}U and ^{40}K geoneutrinos.

$$\frac{\mathrm{d}\Phi(E_\nu, \vec{r})}{\mathrm{d}E_\nu} = A \frac{\mathrm{d}n(E_\nu)}{\mathrm{d}E_\nu} \int_\oplus \mathrm{d}^3\vec{r}\,' \frac{a(\vec{r}\,')\rho(\vec{r}\,')P(E_\nu, |\vec{r} - \vec{r}\,'|)}{4\pi|\vec{r} - \vec{r}\,'|^2} \tag{3}$$

where A is the decay rate per unit mass, $\mathrm{d}n(E_\nu)/\mathrm{d}E_\nu$ is the energy spectrum of neutrinos per decay, $\rho(\vec{r}\,')$ is the rock density, and $P(E_\nu, |\vec{r} - \vec{r}\,'|)$ is the neutrino survival probability after traveling from the source position $\vec{r}\,'$ to the detector position $\vec{r}$. The neutrino survival probability is given by a well-established formula,

$$P(E_\nu, L) \cong 1 - \sin^2 2\theta_{12} \sin^2\left(\frac{1.27\Delta m_{12}^2[\mathrm{eV}^2]L[\mathrm{m}]}{E_\nu[\mathrm{MeV}]}\right) \tag{4}$$

where $L = |\vec{r} - \vec{r}\,'|$. In order to evaluate the geoneutrino rate detectable by KamLAND, we constructed a model of the Earth (Enomoto, 2005a, b). The model is based on the BSE composition given by McDonough (McDonough and Sun, 1995) and a crustal composition model given by Rudnick and Fountain (1995). Following common practice, we assume that the core does not contain U and Th. The mantle composition is assumed to be uniform and is obtained by subtracting the crustal composition from the BSE composition. With this Earth model, the geoneutrino flux at the KamLAND site (36° 25′36″ N, 137°18′43″ E, 358 m elevation from the sea level) is calculated to be 2.34×10^6 cm^{-2}sec^{-1} from ^{238}U decay, 0.07×10^6 cm^{-2}sec^{-1} from ^{235}U decay, 1.99×10^6 cm^{-2}sec^{-1} from ^{232}Th decay, and 1.48×10^6 cm^{-2}sec^{-1} from ^{40}K decay. As shown in Figure 2, about half of the total geoneutrino flux originates within 500 km radius, and about a quarter comes from the mantle. By modifying the Earth model, we found that re-distribution of the sources between the upper and lower mantle makes 3% variation to the total

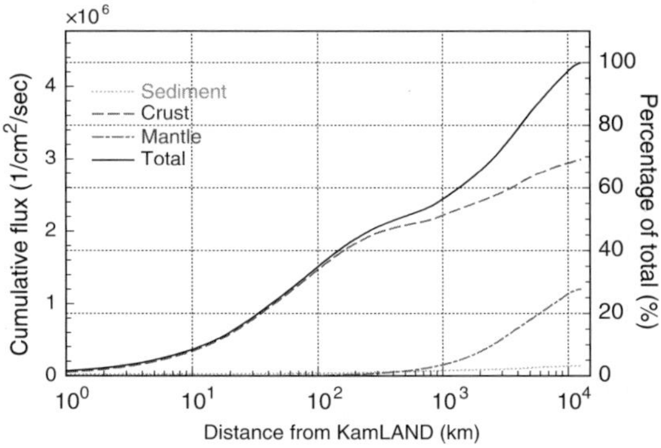

Figure 2. The expected ^{238}U and ^{232}Th geoneutrino flux within a given distance from KamLAND. Approximately 25 and 50% of the total flux originates within 50 km and 500 km of KamLAND, respectively. The mantle contributes 27% of the total flux.

flux, and alternation of the crustal models (i.e., adopting other crustal models (McLennan and Taylor, 1999; Wedepohl, 1995) with keeping the BSE composition by adjusting the mantle composition) makes 8% variation in the total flux (for details, see Enomoto, 2005a, b). The model parameters and the calculated flux contribution from each part of the Earth are summarized in Table II.

Due to large contribution from the near-field region, regional geological effect may considerably influence the total geoneutrino flux, while not significantly affecting the global geophysics. An estimation of the Japan Island Arc crustal composition shows depletion of U and Th in the Japan crust, which leads to reduction of U and Th geoneutrino flux at the KamLAND site by 6.4% and 8.4%, respectively. Based on a large scale geochemical study, the near-field surface geological non-homogeneity might affect the total flux at about 3–10%. Other regional geological effects, such as subducting plate and stagnant slab beneath Japan, Sea of Japan geology, Kamioka mine (where the KamLAND detector is located) geology, possible undiscovered uranium deposits, are estimated and found to affect the total flux at less than few percent (Enomoto, 2005a, b). In addition to the geological uncertainties, the uncertainty of the neutrino oscillation parameters makes 6% error in the flux estimation.

KamLAND detects electron antineutrinos via neutron inverse beta-decay,

$$\bar{\nu}_e + p \rightarrow e^+ + n \tag{5}$$

which has a 1.8 MeV neutrino energy threshold, and a well-established cross-section (Vogel and Beacom, 1999). Only antineutrinos emitted from the ^{238}U and ^{232}Th decay series have energy higher than this threshold energy. The number of geoneutrino events from the reaction for the ^{238}U and ^{232}Th antineutrinos are calculated by

TABLE II

Reference Earth model and geoneutrino flux at Kamioka. U and Th concentrations, a_{element}, represent the geochemical model of the Earth, and the values $\Phi^{\mathrm{Kamioka}}/L_{\bar{\nu}}^{\mathrm{Reservoir}}$ are determined only from the geometrical shape of the reservoirs

Reservoir		a_{element} (ppm)		$\Phi^{\mathrm{Kamioka}}/L_{\bar{\nu}}^{\mathrm{Reservoir}} \times 10^{-19}(1\,\mathrm{cm}^{-2})$	$\Phi^{\mathrm{Kamioka}} \times 10^{5}(1\ \mathrm{cm}^{-2}\mathrm{s}^{-1})$	
		U	Th		^{238}U	^{232}Th
Sediment	Continental	2.8	10.7	3.16	0.61	0.51
	Oceanic	1.7	6.9	2.61	0.14	0.12
Continental Crust	Upper	2.8	10.7	8.34	11.5	9.57
	Middle	1.6	6.1	4.93	4.31	3.57
	Lower	0.2	1.2	4.97	0.53	0.69
Oceanic Crust		0.10	0.22	2.25	0.09	0.04
Mantle	Upper	0.012	0.048	2.31	2.20	1.91
	Lower	0.012	0.048	1.53	4.03	3.51
Core	Outer	0	0	1.24	0	0
	Inner	0	0	1.17	0	0
Bulk Silicate Earth (BSE)		0.02	0.08	3.89	23.4	19.9

$$N = N_{\mathrm{proton}} \cdot \tau \cdot \varepsilon \cdot \int \mathrm{d}E_\nu \sigma(E_\nu) \frac{\mathrm{d}\Phi(E_\nu)}{\mathrm{d}E_\nu} \tag{6}$$

where N_{proton} is the number of target protons, τ is the detector exposure time, ε is the detection efficiency, and $\sigma(E_\nu)$ is the cross-section of the reaction. With 1-year exposure of 10^{32} target protons assuming 100% efficiency, one ^{238}U-neutrino event corresponds to a flux of 7.67×10^4 cm^{-2}s^{-1}, and one ^{232}Th-neutrino event corresponds to a flux of 2.48×10^5 cm^{-2}s^{-1}.

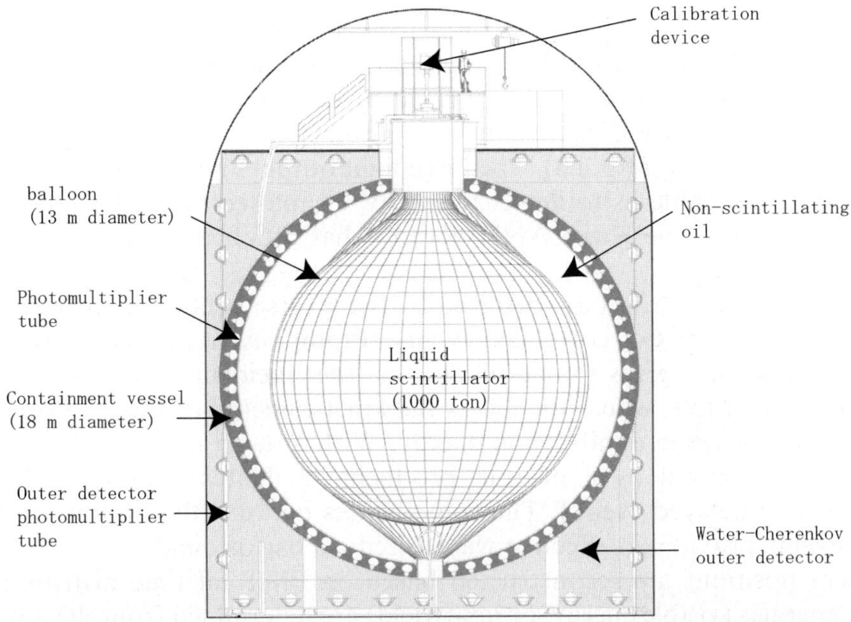

Figure 3. Schematic diagram of the KamLAND detector. The neutrino target/detector is 1000 ton of ultra-pure liquid scintillator (LS, 1,2,4-trimethylbenzene 20%, dodecane 80%, PPO 1.52 g/l) contained in a 13-m-diameter spherical plastic balloon (135-μm-thick transparent nylon/EVOH composite film). The balloon is supported and constrained by a network of Kevlar ropes. A buffer of dodecane and isoparaffin oils between the balloon and an 18-m-diameter spherical stainless-steel containment vessel shields the LS from external radiation. The specific gravity of the buffer oil is adjusted to be 0.04% lower than that of the LS. An array of 1879 photomultiplier tubes (PMT's) are mounted on the inner surface of the containment vessel. The array includes 1325 specially developed fast PMT's with 17-inch-diameter photocathodes and 554 older Kamiokande 20-inch PMT's. While the total photo-cathode coverage is 34%, only the 17-inch PMT's corresponding to 22% coverage are used for this geoneutrino analysis. A 3-mm-thick acrylic barrier at 16.6-m diameter helps prevent radon emanating from PMT grass from entering the LS. The containment vessel is surrounded by a 3200 ton water-Cherenkov detector with 225 20-inch PMT's. This outer detector absorbs gamma-rays and neutrons from the surrounding rock and provides a tag for cosmic-ray muons.

2. The KamLAND Detector

KamLAND is located in the Kamioka mine, 1000 m below the summit of Mt. Ikenoyama, Gifu prefecture, Japan (36°25′36″ N, 137°18′ 43″ E, 2700 m.w.e. underground) (http://www.awa.tohoku.ac.jp/KamLAND). The detector consists of 1000 ton of ultra-pure liquid scintillator (1,2,4-trimethylbenzene 20%, dodecane 80%, and PPO 1.52 g/l) surrounded by 1879 photo-multiplier tubes (PMT), 1325 of which are specially developed fast 17-inch PMT's and 554 of which are older Kamiokande 20-inch PMT's. Figure 3 illustrates the detector construction. The light output of the liquid scintillator is about 8000 photons/MeV, and the photo-coverage of the 1879 PMT's is 34%, providing ~500 p.e./MeV. For the geoneutrino analysis presented in this paper, only 17-inch PMT's are used, corresponding to 22% photo-coverage. The liquid scintillator was purified by the water extraction technique, resulting in very low radioactive impurity level ($(3.5 \pm 0.5) \times 10^{-18}$g/g for U, $(5.2 \pm 0.8) \times 10^{-17}$g/g for Th). The large light output and the low radioactive impurity level enable us to detect few MeV of antineutrinos. Together with the large volume, the KamLAND detector has attained the first sensitivity for geoneutrinos.

Electron antineutrinos are detected by the inverse beta-decay reaction (5). The reaction makes two correlated events. 'Prompt event', which is produced by the position, gives an estimate of the incident neutrino energy, $E_\nu \approx E_{e^+} + 0.8\text{MeV}$, where the E_{e^+} is the kinetic energy of the positron plus the electron-positron annihilation energy. With a mean time of ~200 μs, the neutron is captured by a proton, producing a deuteron and a 2.2 MeV gamma-ray ('delayed event'). The coincidences between the prompt and delayed events provide an effective way to reduce background.

Event positions are reconstructed based on PMT hit time distributions. Event energies (visible energy, or light yield) are determined from the amount of detected light after correcting for spatial dependences of the detector response. Particle energies are then calculated from the visible energy considering scintillator response (quenching) and Cherenkov light emission. The event position and energy determination are calibrated with gamma-ray sources deployed along the vertical center axis of the detector. Within the geoneutrino energy range ($0.9\text{ MeV} < E_{\text{visible}} < 2.6\text{ MeV}$, where E_{visible} roughly corresponds to E_{e^+}), positions and energies are determined within 5 cm and 2% accuracy, respectively, throughout the spherical volume of 5 m radius from the detector center. The scintillator mass in the 5 m radius spherical fiducial volume is estimated using reconstructed vertices of uniformly distributed cosmogenic events (^{8}B and ^{9}Li) and the measured total scintillator mass. The number of target protons in the fiducial volume is estimated at $(3.46 \pm 0.17) \times 10^{31}$.

2.1. Data analysis

The data presented here are based on a total detector live-time of 749.1 days after basic data quality cuts. The total exposure is $(7.09 \pm 0.35) \times 10^{31}$ target-proton years (Araki et al., 2005b). Neutrino event candidates are selected by searching for prompt events accompanied with a single 2.2 MeV delayed event. The time coincidence between the prompt and delayed events (ΔT) is required to satisfy 0.5μ s $\leq \Delta T < 500\mu$ s, and the spatial coincidence between the prompt and delayed (Δr) is required to satisfy $\Delta r < 100$ cm. The delayed event energy (E_d) is required to satisfy $| E_d - 2.2\text{MeV}| < 0.4\text{MeV}$, taking account of the detector energy resolution. The energy window for the prompt events of the geoneutrino candidates is set to $0.9\text{MeV} \leq E_{e+} < 2.6$ MeV, which corresponds to the neutrino energy of $1.7\text{MeV} \leq E_\nu < 3.4$ MeV (the energy range reaches below the inverse beta-decay threshold owing to the detector energy resolution). The overall efficiency for detecting geoneutrino candidates within this energy window in the 5 m radius spherical fiducial volume is estimated to be 0.687 ± 0.007. To reject cosmogenic backgrounds, whole volume or partial volume vetoes are applied following the passage of every cosmic muon. The deadtime due to the vetoes, 10.5% of the total run time, is subtracted from the live-time.

Background for geoneutrino candidates is dominated by antineutrinos from nuclear reactors in the vicinity of the KamLAND detector, and by alpha-particle induced neutrons due to radioactive contamination within the detector. As shown in Figure 4a, reactor neutrinos reach substantially higher energies than geoneutrions, and the properties of the reactor antineutrinos have been studied in detail using this energy region (Eguchi, 2003; Araki et al., 2005). Using the neutrino oscillation parameters determined with antineutrinos in the $E_\nu > 3.3$ MeV region, together with the global analysis of the solar neutrinos (assuming CPT invariance), the reactor antineutrino background in the geoneutrino analysis window is estimated to be 80.4 ± 7.2. The error is dominated by the neutrino oscillation parameter uncertainties, and the initial reactor neutrino flux was calculated within a 3% error.

The alpha-particle-induced neutron background is due to the $^{13}C(\alpha, n)^{16}O$ reaction, where the alpha particle is produced in the ^{210}Po decay. The initial energy of the alpha particle emitted by the ^{210}Po decay is 5.3 MeV, and the $^{13}C(\alpha, n)^{16}O$ reaction produces neutrons with kinetic energy up to 7.3 MeV. The neutrons scatter protons as they thermalize, and the scattered protons yield scintillation light with a total visible energy of few MeV. Here the visible energy is much smaller than the initial neutron kinetic energy owning to scintillation light quenching for high ionization density. The thermal neutrons are captured by protons and produce a 2.2 MeV delayed gamma signal, in exactly the same way as neutrons produced by the inverse beta decay reaction.

^{210}Po's is produced by decay of ^{210}Pb, which has a half-life of 22 years. ^{210}Pb's is populated throughout the whole scintillator volume from the decay of ^{222}Rn, which might have been injected into the liquid scintillator during the detector construction process. Non-equilibrium of the ^{210}Po decay rate implies that ^{210}Pb could be populated during a short period of scintillator filling into the detector, and an estimation of the initial amount of ^{222}Rn based on this assumption resulted in about 100 Bq/m^3, which is more than the maximum amount under contact of the liquid scintillator with atmospheric air (Note that Rn concentration in the mine air is much higher than atmospheric air).

On the basis of the ^{13}C(α, n)^{16}O reaction cross-section, the alpha-particle energy loss in the scintillator, and the number of ^{210}Po decays, the total number of neutrons produced is calculated to be 93 ± 22, where the error is dominated by 20% uncertainty in the ^{13}C(α, n)^{16}O reaction cross-section and 14% uncertainty on an estimation of the ^{210}Po decay rate. The neutron energy distribution is calculated using a ^{13}C(α, n)^{16}O neutron angular distribution data. The visible energy spectrum is then calculated with neutron–proton scattering simulation and an estimated quenching factor for protons, assigning reasonably large error (10% in energy determination) for uncertainties in the quenching factor estimation. Including the efficiency for passing the neutrino selection cuts, the number of ^{13}C(α, n)^{16}O events is estimated to be 42 ± 11.

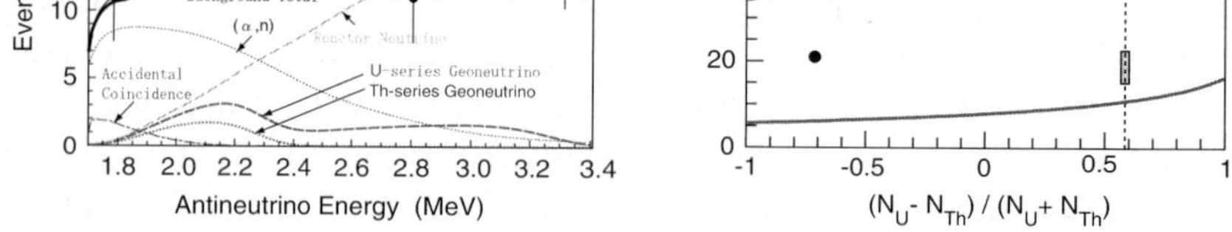

Figure 4. Neutrino energy spectra in KamLAND (left) and spectrum shape analysis result (right). Panel (a) shows the experimental points together with the total expectation (thin solid line), the total expected spectrum excluding the geoneutrino signal (thick solid line), and the expected spectra of the background signals. The geoneutrino spectra are calculated from our reference Earth model, which is based on the BSE model. In panel (b), the 68.3, 95.4, and 99.7% confidence level contours for detected ^{238}U and ^{232}Th geoneutrinos are shown. The small shaded area represents the prediction from the reference Earth model. The vertical dashed line represents the value of $(N_{\rm U}-N_{\rm Th})/(N_{\rm U}+N_{\rm Th})$ assuming the mass ratio of Th/U = 3.9. The dot represents our best fit point, favoring 3 ^{238}U geoneutrinos and 18 ^{232}Th geoneutrinos.

The contribution from accidental coincidences is 2.38 is eliminated as small as 2.38 ± 0.01 events, by adopting tighter neutrino-selection criteria with a smaller radius fiducial volume than used for the reactor neutrino analysis (Araki et al., 2005a). There is a small contribution to the background from decays of cosmic-muon-induced unstable nuclei such as ^{9}Li (which has a neutron in the final state), and neutrinos from spent reactor fuel. There is a negligible contribution from fast neutrons generated by cosmic ray interactions and spontaneous fission of ^{238}U. Other background sources considered and found to be negligible include neutron emitters and correlated decays in the radioactive impurities, (γ, n) reactions, solar neutrino induced break-up of ^{2}H, and cosmic neutrino interaction with scintillator-composing nuclei. The total number of background events is estimated to be 127 ± 13.

The total number of observed antineutrino candidates within the energy range of 1.7MeV $\leq E_{\nu} <$ 3.3MeV is 152. The error on the geoneutrino detection efficiency is estimated at 5.0%, which is dominated by the fiducial volume determination uncertainty (4.9%). Including the detection systematic errors, part of which are correlated with the background estimation errors, a 'rate-only' analysis gives 25^{+19}_{-18} geoneutrino candidate events. Dividing by the detection efficiency, live-time, and number of target protons, the total geoneutrino rate obtained is 51^{+39}_{-36} per 10^{32}-protons per year.

Figure 4a shows the observed neutrino candidate event energy spectrum, the estimated background spectra, and the expected geoneutrino spectra calculated based on our Earth model. The confidence intervals for the number of observed geoneutrinos from an un-binned maximum likelihood analysis is shown in Figure 4b. As shown in the figure, KamLAND has sensitivity in determination of the total geoneutrino flux (vertical axis in the figure), while KamLAND is less sensitive in discrimination between the U-series and Th-series geoneutrinos (horizontal axis in the figure). On the other hand, the Th/U mass ratios in the chondritic meteorites, the solar photosphere, and surface rock samples show stable value at around 3.9, and model estimations of the Th/U mass ratios in the bulk Earth, the crusts and the mantle, are relatively more reliable than the absolute concentrations of U and Th. Assuming a Th/U mass ratio of 3.9, the 90% confidence interval for the total number of geoneutrino candidates is estimated to be 4.5–54.2, with a central value of 28.0. The 99% C.L. upper limit is given at 70.7 events, which corresponds to 145 geoneutrinos per 10^{32}-protons per year and a flux of 1.62×10^{7} cm^{-2}s^{-1}.

3. Discussion

One of the primary interests of geoneutrino observation is to determine the Earth's global heat budget (i.e., direct test of the BSE model). As mentioned previously, the dependence of the flux on the crustal and mantle modeling is about the same order of magnitude as the current uncertainty of the BSE composition. Figure 5 shows the relation between the U and Th amount in each reservoir of the Earth and the geoneutrino flux from each reservoir, calculated with our reference Earth model. In the figure, alternation of reservoir's composition moves the corresponding point along the diagonal line (the slope of which is determined by the geometrical shape of the reservoir and the detector position), and the total geoneutrino flux from all of the reservoirs is obtained as a simple vector sum of each reservoir point. Assuming that we scale concentrations of U and Th in all reservoirs equally, the total geoneutrino flux observed with KamLAND F_{U+Th} is related to the total ^{238}U and ^{232}Th mass in the Earth M_{U+Th} and the heat generation H_{U+Th} as

$$H_{U+Th}[\mathrm{TW}] = \frac{M_{U+Th}[\mathrm{kg}]}{2.49 \times 10^{16}} = \frac{F_{U+Th}[\mathrm{cm}^{-2}\mathrm{s}^{-1}]}{2.70 \times 10^{5}}. \tag{7}$$

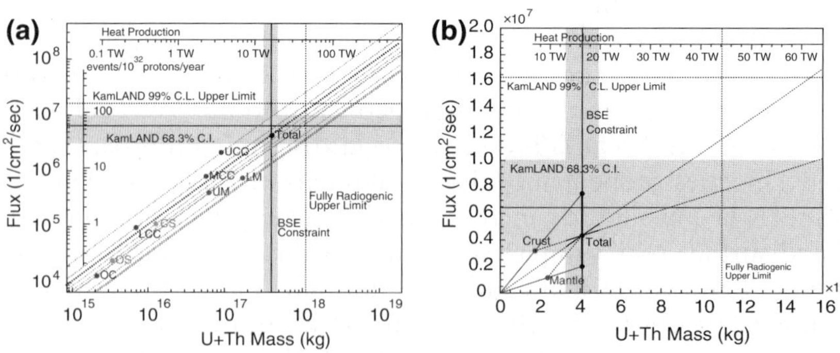

Figure 5. Comparison with Earth model predictions. Both figures show the same values but in a different axis scaling and different reservoir classification. The horizontal line is the KamLAND best-fit flux and the horizontal shaded band shows the interval of 68.3% C.L. The horizontal dotted line is the 99% C.L. upper limit. The points represent the expected neutrino flux at the KamLAND site in the reference Earth model; Upper Continental Crust (UCC), Middle Continental Crust (MCC), Lower Continental Crust (LCC), Oceanic Crust (OC), Continental Sediment (CS), Oceanic Sediment (OS), Upper Mantle (UM), and Lower Mantle (LM). This assumes a Th/U ratio of 3.9. The auxiliary axis on the top shows heat generation from U and Th for a given U + Th mass.

Therefore the observed geoneutrino flux corresponds to 23^{+13}_{-12} TW of the total radiogenic heat by ^{238}U and ^{232}Th decays. If we fix the crustal composition and parameterize the mantle composition as another example, the relation becomes

$$H_{\mathrm{U+Th}}[\mathrm{TW}] = \frac{M_{\mathrm{U+Th}}[\mathrm{kg}]}{2.49 \times 10^{16}} = 6.58 + \frac{(F_{\mathrm{U+Th}}[\mathrm{cm}^{-2}\mathrm{s}^{-1}] - 31.7 \times 10^{5})}{1.22 \times 10^{5}}. \quad (8)$$

In this case the observed geoneutrino flux corresponds to 33^{+30}_{-28} TW of the total radiogenic heat. Relations under other constraints can be directly obtained from the points and lines in Figure 5 and/or the numbers listed in Table II in the same way.

The 99% C.L. flux upper limit is 3.8 times higher than the predicted flux based on the reference Earth model, which gives 16 TW of total radiogenic heat from U and Th. Since the 99% C.L. flux upper limit is far above the predictions from Earth models due to limited statistics, we do not expect that any realistic Earth models can be used to translate this large flux to heat production. However, if we simply scale the heat production with relation (7) just to show the possible effect that geoneutrino data would have on geophysical thermal constraints, then the 99% C.L. flux upper limit is translated to a heat production of 60 TW.

The current KamLAND observation (confidence interval and 99% upper limit) is shown in Figure 5. The observation is in agreement with our Earth model prediction based on the BSE composition, although the measurement errors are relatively large. Even if the current KamLAND observation is not as precise as predictions or limits by Earth models, one can see that geoneutrino observation is approaching the point where we can gain fruitful geophysical information with geoneutrinos.

3.1. Future prospects

The current KamLAND observation suffers from large background originating from the reactor antineutrinos and (α, n) reactions. In particular, the contribution from the (α, n) reaction background is poorly estimated due to large uncertainties of the ^{210}Po decay rate, (α, n) cross-section σ (E), (α, n) partial cross-section $d\sigma(E, \theta)/d\Omega$ where θ is the neutron scattering angle, and the neutron (or proton) quenching factor.

In order to reduce the uncertainty on ^{210}Po decay rate, we have improved the event reconstruction tool, resulting in a small bias in the low energy and large R (i.e., near the boundary of the fiducial volume) region. A new measurement of the (α, n) total cross-section is now available (Harissopulos,

2005), which reduces the error on the total cross-section from 20% to 4%. In order to understand the liquid scintillator response to protons, we have performed a measurement with a mono-energetic neutron beam. A new calibration source, which consists of an α source and ^{13}C, is being prepared. In summary, we will have a better understanding of the (α, *n*) background process and will be able to reduce one of the largest uncertainties of geoneutrino observation.

For the next phase of the KamLAND experiment, which primarily aims for real-time observation of ^{7}Be solar neutrinos, a new purification system is being constructed. This system is beneficial for geoneutrions measurement, since it reduces background due to accidental coincidences and (α, *n*) to a negligible level. Reduction of accidental coincidences may allow extending the fiducial volume and improving the detection efficiencies by relaxing the delayed coincidence event selection criteria.

We estimated prospects of future KamLAND geoneutrino observation after purification by applying the same analysis method as used for the current geoneutrino analysis (Araki et al., 2005b) to software-generated neutrino event candidates. An extended fiducial volume of 5.5 m radius together with an improved detection efficiency of 99% are used, which are the values currently implemented in the KamLAND reactor neutrino analysis (Araki et al., 2005a). Figure 6 shows the expected analysis results after 750 days exposure of the detector, which is the same period as used in the current KamLAND geoneutrino result. KamLAND can determine the geoneutrino flux within 35% accuracy, which is a great improvement from

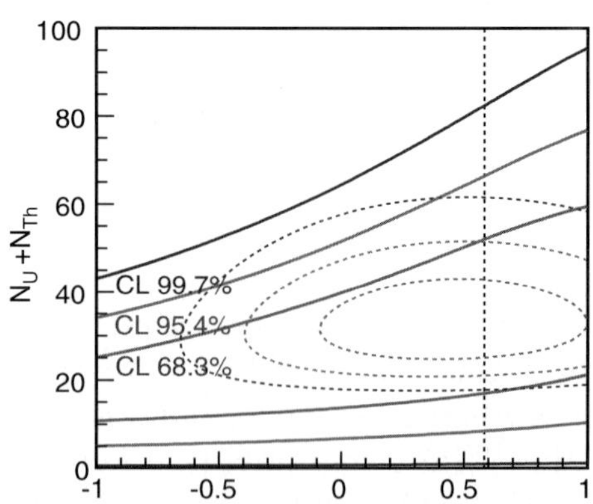

Figure 6. Prospects of future KamLAND geoneutrino observation after purification. The same analysis method as used for the current analysis is applied, except for reduced radioactive background estimation and consequent selection criteria relaxation. The solid lines show the 68.3, 95.4, and 99.7% confidence level contours. The dotted lines show the same analysis of prospects, except for absence of reactor neutrino background.

the current 54% accuracy. The significance of positive geoneutrino signal may reach 99.96%. If we combine it with the current data, the accuracy becomes 28%, which is comparable with the model prediction uncertainty. The 99% flux upper limit may be given at around 30 TW heat-generation-equivalent, which is below the surface heat-flow measurement results (44.2 ± 1.0 TW (Pollack, 1993)). Separation of ^{238}U and ^{232}Th neutrinos seems to be still difficult due to large backgrounds from the surrounding nuclear power reactors, as shown in the figure by the contrast between the contours with and without the reactor background.

4. Conclusion

The measurement of antineutrinos with an exposure of $(7.09 \pm 0.35) \times 10^{31}$ target-proton years of KamLAND is the first experimental study of geoneutrinos. The present measurement is consistent with our model predictions based on the bulk silicate Earth composition, and sets an 99% C.L. upper limit of the geoneutrino flux from ^{238}U and ^{232}Th decays within the Earth at 1.62×10^7 cm^{-2}s^{-1}, which scales to 60 TW of radiogenic heat production equivalent. Further exposure with improved accumulation of statistics with improved analysis method, reduced systematic error, and removal of radioactive background sources by the newly installed purification system will provide more precise determination of the geoneutrino flux, which might be comparable with uncertainties in the Earth models. The current result of KamLAND demonstrates that geoneutrinos are practical new tool to study the Earth interior, and paves the way to future, more accurate measurements.

References

Aleksankin, V. G. et al. 1989, in P. M. Rubzov (ed.), *Beta and Antineutrino Radiation of Radioactive Nuclei*, Energoatomizdat, Moscow, 719 pp.

Araki, T. et al. [KamLAND Collaboration]: 2005a, *Phys. Rev. Lett.* **94**, 081801.

Araki, T. et al. [KamLAND Collaboration]: 2005b, *Nature* **436**, 499–503.

Avilez, C., Marx, G. and Fuentes, B.: 1981, *Phys. Rev. D* **23**, 1116–1117.

Eder, G.: 1966, *Nucl. Phys.* **78**, 657–662.

Eguchi, K. et al. [KamLAND Collaboration]: 2003, *Phys. Rev. Lett.* **90**, 021802.

Enomoto, S.: 2005a, Neutrino Geophysics and Observation of Geo-Neutrinos at KamLAND, Thesis, Tohoku Univ.; available at http://www.awa.tohoku.ac.jp/KamLAND/publications/Sanshiro_thesis.pdf.

Enomoto, S. et al.: 2005b, Neutrino Geophysics with KamLAND and future prospects, submitted, arXiv:hep-ph/0508049.

Firestone, R. B. and Shirley, V. S.: 1996. *Table of Isotopes* (8th ed.). John Wiley, New York.

Harissopulos, S. et al.: 2005, *Phys. Rev. C* **72**, 062801.

http://www.awa.tohoku.ac.jp/KamLAND/.

Kobayashi, M. and Fukao, Y.: 1991, *Geophys. Res. Lett.* **18**, 633–636.
Krauss, L. M., Glashow, S. L. and Schramm, D. N.: 1984, *Nature* **310**, 191–198.
Mantovani, F. et al.: 2004, *Phys. Rev. D* **69**, 013001.
Marx, G.: 1969, *Czech. J. Phys. B* **19**, 1471–1479.
McDonough, W. F. and Sun, S.-s.: 1995, *Chem. Geol.* **120**, 223–253.
McLennan S. M. and Taylor S. R.: 1999, in Marshall C. P. and Fairbridge R. W. (eds.), *Encyclopedia of geochemistry*, Kluwer Academic Publishers, Dordrecht, The Netherlands, pp. 145–150.
Pollack, H. N. et al.: 1993, *Rev. Geophys.* **31**, 267–280.
Raghavan, R. S. et al.: 1998, *Phys. Rev. Lett.* **80**, 635–638.
Rothschild, C. G., Chen, M. C. and Calaprice, F. P.: 1998, *Geophys. Res. Lett.* **25**, 1083–1086.
Rudnick, R. L. and Fountain, D. M.: 1995, *Rev. Geophys.* **33**, 267–309.
Vogel, P. and Beacom, J. F.: 1999, *Phys. Rev. D*, **60**, 053003.
Wedepohl, K. H.: 1995, *Geochem. Cosmochem. Acta* **59**, 1217–1232.

Earth, Moon, and Planets (2006) 99:147–153
DOI 10.1007/s11038-006-9123-5

Experimental Status of Geo-reactor Search with KamLAND Detector

JELENA MARICIC for KAMLAND COLLABORATION
University of Hawaii at Manoa, Honolulu, HI 96822, USA
(E-mail: jelena@physics.drexel.edu)

Drexel University, Philadelphia, PA 19104, USA

(Received 28 February 2006; Accepted 9 September 2006)

Abstract. A natural nuclear fission reactor operating in the center of the Earth has been proposed by Herndon (Hollenbach and Herndon, 2001) as the energy source that powers the geo-magnetic field. The upper limit on the expected geo-reactor power is set by the estimated 12 TW (Buffett, 2003) heat flow from the Earth's core. If it exists, a nuclear reactor of that size emits a strong anti-neutrino flux. Emitted electron anti-neutrinos can be detected by the Kamioka liquid scintillator anti-neutrino detector (KamLAND) (Raghavan, 2002), and the geo-reactor power level is proporional to the anti-neutrino emission rate. KamLAND measures the geo-reactor power as a constant positive offset in detected anti-neutrino rate on top of the varying anti-neutrino rate coming from man-made reactors. Here we present the first attempt to measure the geo-reactor power. Based on a 776 ton-year exposure of KamLAND to electron anti-neutrinos, the detected flux corresponds to (6 ± 6) TW. The upper limit on the geo-reactor power at 90% confidence level is 18 TW, which is below the lower limit of the total Earth's radiogenic heat, estimated to be between 19 and 31TW (Anderson, 2003).

Keywords: Anti-neutrinos, geo-reactor, geo-magnetic field, natural nuclear reactor

1. Introduction

J.M. Herndon suggests that a nuclear fission reactor operates in the Earth's core (Hollenbach and Herndon, 2001), energizing the Earth's magnetic field. Confirmation of the existence of the geo-reactor would provide very important and revolutionary information about the Earth's dynamics, inner core composition and would require rethinking of current theories of planet formation. Lack of geo-reactor existence, although less revolutionary, would nevertheless present significant independent confirmation of the mainstream models of the Earth's interior.

Existence of the geo-reactor can be confirmed or denied by the measurement of its anti-neutrino flux. Being a fission reactor, the geo-reactor must emit a strong anti-neutrino flux. We assume that the flux is constant over the data-taking period of a few years. The constant intensity of the geo-magnetic field implies that the geo-reactor, being its power source, has a steady power output as well. Because anti-neutrino flux is proportional to reactor power, constant power level implies constant anti-neutrino flux emission. The KamLAND detector started its operations in January 2002 to study phenomenon of anti-neutrino oscillations by observing anti-neutrino flux coming from nuclear reactors around Japan and the rest of the world. This resulted in the measurement of neutrino oscillation parameters (Araki et al., 2005a). The KamLAND detector is also sensitive to anti-neutrino flux coming from the geo-reactor and can observe its flux as a fixed increase in the detected rate on top of the time varying flux coming from man-made reactors (their power levels and thus their anti-neutrino emission rates change over time). Figure 1 illustrates the variation of the unoscillated anti-neutrino rate from man-made reactors and the level of the geo-reactor contribution for two arbitrary values of geo-reactor power.

2. A Nuclear Reactor in the Earth's Core?

The geo-reactor hypothesis requires a radically different inner core composition (Herndon, 1979, 1980, 1996) from the composition predicted by the traditional BSE model (McDonough and Sun, 1995) which makes it controversial. However, there is some indirect evidence that support existence of the geo-reactor in the Earth's core (Herndon, 2003). Details of the geo-reactor model are described by the hypothesis author (Hollenbach and Herndon, 2001; Herndon, 2006).

KamLAND can detect the hypothetical geo-reactor via its anti-neutrino flux. The expected anti-neutrino rate is assumed to be constant and scales

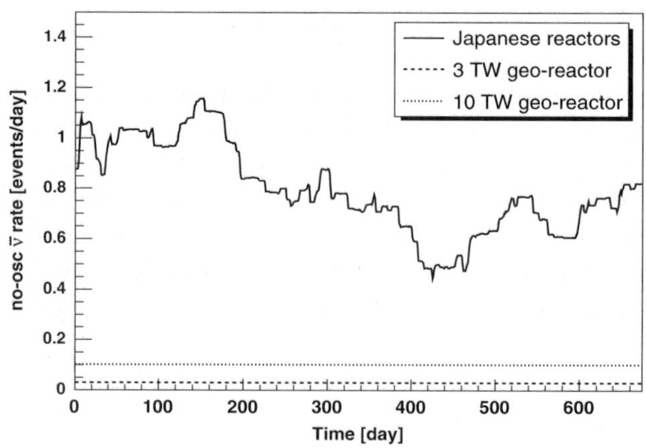

Figure 1. Unoscillated expected rate of anti-neutrinos from commercial nuclear reactors vary on daily basis, while the geo-reactor provides a small, albeit constant contribution to the overall expected rate.

with: geo-reactor power, inverse square of the Earth radius (assuming that geo-reactor is located in the Earth's center) and survival probability of anti-neutrinos due to the oscillation effects.

Anti-neutrino survival probability is described with the following formula:

$$P(\bar{\nu}_e \rightarrow \bar{\nu}_e) \approx 1 - \sin^2 2\theta_{12} \sin^2\left(\frac{1.27\Delta m^2_{12}[\mathrm{eV}^2]}{E_{\bar{\nu}_e}[\mathrm{MeV}]} L[m]\right), \tag{1}$$

where $\sin^2 2\theta_{12} = (0.816^{+0.054}_{-0.037})(\theta_{12}$ is the mixing angle between $\bar{\nu}_e$ and $\bar{\nu}_\mu$ from the combined global solar solution (Ahmed, 2004)). The value of mixing mass squared difference in the same solution is $\Delta m^2_{12} = (6.45^{+2.2}_{-1.6} \cdot 10^{-5})\mathrm{eV}^2$. Although Δm^2 is better constrained by a measurement performed by KamLAND (Araki et al., 2005a), this cannot be employed herein since that solution was obtained under the assumption of negligible geo-reactor signal and would therefore bias our analysis toward zero geo-reactor power. Since the geo-reactor is many oscillation lengths distant, the survival probability formula can be replaced by $P(\bar{\nu}_e \rightarrow \bar{\nu}_e) \approx 1 - 0.5 \cdot \sin^2 2\theta_{12}$, with negligible error. However this approximation can not be used for man-made reactors that are several hundred kilometers away from KamLAND.

3. Anti-neutrino Detection with KamLAND

KamLAND is a 1 kton liquid scintillator detector located 1 km undeground, below the Mt. Ikenoyama, Gifu perfecture, Japan. Details of the KamLAND detector are given in Eguchi et al. (2003). Electron anti-neutrinos in KamLAND are detected via the inverse β-decay reaction,

$$\bar{\nu}_e + \mathrm{p} \rightarrow \mathrm{e}^+ + \mathrm{n}. \tag{2}$$

The inverse beta reaction consists of a prompt and a delayed event. The prompt event takes place when the created positron annihilates with a nearby electron emitting two gamma rays in opposite direction. The visible energy of the prompt event is 0.8 MeV lower than the energy of the incoming anti-neutrino. The delayed event takes place when the thermalized neutron gets captured on a proton, emitting a gamma ray with energy of 2.2 MeV. The delayed event typically happens within ~2 m of the prompt event and about 200 μs later (on average). This delayed coincidence detection scheme makes anti-neutrino interactions distinguishable from the majority of background sources, which are characterized by single events. Since the energy threshold of the inverse β decay is 1.8 MeV, only anti-neutrinos above this energy can be detected.

The search for the geo-reactor has been performed in the energy range above 3.4 MeV to exclude geological anti-neutrinos coming from decay chains of ^{238}U and ^{232}Th, whose flux is not well known (Araki et al., 2005b) and whose maximum energy is 3.4 MeV. The data sample used for this analysis corresponds to 515.1 days of livetime with a total detection efficiency of (0.898 ± 0.015). The target volume had a radius of 5.5 m and included 4.61×10^{32} target protons.

Anti-neutrinos coming from man-made nuclear reactors around Japan and the rest of the world are the background in the search for the geo-reactor signal. The expected number of background events depends on the choice of oscillation parameters. The expected number of anti-neutrinos before inclusion of neutrino oscillation effect is (365.2 ± 23.7), but after neutrino oscillations are applied the actual expected number is (243^{+6}_{-65}), using values of oscillation parameters taken from SNO experiment (Ahmed, 2004).

There are background sources in the detector that mimic the anti-neutrino signal. Accidental background is caused by coincidence of single events that pass the anti-neutrino event selection criteria by accident. The accidental background in the chosen data sample is estimated to be (2.69 ± 0.02) events. Another source of background is caused by cosmic rays traversing the detector. Although the detector is vetoed after each muon via a set of cuts (so called muon spallation cuts), long-lived spallation products like $^{9}Li/^{8}He$ still contribute to the background. Their contribution is estimated to be (4.8 ± 0.9) events. The largest non-anti-neutrino background comes from α particles from the ^{222}Rn decay chain that engages the reaction $^{13}C(\alpha, n)^{16}O$ contributing (10.3 ± 7.1) events. Thus, the total non anti-neutrino background in the chosen data sample is estimated to be (17.8 ± 8.0) events.

The expected geo-reactor signal is estimated under the assumptions of stable power output and an anti-neutrino energy spectrum resembling that from man-made nuclear reactors whose anti-neutrino production is described in Bemporad et al. (2002). The expected geo-reactor rate in KamLAND above 3.4 MeV and with 100% detection efficiency is 0.0102 events/day · TW_{th} before inclusion of oscillation effects. A comparison of signals is shown in Figure 2.

Anti-neutrino candidates are selected from the data sample based on the following criteria:

- time between the prompt and delayed event is $0.5\mu s < \Delta T < 1000\mu s$
- distance between prompt and delayed event is $\Delta R < 2$ m
- energy of the prompt event is 2.6 MeV $< E_{prompt} <$ 8.5 MeV
- energy of the delayed event is 1.8 MeV $< E_{delayed} <$ 2.6 MeV
- both prompt and delayed event must take place within a fiducial volume with radius $R < 5.5$ m.

The total number of observed anti-neutrino candidates that pass the above selection cuts is 258. The energy spectrum of candidates together with background is shown in Figure 2.

The unbinned Maximum Likelihood Method (Bevington and Robinson, 2003) was used to set an upper limit on the geo-reactor power. A six parameter fit was performed, where geo-reactor power, detection efficiency, $^{13}C(\alpha, n)^{16}O$ background, $^{9}Li/^{8}He$ background, and neutrino oscillation parameters were varied. The fit included daily rate information, anti-neutrino energy spectrum shape information as well as time variations of the incoming anti-neutrino flux on a daily basis. This information was especially important for the geo-reactor search, since the incoming anti-neutrino flux varied significantly over time (as shown in Figure 1), providing greater leverage for the analysis.

The best fit for the likelihood function shown in Figure 2 yields unoscillated geo-reactor rate in KamLAND of (0.06 ± 0.06) events/day or (16 ± 16) geo-reactor anti-neutrinos in our data sample. This corresponds to a geo-reactor power of (5.9 ± 5.9) TW_{th}. The simultaneous best fit solution for the mixing parameters gives $\Delta m^2 = 8.2 \cdot 10^{-5}$ eV^2 and $\sin^2 2\theta = 0.81$ which is consistent both with solar neutrino solution and the KamLAND stand-alone solution. The 90% C.L. upper limit for the putative geo-reactor power is 18 TW_{th} as shown in Figure 3. As a cross-check, an analysis with a lower energy threshold of 2.4 MeV has been performed. This increased the number of anti-neutrino candidates to 362 events or by about 40%, but background sources, as well as their uncertainties, were significantly larger as well. Nevertheless, the best fit solution for the geo-reactor power agrees within experimental error with the higher energy threshold analysis.

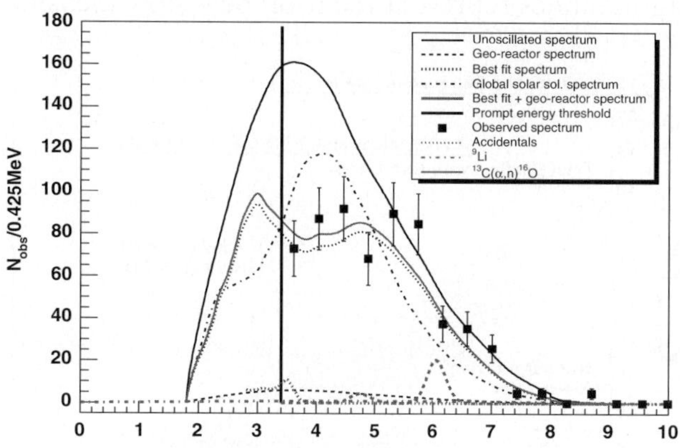

Figure 2. Observed energy spectrum (squares) together with the best fit solution for the geo-reactor spectrum with background sources. The best fit solution with contribution from the geo-reactor is also shown.

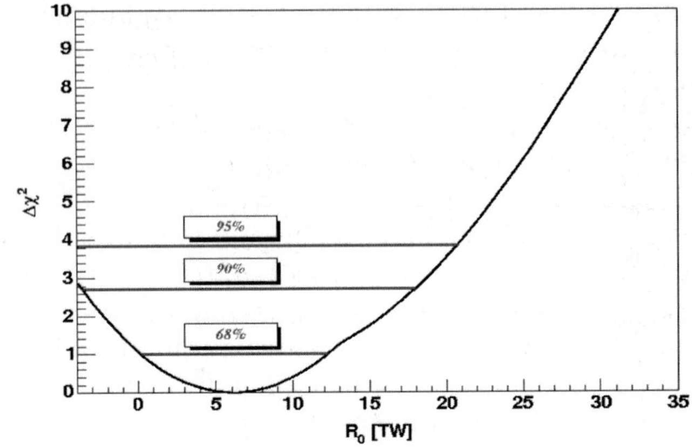

Figure 3. $\Delta\chi^2$ as a function of geo-reactor power. R_0 meaning best fit for measured geo-reactor power based on the experimental data.

4. Discussion of Results and Conclusion

Limits on the power of the hypothetical Earth-centered geo-reactor have been set for the first time. The mean value is within 1σ from zero, making the best fit result inconclusive with respect to geo-reactor existence. The mean value of 6 TW is in agreement with the 10 TW maximum heat flow expected from the Earth's core. The 90% C.L. of 18 TW is below the lower limit of the total radioactive heat flow (Anderson, 2003). The situation is summarized in Figure 4 which shows the range of power attributed to radioactive decay throughout the Earth (19–31 TW) versus the reactor power "measured" herein. The diagonal lines represent the total heat flow measurement from the

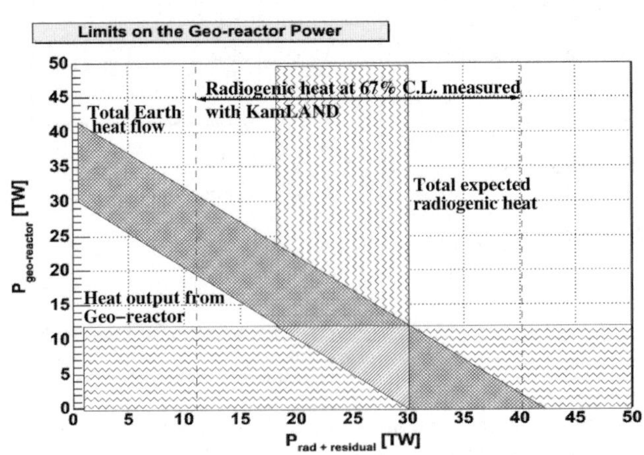

Figure 4. Comparison of geo-reactor result with known geological constraints.

Earth (31-44 TW). Figure 4 shows that geo-reactor is not ruled out by geological constraints. As KamLAND acquires more data, the limits on the geo-reactor power will improve and errors will be reduced. However, measurement of the geo-reactor power that would definitely confirm or reject geo-reactor hypothesis can only be achieved by a KamLAND-sized detector located far from commercial nuclear reactors. Hawaii is among the best locations in the world for this type of measurement. There are efforts to develop such a detector with the Hanohano project, described in these proceedings (Dye et al. 2006).

References

Ahmed, S. N. et al.: 2004, *Phys. Rev. Lett.* **92**, 181301.

Anderson, D.: 2003, Energetics of the Earth and the Missing Heat Source Mystery, available at http://www.mantleplumes.org/Energetics,html.

Araki, T. et al.: 2005a, *Phys. Rev. Lett.* **94**, 081801.

Araki, T. et al.: 2005b, *Nature* **436**, 499–503.

Bemporad, C. et al.: 2002, *Rev. Mod. Phys.* **74**, 297.

Bevington, P. and Robinson, D. K.: 2003, *Data Reduction and Error Analysis for the Physical Sciences*, McGrawHill.

Buffett, B. A.: 2003, *Science* **299**, 1675–1677.

Dye, S. T., et al.: 2006, this volume, doi: 10.1007/s11038-006-9129-z.

Eguchi, K. et al.: 2003, *Phys. Rev. Lett.* **90**, 021802.

Herndon, J. M.: 1979, *Proc. R. Soc. London A* **368**, 495–500.

Herndon, J. M.: 1980, *Proc. R. Soc. London A* **372**, 149–154.

Herndon, J. M.: 1996, *Proc. Natl. Acad. Sci. USA* **93**, 646–648.

Herndon, J. M.: 2003, *Proc. Natl. Acad. Sci. USA* **100**, 3047–3050.

Herndon, J. M.: 2006, this volume, doi: 10.1007/s11038-006-9121-7.

Hollenbach, D. F. and Herndon, J. M.: 2001, *Proc. Nat. Acad. Sci.* **98**, 11085.

McDonough, W. F. and Sun, S.-S.: 1995, *Chem. Geol.* **120**, 223.

Raghavan, R. S.: 2002, *hep-ex/0208038*.

Earth, Moon, and Planets (2006) 99:155–181
DOI 10.1007/s11038-006-9132-4

Imaging the Earth's Interior: the Angular Distribution of Terrestrial Neutrinos

BRIAN D. FIELDS
Department of Astronomy, Center for Theoretical Astrophysics, University of Illinois at Urbana-Champaign, Urbana, IL, USA
(E-mail: bdfields@uiuc.edu)

Department of Physics, University of Illinois at Urbana-Champaign, Urbana, IL, USA

KATHRIN A. HOCHMUTH
Max-Planck-Institut für Physik (Werner-Heisenberg-Institut), Föhringer Ring 6, 80805 München, Germany
(E-mail: hochmuth@ph.tum.de)

(Received 24 August 2006; Accepted 31 October 2006)

Abstract. Decays of radionuclides throughout the earth's interior produce geothermal heat, but also are a source of antineutrinos; these geoneutrinos are now becoming observable in experiments such as KamLAND. The (angle-integrated) geoneutrino flux has been shown to provide a unique probe of geothermal heating due to decays, and an integral constraint on the distribution of radionuclides in the earth. In this paper, we calculate the angular distribution of geoneutrinos, which opens a window on the *differential* radial distribution of terrestrial radionuclides. We develop the general formalism for the neutrino angular distribution. We also present the inverse transformation which recovers the terrestrial radioisotope distribution given a measurement of the neutrino angular distribution. Thus, geoneutrinos not only allow a means to image the earth's interior, but offer a direct measure of the radioactive earth, both revealing the earth's inner structure as probed by radionuclides, and allowing a complete determination of the radioactive heat generation as a function of radius. Turning to specific models, we emphasize the very useful approximation in which the earth is modeled as a series of shells of uniform density. Using this multishell approximation, we present the geoneutrino angular distribution for the favored earth model which has been used to calculate the geoneutrino flux. In this model the neutrino generation is dominated by decays of potassium, uranium, and thorium in the earth's mantle and crust; this leads to a very "peripheral" angular distribution, in which 2/3 of the neutrinos come from angles $\theta \gtrsim 60°$ away from the nadir. We note that a measurement of the neutrino intensity in peripheral directions leads to a strong lower limit to the central intensity. We briefly discuss the challenges facing experiments to measure the geoneutrino angular distribution. Currently available techniques using inverse beta decay of protons require a (for now) unfeasibly large number of events to recover with confidence the forward scattering signal from the background of subsequent elastic scatterings. Nevertheless, it is our hope that future large experiments, and/or more sensitive techniques, can resolve an image of the earth's radioactive interior.

Keywords: Geoneutrinos

1. Introduction

The decays of radioactive species within the earth generate an important component of geothermal heat. However, a quantitative accounting of the radioactive energy generation of the earth requires a detailed knowledge of the abundance distribution of the key long-lived radioisotopes – uranium, thorium and potassium – inside the earth. Because our knowledge of these abundance distributions is incomplete, the radiogenic heat output is consequently model-dependent and thus uncertain. A fundamental diagnostic is the "Urey ratio" which measures the ratio $P_{\rm rad}/P_{\rm lost}$ of total radioactive heat production to the surface heat loss. Current estimates span the range e.g. (McKenzie and Richter, 1981; Stein, 1995)

$$P_{\rm rad}/P_{\rm lost} \sim 0.5 - 0.6 \tag{1}$$

which seems to suggest that radiogenic heating is a dominant heat source, but not the only one. It is even possible that the Urey ratio is closer to 1, as the primordial heat from earth's formation should have radiated away a long time ago (Albarede and van der Hilst, 2002), and so the earth's heat could be fully radiogenic. This would imply that more radioactive material is hidden in the earth than presently suspected. The strength of these qualitative conclusions thus hangs on the strength of the quantitative measurements of the radiogenic heat production and total heat loss.

A beautiful if challenging means of measuring the radiogenic heat production of the earth follows from the realization that β-decays not only are a heat source but also produce neutrinos. Decades ago, the prescient work of Eder (1966) and later Krauss et al. (1984) pointed out that the radioactive heat flux and the neutrino flux from the earth are tightly linked. A measurement of the neutrino flux would constrain the radiogenic heat production, and thus offer a new and direct measure of the Urey ratio. Clearly, a precise measurement of the Urey ratio would provide important insight into the interior structure and dynamics of the earth.

This vision of neutrino geophysics has enjoyed a major advance recently, with the first detection of geoneutrinos by the KamLAND Collaboration (Eguchi et al., 2003; Araki et al., 2005a). In fact, the detection of these geoneutrinos was only a side product of the KamLAND detector, as it is primarily designed to detect the flavor change of reactor antineutrinos produced by the Japanese nuclear power plants. The KamLAND Collaboration reported an excess of $\bar{\nu}_e$ events above background. They estimate at 90%CL that between 4.5 and 52.4 geoneutrino events were detected (Araki et al., 2005a), which gives a geoneutrino flux at KamLAND of about $F_\nu \sim 5 \times 10^6$ neutrinos $\mathrm{cm^{-2}\,s^{-1}}$. The corresponding limits on terrestrial radiogenic heat production are weak (<60 TW at 99%CL) but consistent with geophysical estimates. Obviously these are early days, but we are encouraged that, even in the presence of a dominant anthropogenic background, KamLAND has demonstrated that geoneutrinos exist at observable levels.

Several groups have taken interest in these neutrinos and have in conjunction with geophysicists and geologists worked on models for the distribution of radioactive elements in the earth and predicted the (angle-integrated) geoneutrino flux at different detector sites (Rothschild et al., 1998; Fiorentini et al. 2003a, b; Nunokawa et al. 2003; Mantovani et al., 2004). The flux calculations have already become rather sophisticated, and are based on detailed geophysical models, in some cases even including the effects of the anisotropic radioisotope densities in the crust. These models confirm in detail that indeed the geoneutrino flux is proportional to radiogenic heat flux, but with the important caveat that the exact scaling between the two depends on the detailed abundance and density distributions within the earth.[1] In this sense, the geoneutrino flux measurement at any given site is an *integral* measure of the radioisotope distribution. Combining measurements from multiple sites can constrain the distribution of geoneutrino sources.

In this paper we consider the angular distribution of the geoneutrino flux. We show that, in the idealized futuristic case in which the angular distribution is well measured, it can be inverted to recover the full density distribution of radionuclides – a tomography of the structure and radiogenic heat generation of the earth. In addition, we come to the conclusion that with a future low-energy antineutrino detector with even crude angular resolution, it will be possible to distinguish between the different earth models and solve the problem as to how much radioactive material is contained in the earth and where it is located. Thus, the angular distribution provides a *differential* measurement of the radioisotope distribution, and can reveal a wealth of new information about the structure and content of the earth.

We first present the formal calculations in Section 2. After outlining the general formalism, we consider the useful approximation of a uniform density shell, from which a multiple-shell model of the earth can be constructed. Our formalism can allow for a more realistic, aspherical distribution of radioisotopes; however, this extra detail would complicate our results and obscure some of the simple trends we wish to point out. We can look forward to the day when it becomes possible to measure the angular distribution with

[1] It is true that, at least in a spherically symmetric earth, the surface heat flux is directly related, via Gauss' law, to the total mass of radionuclides (c.f. Equation 24). However, a simple Gauss' law argument fails in the case of neutrinos, because at each point they are emitted isotropically, not just radially, and as we will see, the "sideways" neutrinos make a large contribution to the total surface geoneutrino flux.

significant precision, at such time more realistic model-building (such as already has been accomplished for the geoneutrino flux) would be justified.

We then review in Section 3 models for the radioisotopic content of the earth. Using these models we construct physically motivated spherically symmetric illustrations of plausible geoneutrino angular distributions (Section 4). We present several plots for different geological abundance predictions, finding in general that the high radioisotope content of the crust leads to a large "peripheral" geoneutrino signal. We investigate the possibility of ^{40}K in the core, as recently suggested by several groups (Gessmann and Wood, 2002; Lee and Jeanloz, 2003; Rama Murthy et al., 2003); if core ^{40}K is at the high end of these predictions, then the resulting neutrino signal could be quite large and should lead to a readily observable central intensity peak, though the very low beta-decay endpoint of ^{40}K presents even more severe experimental challenges for the potassium neutrinos than for those from uranium and thorium. Indeed, as we discuss in Section 5, antineutrinos are presently detected via the reaction $\bar{\nu}_e + p \rightarrow e^+ + n$, which allows for some angular resolution due to the forward scattering of the neutron. The threshold of this reaction is 1.8 MeV, implying that all ^{40}K neutrinos and most of the uranium and thorium flux is lost for detection. Thus, low-resolution imaging of the thorium and uranium earth is beyond the reach even of next-generation neutrino experiments, although we note that these should be the first to confirm the anisotropy of the geoneutrino signal (Hochmuth et al., 2005). It remains for even more advanced experiments to achieve an angular sensitivity fit for the challenges of geoneutino detection. With this in mind, we present conclusions in Section 6, and discuss the exciting possibilities that will arise when we are finally able to use neutrinos to image the interior of the earth.

Before we begin the formal development, a word of clarification seems in order. Note that for brevity, we will refer to the emitted particles as neutrinos, although they are of course antineutrinos. In fact, there is some regular ν_e production due to ^{40}K electron captures. However, the branching here is 10.72% of all ^{40}K decays, and thus the majority of the ^{40}K nuclei β-decay and yield $\bar{\nu}_e$ with a continuous energy spectrum. We look forward to the day when the monoenergetic ^{40}K electron capture ν_e flux (and angular distribution!) can be measured and compared against the $\bar{\nu}_e$ signal. However, in this paper we will consider only the dominant, β-decay $\bar{\nu}_e$ signals from K, Th, and U.

2. Formalism

The fundamental quantity we wish to calculate is the differential intensity I, or surface brightness, of geoneutrinos at the surface of the earth. This is just the distribution of neutrino flux F versus solid angle: $I(\theta, \phi) = dF/d\Omega$. Both angular coordinates are local and observer-centered: $\theta \in [0, \pi/2]$ is the nadir angle, i.e., the angle measured from the downward vertical (so that the center of the earth is at $\theta = 0$, and the horizontal is at $\theta = \pi/2$; see Figure 1). The angle ϕ is an azimuth. These angles thus cover the "sky" underfoot (or rather, the terrestrial hemisphere) which we wish to image. In this paper we will consider only the case of a spherically symmetric earth (including the outermost layers). This guarantees that the intensity $I = I(\theta)$ is azimuthally symmetric and so only depends on the nadir angle. In this case, we have $I(\theta) = (2\pi)^{-1}\, dF/d\cos\theta$.

The neutrino intensity in any direction depends on the distribution of sources along that line of sight. The governing equation is that of radiation transfer for neutrinos, which is formally identical to the usual expression for photons (Chandrasekhar, 1950), although of course the microphysics is quite different (the earth is optically thin to neutrinos, but neutrinos do undergo oscillations). Then over a line of sight $\vec{s}$, ignoring scattering and absorption, the intensity changes according to

$$dI/ds = q(\vec{s})/4\pi \tag{2}$$

where q is the source function at point $\vec{s}$, which measures the local neutrino production rate per unit volume.

For each radioisotope species i, this takes the form

$$q_i = \frac{n_i}{\tau_i} = \frac{\rho_i}{\tau_i m_i}. \tag{3}$$

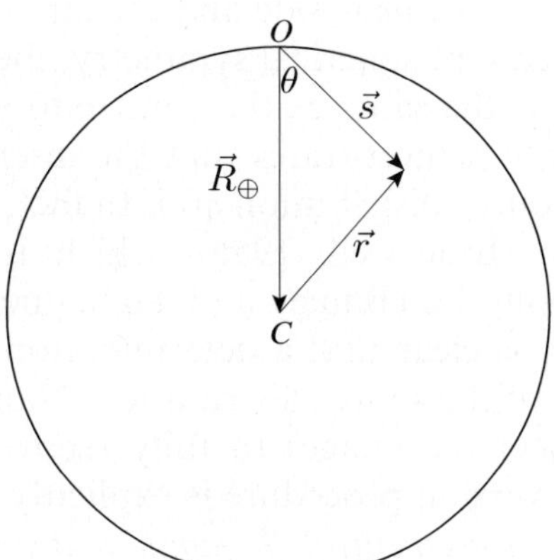

Figure 1. The basic geometry of the problem. An observer at point O measures an intensity, which sums the emission along a line of sight $\vec{s}$. The nadir $\vec{R}_\oplus$ points to the center C of the earth, which makes an angle θ with the line of sight. A given point along $\vec{s}$ is at geocentric distance $\vec{r}$.

Here n_i and ρ_i are the local number and mass densities, respectively, of species i in the earth. The τ_i is the mean lifetime and m_i is the mass of an i nucleus. The total source is just a superposition of all species: $q = \sum_i q_i$. The effects of neutrino oscillations are not yet included; we will address this below (Section 2.2).

Integrating Equation (2) over a line of sight at nadir angle θ, we find the intensity

$$I(\theta) = \int_0^{2R_\oplus \cos\theta} q(\vec{s})ds \tag{4}$$

where $\vec{s}$ is centered on the observer, as seen in Figure 1.

Since models of the earth's structure and composition are expressed in terms of the radius $\vec{r}$, it is very useful to rewrite Equation (4) in these geocentric coordinates. From Figure 1 we see that $\vec{s} = \vec{r} + \vec{R}_\oplus$, so that $\vec{r} = \vec{s} - \vec{R}_\oplus$ and thus

$$s(r,\theta) = R_\oplus \cos\theta \pm \sqrt{r^2 - (1 - \cos^2\theta)R_\oplus^2} \tag{5}$$

where $+(-)$ corresponds to the far side (near side) of the midpoint $s = R_\oplus \cos\theta$ of the line of sight. Therefore, for a fixed nadir angle θ, we can transform the integral to the geocentric coordinate system:

$$I_i(\theta) = 2\int_{R_\oplus \sin\theta}^{R_\oplus} dr \frac{q_i(r)r}{\sqrt{r^2 - (1-\cos^2\theta)R_\oplus^2}} \tag{6}$$

where $q_i(r)$ can in a non-spherically symmetric case also be a function of (geocentric) latitude and longitude. In this case, the factor of two is replaced by the sum of integrals for the near-side and the far-side of the midpoint $s = R\cos\theta$. In our special case of spherical symmetry, the contribution from the near half of the sphere is the same as the contribution from the far side.

Equation (6) explicitly demonstrates that the intensity $I(\theta)$ is an integral transformation of the source distribution $q(r)$. In fact, this mapping is a form of the Abel transform (Bracewell, 1986), which is used in deprojection problems in both astrophysics (Binney and Tremaine, 1987) and geophysics (Dahlen, 2004). Thus it is clear that a determination of the intensity distribution offers a measure of the source distribution. Namely, one can invert the transformation (deproject the image) to fully recover the complete source distribution $q(r)$; this inversion procedure is explicitly presented in Appendix A. In other words, *a measurement of the angular distribution of geoneutrinos not only yields an image of the earth's radioactive interior, but this image can also be inverted to give a tomography of the terrestrial radioisotope distribution.* Clearly, the geoneutrino angular distribution offers a unique and powerful probe of the interior of the earth. This power will be further illustrated below, where we will see that even a partial (low-resolution) determination of $I(\theta)$ offers important geophysical information.

In evaluating Equation (6), it will be convenient to introduce dimensionless scaled variables: a radial fraction $x = r/R_\oplus \in [0, 1]$, a local mass fraction $a_i = \rho_i/\rho$, and a local density measure $\tilde{\rho} = \rho/\bar{\rho}$ (with $\bar{\rho} = 3M/4\pi R_\oplus^3$ the mean earth density). It will also be useful to denote the nadir angle cosine as $\mu = \cos\theta$. We then rewrite Equation (6) as a product of two terms

$$I_i(\theta) = I_{i,0} g_i(\mu). \tag{7}$$

Here the dimensionful overall magnitude is set by

$$I_{i,0} = 2\frac{N_i \bar{a}_i \bar{\rho} R_\oplus}{4\pi A_i \tau_i m_u}, \tag{8}$$

where N_i is neutrino multiplicity, i.e., the number of geoneutrinos released per decay chain and m_u the atomic mass unit. Values of $I_{i,0}$ appear in Table I; the radioisotope abundances are taken from Table II. The dimensionless angular distribution (akin to the "phase function" of radiation transfer (Chandrasekhar, 1950)) is the heart of this paper, and is contained in the function

$$g_i(\mu) = \int_{\sqrt{1-\mu^2}}^{1} dx \frac{\tilde{\rho}_i(x) x}{\sqrt{x^2 - (1-\mu^2)}} \tag{9}$$

which requires knowledge of the density distribution of each radioisotope, usually presented in terms of mass fractions a_i and a total density profile ρ via $\tilde{\rho}_i = \rho_i/\bar{\rho}_i = (a_i/\bar{a}_i)\tilde{\rho}$.

The flux of geoneutrinos integrated over different annuli is also of interest. We quantify this in terms of the flux *exterior* to the nadir angle θ, via

$$F_i(>\theta) = F_i(<\mu) = \int_{\Omega_{\rm shell}} d\Omega I_i(\mu, \phi) \tag{10}$$

TABLE I

Properties of the principle geoneutrino source nuclei

Radioisotope species	Mean life τ (Gyr)	$\bar{\nu}_e$ multiplicity N_i	Isotopic abundance (%)	Mean terrestrial abundance $\bar{a}_i$	Intensity normalization $I_{i,0}$ [neutrinos cm^{-2} s^{-1} sr^{-1}]
^{40}K	1.84	1	0.0117	1.8×10^{-8}	2.4×10^6
^{238}U	6.45	6	99.2745	5.3×10^{-8}	0.48×10^6
^{232}Th	20.3	4	100	1.35×10^{-8}	0.56×10^6

$$F_i(>\theta) = F_i(<\mu) = 2\pi I_{i,0} \int_0^\mu d\mu g(\mu) \tag{11}$$

$$F_i(>\theta) = F_i(<\mu) \equiv 2\pi I_{i,0} H(\mu) \tag{12}$$

where second and third expressions assume spherical symmetry, and where the dimensionless quantity H encodes the angular dependence. A consequence of this definition is that the total neutrino flux is given by $F_i(\theta > 0) = F_i(\mu < 1) = 2\pi I_{i,0} H(1)$. This can be compared with existing calculations (Krauss et al., 1984; Fiorentini et al., 2003b; Mantovani et al., 2004).

2.1. The uniform shell approximation

We want to consider a single shell with a density $\tilde{\rho}_{i,0}$ in species i which is constant between $r_{\text{in}} = x_{\text{in}} R_\oplus$ and $r_{\text{out}} = x_{\text{out}} R_\oplus$, and zero otherwise:

$$\tilde{\rho}_i(x) = \begin{cases} \tilde{\rho}_{i,0} & x_{\text{in}} < x < x_{\text{out}}, \\ 0, & \text{otherwise} \end{cases} \tag{13}$$

This form allows us to simplify the integral and solve it analytically. In particular, owing to the constant density, the intensity at each line of sight is just proportional to the shell path length $\Delta s(\theta)$ along that sightline.

There are three cases to be distinguished, as one can see in Figure 2:

$$g(\mu) = \begin{cases} \tilde{\rho}_{i,0}\left(\sqrt{\mu^2 - \mu_{\text{out}}^2} - \sqrt{\mu^2 - \mu_{\text{in}}^2}\right), & \mu > \mu_{\text{in}} \\ \tilde{\rho}_{i,0}\sqrt{\mu^2 - \mu_{\text{out}}^2}, & \mu_{\text{out}} \le \mu \le \mu_{\text{in}} \\ 0, & \mu < \mu_{\text{out}} \end{cases} \tag{14}$$

TABLE II

Mantle and crust elemental abundance distribution (Mantovani et al., 2004)

Region	Radii (km)	a (U)	a (Th)	a (K)
Lower mantle	3480–5600	13.2×10^{-9}	52×10^{-9}	1.6×10^{-4}
Upper mantle	5600–6291	6.5×10^{-9}	17.3×10^{-9}	0.78×10^{-4}
Oceanic crust	6291–6368	0.1×10^{-6}	0.22×10^{-6}	0.125×10^{-2}
Lower crust	6291–6346.6	0.62×10^{-6}	3.7×10^{-6}	0.72×10^{-2}
Middle crust	6346.6–6356	1.6×10^{-6}	6.1×10^{-6}	1.67×10^{-2}
Upper crust	6356–6368	2.5×10^{-6}	9.8×10^{-6}	2.57×10^{-2}
Sediments	6368–6371	1.68×10^{-6}	6.9×10^{-6}	1.7×10^{-2}
Oceans	6368–6371	3.2×10^{-9}	0	4.0×10^{-4}

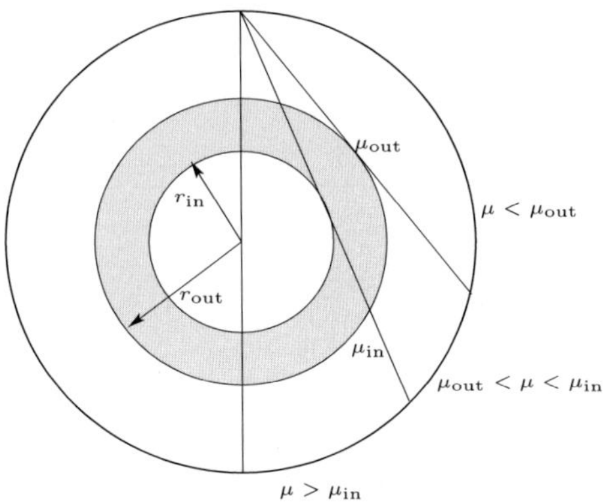

Figure 2. Geometry for the uniform shell model. The inner and outer shell radii are r_{in} and r_{out}, respectively. The lines of sight tangent to the inner and outer edges of the shell are shown, and are at nadir angle cosines μ_{in} and μ_{out}, respectively. These lines of sight divide the terrestrial hemisphere into three distinct regions, which are labeled.

where

$$\mu_{in} = \sqrt{1 - x_{in}^2}, \quad \mu_{out} = \sqrt{1 - x_{out}^2}. \tag{15}$$

These are integrable, and give

$$H(\mu) = \begin{cases} \tilde{\rho}_{i,0}\left[\mu_{out}^2 h(\mu/\mu_{out}) - \mu_{in}^2 h(\mu/\mu_{in})\right], & \mu > \mu_{in} \\ \tilde{\rho}_{i,0}\mu_{out}^2 h(\mu/\mu_{out}), & \mu_{out} \leq \mu \leq \mu_{in} \\ 0 & \mu < \mu_{out} \end{cases} \tag{16}$$

where

$$h(u) = \frac{1}{2}u\sqrt{u^2 - 1} - \frac{1}{2}\ln\left(u + \sqrt{u^2 - 1}\right). \tag{17}$$

Note that Equation (16) calls $h(u)$ only in the domain $u \geq 1$, and that $h(1) = 0$. The total flux for a uniform shell was first calculated in Krauss et al. (1984), and one can easily show that $F(\mu < 1)$ reproduces their result.

There are four cases we want to illustrate; these are plotted in Figure 3. In the "uniform earth" model, the density is the same throughout the whole planet ($x_{in} = 0$, $x_{out} = 1$). This gives an angular distribution which grows linearly with μ, $g(\mu) = \tilde{\rho}_{i,0}\mu$, and hence $I(\theta) \propto \cos(\theta)$. The integrated flux which increases quadratically as $H(\mu) = \tilde{\rho}_{i,0}\mu^2/2$, i.e., $H(\theta) \propto \cos^2\theta$. This therefore gives a very centrally bright distribution. In the "uniform core" model, the density is non-zero only in a central region which extends from $x_{in} = 0$ to x_{out}. As seen in Figure 3, this gives an inner intensity distribution which is similar to the uniform earth model, as one would expect, but which goes to zero, as it should, at the outer tangent μ_{out}.

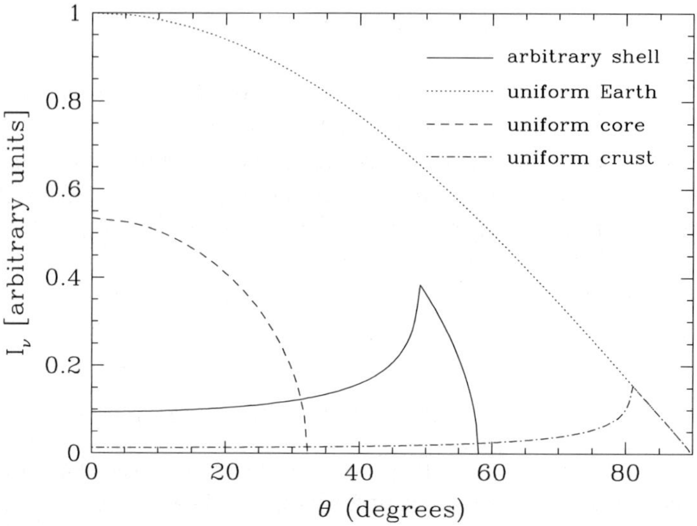

Figure 3. The angular distribution of intensity for different cases of a 1-zone, uniform density model, shown as a function of nadir angle θ. The units in the vertical scale are arbitrary, and the normalizations of the different curves are chosen for clarity.

As we will see below, it turns out that a more physically motivated case is the "uniform crust" model, where only the layers of the earth (from $x_{\rm in}$ to $x_{\rm out} = 1$) contribute. This yields an intensity distribution

$$g(\mu) = \begin{cases} \tilde{\rho}_{i,0}\left[\mu - \sqrt{\mu^2 - \mu_{\rm in}^2}\ \right], & \mu > \mu_{\rm in} \\ \tilde{\rho}_{i,0}\mu, & \mu \leq \mu_{\rm in} \end{cases} \tag{18}$$

$$H(\mu) = \begin{cases} \frac{\tilde{\rho}_{i,0}}{2}\left[\mu_{\rm out}^2 h(\mu/\mu_{\rm out}) - \mu_{\rm in}^2 h(\mu/\mu_{\rm in})\right], & \mu > \mu_{\rm in} \\ \frac{\tilde{\rho}_{i,0}}{2}\mu^2, & \mu \leq \mu_{\rm in} \end{cases} \tag{19}$$

which is also linear in μ, and thus scales as $I \propto \cos(\theta)$, for large angles. The intensity peaks at inner tangent point $\mu_{\rm in}$, where the line of sight is longest. A measurement of this peak would thus give the position of the inner edge. Interior to $\mu_{\rm in}$, the intensity drops to a minimum at the nadir, where the column density is the smallest. But the central intensity is nonzero, and is simply related to the peak intensity via

$$\frac{I_{\rm peak}}{I_{\rm center}} = \frac{I(\mu_{\rm in})}{I(1)} = \frac{\Delta s(\mu_{\rm in})}{\Delta r_{\rm shell}} = \sqrt{\frac{1 + x_{\rm in}}{1 - x_{\rm in}}} \tag{20}$$

This useful relation allows one to use the peripheral flux, due to emission from earth's outer layers, to put a lower limit on the contribution of these layers to the central flux. Any observed central flux in excess of this limit must be due to emission from the inner earth.

The last model illustrated in Figure 3 is a single shell with arbitrary inner and outer radii. The intensity distribution is an amalgam of the features seen in the special cases of the crust and core models. As with the crust model, the intensity peaks at the inner tangent point, where the line of sight is longest. As with the core model, the intensity goes to zero at the outer tangent. Thus, a measurement of the peak would give the position of the shell's inner edge, while a measurement of the cutoff would give the outer edge.

2.2. Towards a realistic multishell model

Based on the idealized calculations in the last section we want to build now a model that is applicable for the earth's Interior. The earth of course does not have a constant density. However, we can approximate the density structure as a series of uniform density shells. Indeed, earth models are in practice typically expressed in this manner. Formally, the generalization is trivial, thanks to the lack of neutrino absorption and scattering, which guarantees that superposition holds:

$$I_i(\theta) = \sum_{\text{shells}\, j} I_i^{(j)}(\theta) \tag{21}$$

$$F_i(>\theta) = \sum_{\text{shells}\, j} F_i^{(j)}(>\theta) \tag{22}$$

where $I_i^{(j)}(\theta)$ and $F_i^{(j)}(\theta)$ are the intensity and flux, respectively, from shell j.

Moreover, we must consider the effect of neutrino oscillations (Fiorentini et al., 2003a; Nunokawa et al., 2003). As the mass eigenstates of the neutrinos differ from the flavor eigenstates, some of the initial $\bar{\nu}_e$ signal will be converted into other flavors during propagation, and thus reduce the $\bar{\nu}_e$ signal in the detector. In general, the oscillations will have a "vacuum" contribution, but will be modified due to the earth matter effects, which are density- and energy-dependent. If matter effects are important, then it complicates the determination of the angular distribution, introducing a density- and path-dependent oscillation factor into Equation (6) and its descendants. This factor would complicate the inversion of the angular distribution $I(\theta)$ to recover the radioisotope source distribution $q(r)$.

However, we do not expect matter effects to play any greater role, due to the low energy of the geoneutrinos and the low density of the earth. This can be justified if the vacuum oscillation length L_v is much smaller than the oscillation length in matter L_m. If we take the values of Barger et al. (1994) for the matter in the earth and a value for $\Delta m^2 \sim 10^{-4}$ eV^2, we obtain $L_v \sim$ 10 km and $L_m \sim$ 1000 km. Therefore $|L_v| \ll |L_m|$, and we see that, roughly speaking, the vacuum oscillation length is short compared to both the matter oscillation length and to the changes in density; thus the vacuum oscillations will wash out any matter effect and average out the pathlength dependence. We thus follow Mantovani et al. (2004) and introduce only a density-independent survival probability of $(1 - \frac{1}{2}\sin^2 2\vartheta_{12}) = 0.59$ in our equations. Here ϑ_{12} is the dominant electron-neutrino $\nu_e - \nu_x$ mixing angle; solar neutrino experiments and KamLAND are best fit (Araki et al., 2005b) by $\tan^2\vartheta_{12} = 0.40^{+0.10}_{-0.07}$ (and hence $\sin^2 2\theta_{12} = 0.82$) and $\Delta m^2 = (7.9^{+0.6}_{-0.5})\times$ 10^{-5} eV^2. Therefore the intensity scaling of Equation (8) changes to:

$$I^{\text{eff}}_{i,0} = \left(1 - \frac{1}{2}\sin^2 2\vartheta_{12}\right) I_{i,0}. \tag{23}$$

Table I sums up the important properties of the radioactive elements in question.

In the next section we want to use this and discuss different earth models.

3. Earth Models

The geoneutrino intensity depends on the radioisotope density structure $\rho_i(r)$, which is usually presented as abundances a_i and a total density ρ, where $\rho_i = a_i\,\rho$. The earth's interior structure and total density are primarily probed via the propagation of seismic waves. These results have been synthesized by Dziewonski and Anderson in their Preliminary Reference Earth Model (Dziewonski and Anderson, 1981). Following Mantovani et al. (2004) we will adopt this model. In addition, for our spherically symmetric study we took the outermost 3 km of the earth to be sediments.

The distribution of radioisotope abundances a_i are obtained with different geological measurements which, for most of the earth's interior, are necessarily indirect. Consequently, the abundance distribution remains model-dependent. Indeed, the measurement of the angular distribution of geoneutrinos (as well as the geoneutrino energy spectrum) would provide a powerful new method to *measure* the radioisotope distribution. For the purposes of illustration here, we will adopt the values of a_i given in the reference model of Mantovani et al. (2004), which is very detailed and draws on a wealth of geophysical data. Note that in the following we will refer to the model of Mantovani et al. as (geophysical) reference model. This model takes into account the bulk silicate earth model, which describes the composition of the crust–mantle system (McDonough and Sun, 1995). Table II shows the adopted abundance distribution. Note that these are *elemental* abundances; we assume that the isotopic fractions of Table I hold throughout the earth. This correction is particularly important for ^{40}K.

The composition of the earth's core deserves special attention. The earth's core consists largely of iron. But its density is lower than one would expect if the core were pure iron. Therefore it is assumed that light elements in the form of alloys are present (McDonough, 2002). So far it was generally believed, that the core does not hold any significant amount of radioactive elements, as there was no evidence that the radioactive isotopes in question could alloy with iron. Hence, in the reference model radioactive elements are only placed in the mantle and in the crust, but are absent from the core.

On the other hand, it is puzzling why carbonaceous chondrites have a K/U ratio that is eight times higher than in the crust–mantle system (Wasserburg et al. 1964). In this connection, it is noteworthy that recent experiments demonstrate that potassium does form alloys with iron under high temperature and pressure conditions which likely were present at earth's formation. The maximum possible amounts of potassium in earth's core suggested by experiments and cosmochemical considerations range from 60–130 ppm (Rama Murthy et al., 2003), to 1200 ppm (Gessmann and Wood, 2002), to as high as 7000 ppm (Lee and Jeanloz, 2003). In light of these experiments, we will consider the possibility of an additional radiogenic contribution from the core, and quantify the impact of core ^{40}K on the reference model.

The amount of radioactive material in the core contributes also to the radiogenic heat of the earth. Presently it is assumed that the earth's surface loses about 44 TW or 87 mW/m^2 (Pollack et al., 1993). However, this is not a fixed value and the true heat loss is a matter of intense discussion (Hofmeister and Criss, 2005). Note that the measured heat flow at the surface of the earth is reasonably similar for various locations. The amount of surface radiation depends on convection and conduction properties of the earth's interior. The present day heat production H in units of TW of uranium, thorium and potassium with a total mass in kg of M_{U}, M_{Th}, M_{K}, respectively is

$$H = \sum_i \epsilon_i M_i \approx 10\left(\frac{M_{\mathrm{U}}}{10^{17}\ \mathrm{kg}}\right) + 2.7\left(\frac{M_{\mathrm{Th}}}{10^{17}\ \mathrm{kg}}\right) + 3.4 \times 10^{-4}\left(\frac{M_{\mathrm{K}}}{10^{17}\ \mathrm{kg}}\right)\mathrm{TW} \tag{24}$$

where $\epsilon_i = Q_i/m_i\,\tau_{\beta,i}$ is the specific non-neutrino energy loss per nucleus, and where M_{K} is the *total* potassium mass, for which the ^{40}K isotopic fraction appears in Table I. To obtain the heat production of potassium we used, that in 89.28% of all cases a ^{40}K nucleus β-decays with an average energy of 0.598 MeV (Van Schmus, 1995). The other decay mode of potassium is electron capture with an energy of 1.505 MeV.

With knowledge of the radioisotopic content of the earth from geoneutrinos, Equation (24) can be compared with the global heat flux, and used to determine the Urey ratio Equation (1). While the radioisotope abundances remain uncertain, one can rather turn the problem around, and use the global heat flux with Equation (24) to set an upper limit to the radioisotopic content and thus to the geoneutrino flux. Such an analysis has been carried out by Mantovani et al., who find that this "maximal radiogenic" model leads to fluxes about twice the level of their reference model.

In the next section we will discuss the plots we created for the different earth models.

4. Results

We now combine our general formalism with various earth models to arrive at predictions for the geoneutrino angular distribution. We first consider the reference model, then its variants and its uncertainties, and finally we comment on the effect of anisotropies in the radioisotope distributions.

Since our theoretical discussion is already rather forward-looking compared to present experimental capabilities (see Section 5), we illustrate the angular distributions from all three principal geoneutrino radioisotopes. However, the reader should bear in mind that given the low endpoint energy of ^{40}K, its neutrinos are below threshold for inverse beta decay, and thus cannot be seen by this technique. Hence, the ^{40}K neutrinos are even more difficult to observe; for this reason, we will also show results when only the ^{238}U and ^{232}Th neutrinos are measured.

4.1. The reference model

The reference model serves as our standard and fiducial case. The geoneutrino intensity distribution based on the abundance values of this model appears in Figure 4. We see that the total intensity is peaked near the horizon, at large nadir angles. The strikingly "peripheral" character of this neutrino distribution is a direct consequence of the location of the radioisotopes in the mantle and crust.

The experimental ability to detect this pattern is perhaps best quantified in Figure 5, which displays the cumulative angle-integrated flux F for the reference model (c.f. Equation 10). The change in the normalized flux $F/F_{\rm tot}$ over any angle interval gives the contribution of that interval to the total flux. Figure 5 thus shows that fully 2/3 of the total flux arrives in the outermost nadir angles $\theta \gtrsim 60°$; this result holds whether or not ^{40}K neutrinos are observed.

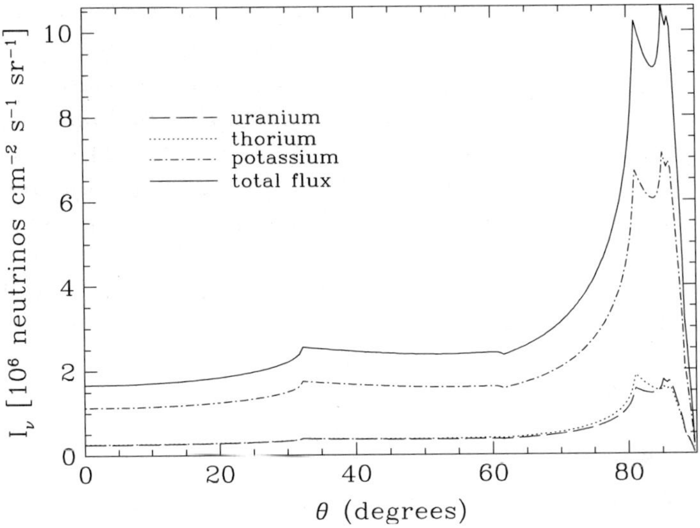

Figure 4. The figure shows the reference model. The dashed-dotted line represents the total expected intensity. No contribution coming from the core has been added yet. The other curves show the intensities from thorium, uranium and potassium separately. It can be seen that the major contribution is coming from potassium.

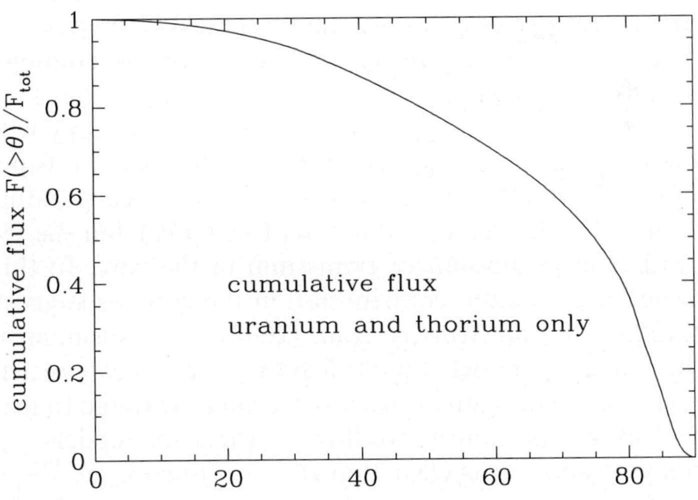

Figure 5. Cumulative flux expected for the reference model with the uranium and thorium fluxes only. The flux is normalized to the total, and thus is dimensionless, spanning the range 0 to 1.

This model can thus be easily tested with an experiment having even very modest angular resolution. For example, an experiment with 30° resolution could test whether the counts in the outer $\theta > 60°$ are a factor ~2 higher than the counts in the inner $\theta < 60°$. The confirmation of this model would vindicate the idea that U, Th, and K are congregate only in the mantle and crust.

Figure 4 also shows the expected angular distribution separately for uranium, thorium, potassium and the cumulative intensity for the reference model. The uranium to thorium ratio stays approximately the same, whereas the amount of potassium increases steeply in the crust. The intensity shows a double peak, which is due to the fact, that the abundance of radioactive elements is lower in the middle crust than in the upper and lower crust.

We note that the reference model gives a geoneutrino distribution qualitatively similar to the uniform crust model presented in Section 2. This of course traces to the positioning of the radioisotopes in the outer earth.

4.2. Models with core potassium

We now turn to the possibility that the earth's core might contain significant amounts of ^{40}K. Based on the experimental evidence that potassium can form alloys with iron (Lee and Jeanloz, 2003; Rama Murthy et al., 2003) we use the fiducial values obtained by different authors and add them to the reference model. We are aware that if the abundance ratios and mass ratios of the reference models are correct and only the total mass estimates of U, Th and K are incorrect, then the intensity of the neutrinos coming from the crust–mantle system should go down, but the overall shape of Figure 4 should stay the same. This decrease in the peripheral intensity will make the core contribution even more dominant. On the other hand, it is possible that the assumptions of the reference model are correct (incorporating the bulk silicate earth model with a heat production of ~20 TW), but the Urey ratio is closer to 1 with a large amount of potassium in the core. In this case it is legitimate to add a potassium contribution in the core, as suggested by Lee and Jeanloz (2003), Rama Murthy et al. (2003) and Gessmann and Wood (2002), to the reference model. Figure 6 shows the angular distribution for different amounts of potassium in the core as could be found in the literature, while Figure 7 shows the cumulative flux for the same models.

From Figures 6 and 7 it is clear that the introduction of ^{40}K in the core can significantly alter the geoneutrino angular distribution. The effect is to enhance the central intensity ($\theta \lesssim 30°$), possibly also raising the overall detected flux. The departure from the reference model depends of course on the core potassium abundance. The value given in Lee and Jeanloz (2003) of 7000 ppm potassium in the core is only an upper limit, but in this case the core would clearly dominate the distribution. For 1200 ppm of potassium in the core (Gessmann and Wood, 2002) the maximum intensities coming from crust and core are approximately the same. But even the much lower value

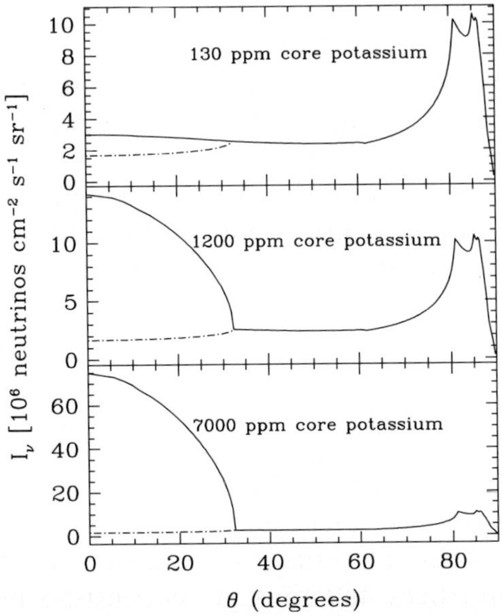

Figure 6. The curves show the possible abundances for potassium in the core as found in Rama Murthy et al. (2003), Lee and Jeanloz (2003) and Gessmann and Wood (2002). The dashed-dotted line is the intensity of the reference model.

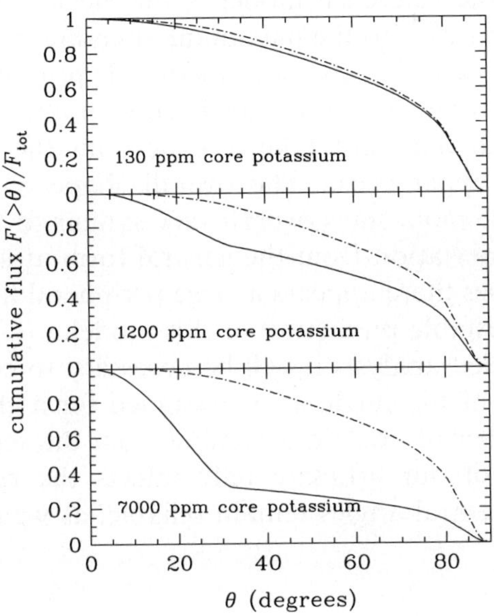

Figure 7. The models are equivalent to Figure 6, showing the cumulative fluxes of models with potassium contribution in the core (normalization as in Figure 5). The dashed line corresponds to the flux of the reference model.

obtained by Rama Murthy et al. (2003) of 60–130 ppm might still be detectable with a future neutrino detector.

A future low energy neutrino detector with angular resolution will be able to distinguish between the opposing models. Again, with only a modest resolution, it will already be possible to make important statements. An experiment with 30° resolution could divide the emission into central, medial, and peripheral bins, and the relative counts would test both the concentration of radioisotopes in the mantle and crust, as well as the possible presence of ^{40}K in the core.

4.3. Uncertainties

We want to investigate the impact of uncertainties of the reference model on the geoneutrino distribution. The uncertainties for crust and mantle are independent of each other. The crust itself can vary by about a factor of two in radioisotope abundances and thus in geoneutrino intensity (Mantovani et al., 2004). The net effect of these variations thus depends on how the crust and mantle uncertainties combine.

Figure 8 shows some of the possible uncertainties. In the plot showing the minimum amount of radioactive elements in the crust–mantle system it can be perceived, that the reference model is on the low side of the possible abundances in comparison to the maximum abundances, where the intensity grows by a factor of two. The absence of a double peak in the intensity is due to the fact, that in Mantovani et al. (2004) only an uncertainty for the whole crust is given, which does not take into account the distinction between lower, middle and upper crust. The overall shape of the maximum and minimum abundance plots stays overall very similar to the reference model, although there is a deviation from the general form at angles $\theta \gtrsim 30°$. Thus we see that in all cases there appears a large peripheral flux, which remains a robust and highly testable prediction of this model.

For a more detailed analysis it will be necessary to construct a model of the outermost layer of the earth, as we assumed for a change in altitude an increase in the number of neutrinos coming from the sediments. But in any case the crudeness of our estimate only infects the results for the outer periphery, and the central angles remain reliable, as we now see.

4.4. Crust anisotropies and observing strategies

We have assumed spherical symmetry throughout, and thus our calculations cannot directly address the effect of anisotropies in the radioisotope distributions. Yet these anisotropies, which reside in the crust, can have a very

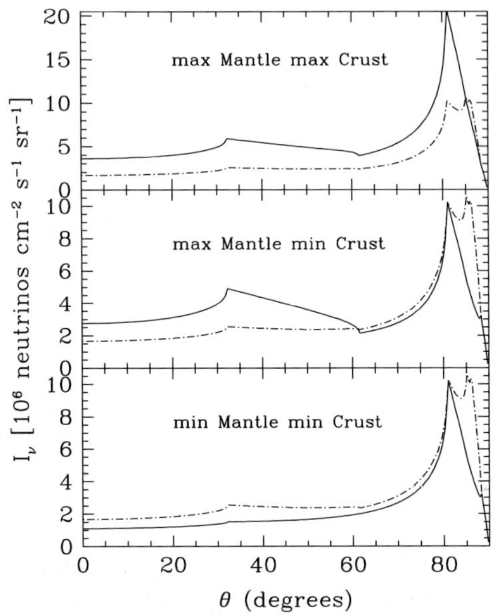

Figure 8. The upper plot shows the intensity with a maximal amount of radioactive elements in crust and mantle. In the lower panel the intensity for a minimal abundance in crust and mantle is plotted. As the uncertainties in the abundances of mantle and crust are independent of each other (Mantovani et al., 2004), the plot in the middle shows a hybrid scenario with a maximal abundance in the mantle, whereas the crust abundance is minimized. The dashed-dotted line in all three plots is the reference model intensity is added as dashed-dotted line in all three plots for comparison. It can be seen that the reference model abundances are at the lower limit of the possible range of values. That the crust is only represented by a single peak is an artifact which arises because the uncertainties given in Mantovani et al. (2004) do not distinguish between the different layers of the crust; this changes the intensity in the outer crust significantly.

significant impact (Mantovani et al., 2004), since the crust is the largest radioisotope reservoir. For example, Mantovani et al. (2004) predict fluxes which differ by more than a factor of 2 between locations above minimum and maximum crust depths. This would lead to changes in the overall geo-neutrino intensity, and possibly to observable azimuthal asymmetry.

We note that emission from the crust will affect the intensity only at the largest nadir angles. Thus we expect our spherical calculation to be reliable at small to median angles. Furthermore, one can get a rough understanding of the effect of different crust depths by adopting spherical models in which the crust layers have the properties appropriate for the experimental location.

As an extreme example, consider a hypothetical model where we assume that the neutrino detector is on a ship. That means that the crust has the abundances of the oceanic crust and the top layer is 3 km of ocean water. Of course in this case there is no sufficient shielding from cosmic rays and

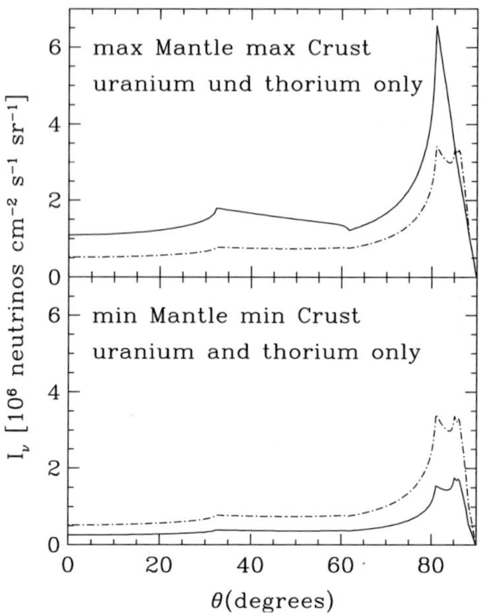

Figure 9. Same as for 8 but only for the uranium and thorium intensities.

atmospheric radiation, which will make the detection of low energy neutrinos hard if not impossible. But nevertheless the effects are interesting, as the contribution coming from large angles is reduced by a factor of three, which should make it easier to detect the effects coming from the core and mantle. We have presently no model for an anisotropic earth, but we expect, that for the real earth in the described case the intensity for lower angles ($\theta \lesssim 60°$) will be very similar to the reference model (in case of its correctness), only the contribution from the angles $\gtrsim 30°$ will be noticeably reduced.

5. Experimental Challenges and Prospects

We briefly want to address the practical aspects of measuring the geoneutrino angular distribution. For a more thorough discussion and Monte Carlo study see Hochmuth et al. (2005); here we summarize the basic results, which are sobering, at least for the present.

Antineutrinos are usually detected via inverse β-decay of protons, $\bar{\nu}_e + p \to e^+ + n$, and the subsequent observation of the rapid e^+ annihilation signal as well as the delayed n capture onto an ambient nucleus. Part of the momentum of the neutrino is transferred to the neutron, which thus contains directional information (Vogel and Beacom, 1999). This directionality is, however, gradually lost as the neutron elastically scatters before it is captured. As noted by Vogel and Beacom (1999), the number of elastic scatterings is large enough that individual neutron directionality is lost; the original forward direction can only be recovered statistically. For the case of a collimated neutrino beam, the resulting neutron distribution is thus a sphere of radius ~10 cm, whose center is offset from the e^+ in the forward direction by only ~2 cm. This mismatch in scales requires a very large number of events to resolve; this is the crux of the (main) difficulty in measuring geoneutrino directionality. Given the "Poisson" blurring of elastic scatting, directional sensitivity is best for liquid scintillators which use materials (such as gadolinium) that have a large neutron capture cross section. At present, no such measurement appears possible: KamLAND is a liquid scintillator experiment, but the neutron capture on protons is slow compared to gadolinium. However, the reactor neutrino experiment CHOOZ did use a gadolinium-loaded scintillator, and was able to reconstruct the source direction to within about ~20° (Apollonio et al., 2000), which as we have emphasized would already be geophysically interesting. While CHOOZ is no longer operating, and was too small to detect the geoneutrino intensity, it nevertheless demonstrated that antineutrinos with an energy spectrum similar to that of geoneutrinos can be already detected with modest angular sensitivity.

Of course, even the crude sensitivity of CHOOZ was only achieved by observing about 2500 events which originated from a well-localized point source. Our situation is more demanding, as we wish to determine a pattern with azimuthal symmetry. One thus would expect that the simplest approach would test for anisotropy, e.g., how does the net upgoing neutron count rate compare to the downgoing count rate, that is, does the geoneutrino signal have a nonzero dipole?

In Hochmuth et al. (2005), it is shown that a large-volume, 50 kt scintillator experiment can indeed detect deviations from an isotropic geoneutrino flux. Indeed, the simple presence of a dipole can be determined with a few thousand events. However, the precision of the detection is very low. Thus, to confidently distinguish between interestingly different geophysical models – i.e., different dipole predictions for, say, a central vs. peripheral distribution – requires a number of events $N \gtrsim 10^5$. Hochmuth et al. (2005) thus conclude that upcoming experiments can hope to detect anisotropy but with little discriminating power. Of course, a precision measurement of the geoneutrino flux, and its spectrum, will of course still offer important geophysical information.

Other experimental difficulties are also present. While the geoneutrino flux is significantly higher than the supernova background, geoneutrino detection will be all the more challenging given that typical geoneutrino energies are relatively low, ~0.5–1.5 MeV (Krauss et al., 1984; Mantovani et al., 2004). Indeed, the low threshold of ^{40}K places its neutrino signal below the 1.8 MeV threshold for inverse beta decay, so that these geoneutrinos will be elusive for the foreseeable future – barring innovative advances in experimental techniques. And of course, a good geoneutrino experiment clearly will need to be as far as possible from reactors. This is particularly important if the radioisotope distribution really is concentrated in the outer earth, as this will lead to a peripheral flux which can be confused with (but possibly calibrated by) the contribution from nearby detectors. A more promising aspect of geoneutrino physics is the detection of the speculative georeactor (Herndon, 2003). We just want to point out that such a reactor would have a similar spectrum to conventional power plants, which sets a detection within reach of a large future detector as noted in e.g. Hochmuth et al. (2005).

6. Discussion

Geoneutrinos offer invaluable and unique information about the energetics and structure of the earth (Krauss et al., 1984; Fiorentini et al. 2003b). The longstanding dream of measuring this neutrino population has now begun to be realized with the first detection of a geophysical signal by the KamLAND experiment (Araki et al., 2005a; Eguchi et al., 2003). This achievement is already a triumph, as the geophysical component is (by design!) dominated by the signal from reactor neutrinos. Nevertheless, we believe it is now worthwhile to look forward to the even more challenging results of determining the angular distribution of geoneutrinos, which offers a wealth of new information.

In this paper we thus have calculated the angular distribution of geoneutrinos which arise in β-decays of potassium, thorium, and uranium. We have developed the general formalism for the neutrino intensity in a spherically symmetric earth. We find that the geoneutrino angular distribution, once known, can be inverted to fully recover the terrestrial radioisotope distribution. Thus the geoneutrino "sky" can provide a tomography of the earth's structure, and yields the full radial dependence of the radiogenic heat production.

Turning to model-building, we explore the idealized case of an arbitrary shell of uniform density. This can be generalized to give an earth model which is a series of concentric uniform shells. We then adopt the radioisotope profile of the reference model and calculate the resulting angular distribution (Figure 4). Because the reference model places all radioisotopes in the mantle and crust, the resulting geoneutrino intensity is highly "peripheral," with 2/3 of the flux coming from nadir angles $\theta \gtrsim 60°$. Thus, even a crude measurement of the angular distribution (say, in three 30° bins) would strongly test this prediction.

We have also investigated the effect of physically plausible variations to the reference model. Mantovani et al. (2004) identify uncertainties in their radioisotope distributions which have the effect of multiplicatively raising or lowering the neutrino intensity, without a significant change to the angular shape. However, both the shape and normalization of the intensity can change strongly if the earth's core contains a significant amount of potassium, contrary to the assumptions of the reference model. If core potassium abundances are near the current upper limits (Lee and Jeanloz, 2003), the resulting geoneutrino signal can dominate the total flux. Measurements of the angular distribution will probe the radioisotope abundances in the core. In particular, measurement of the intensity inside a nadir angle $\theta \lesssim 30°$ can discriminate among possibilities recently suggested in the literature.

Moreover, the central and peripheral intensities are related, since the peripheral neutrinos come from outer shells which also contribute to the central signal. Thus a measurement of the peripheral flux can be used to place a lower limit to the central flux. A difference between this lower limit and the observed central flux then amounts to a detection of some radioisotopes in the core.

A determination of the geoneutrino angular distribution will also solidify the connection between the geothermal heat flux and the geoneutrino flux. As noted by Fiorentini et al. (2003b), the *radial* component of the neutrino flux is directly related, by Gauss' law, to the geothermal heat production (Equation 24). But neutrino emission is locally isotropic and hence contains non-radial components; thus the angle-integrated geoneutrino flux in turn includes the non-radial components, and thus can only be related to the heat flux given a model of the radioisotope density profile. However, a measurement of the geoneutrino angular distribution can be inverted to recover the terrestrial radioisotope density distribution. This will allow for a full calculation not only of the global radiogenic heat production, but also of its radial dependence. Thus one can test in detail models of heat production and transport.

In this way, the neutrino intensity can be used to measure the radioactive contribution to the geothermal heat flux. This can then be compared to geophysical measurements (Stein, 1995) of the total heat flux. A comparison of these results will yield a new and robust measurement of the Urey ratio (Equation 1) (McKenzie and Richter, 1981). This in turn will shed light on the thermal history of the earth, and quantify the importance of any non-radioactive heating, presumably due to residual "primordial" processes during the formation of the earth.

Also, we note that the reference model we have adopted normalizes to the observed terrestrial heat flux and a Urey ratio of 0.5. Thus, if this Urey ratio is correct, but the heat flux contains a significant component from the core, this will reduce the contribution from the outer layers, which will act to redistribute the peripheral intensity to the interior.

These particular results illustrate a larger more general conclusion, that geoneutrinos open a new window to the earth's interior. Measurements of the angular distribution of geoneutrinos will allow us to infer the radioisotope distribution of the earth. This in turn offers a new probe of the earth's structure – for example, allowing a test of how sharp the radioisotope boundaries are in going from crust to mantle to core. If the boundaries are sharp, then the angular distribution will offer unprecedented new measures of the positions of these boundaries and thus will be a general probe of the interior structure of the earth.

Unfortunately, these lofty goals will, to say the least, require great patience (and/or innovation!). For inverse beta decay detection reactions, the isotropizing effect of elastic scatterings presents enormous challenges in recovering the neutrino directionality, particularly for a diffuse signal of the kind we are considering. While next-generation large-volume scintillator experiments can firmly establish the anisotropy of the geoneutrino signal, the dipole sensitivity for any realistic experiment will be too poor to discriminate among even drastically different radionuclide distributions.

In the meantime, there is hope; we note as well that we have so far considered the total, energy-integrated, intensity. However, neutrino energy information is also available, and indeed geoneutrino spectra have been presented (Mantovani et al., 2004). Since the emitted energy spectra take the well-understood β-decay form, with sufficient energy resolution it is possible to separate the U and Th components. If this can be done in conjunction with even the crudest angular resolution, it would be possible to actually distinguish beyond all doubt between neutrinos coming from U, Th and K and to obtain a particularly complete picture of the radioactive earth.

Thus we believe the quest to measure an image of the "geoneutrino sky" is a worthy if challenging and futuristic goal. As we have shown, even the first, crudest attempts at neutrino imaging will yield important results, and will thus impel further improvements.

Acknowledgements

We are grateful to Stuart Freedman for very helpful discussion regarding experimental issues and prospects regarding antineutrino directional sensitivity. We thank Charles Gammie for encouragement and for alerting us to the Abel transform, and V. Rama Murthy for enlightening discussions regarding core potassium. We are particularly thankful to Georg Raffelt for guidance and insight. The work of BDF is supported by the National Science Foundation grant AST-0092939.

Appendix A: Terrestrial Tomography: Inverting the Angular Distribution

We may write the intensity distribution (Equation 6) in dimensionless units as

$$I(\sigma) = 2R_{\oplus} \int_{\sigma}^{1} dx \frac{xq(x)}{\sqrt{x^2 - \sigma^2}} \equiv \int_{0}^{\sigma} dx K(\sigma, x) q(x) \tag{A1}$$

where $\sigma = \sin\theta \in [0, 1]$ and $x = r/R_{\oplus} \in [0, 1]$. Thus both I and q are defined on the interval [0, 1]. Furthermore, for an experiment on the surface of the earth, we expect that $I(1) = 0 = q(1)$ because the earth's density goes to zero at the surface (by definition!). However, a real experiment located slightly under the surface of the earth might have a nonzero horizontal flux $I(1)$.

Clearly, $I(\sigma)$ is an integral transformation, with $K(\sigma, x) = x/\sqrt{x^2 - \sigma^2}$ the kernal. Specifically, Equation (A1) is a version of the Abel transform (Bracewell, 1986). In fact, the usual Abel transform is applied to a function defined over an infinite domain, but fortunately one can show that the key results carry over to our case of a finite domain.

The inverse Abel transform appropriate for our case is

$$q(x) = -\frac{1}{\pi R_{\oplus}} \int_{x}^{1} d\sigma \frac{I'(\sigma)}{\sqrt{\sigma^2 - x^2}} + \frac{I(1)}{\pi R_{\oplus}\sqrt{1 - x^2}} \tag{A2}$$

$$q(x) = -\int_{0}^{\sqrt{1-x^2}} d\mu \frac{I'(\mu)}{\sqrt{1 - x^2 - \mu^2}} + \frac{I(\mu = 0)}{\pi R_{\oplus}\sqrt{1 - x^2}} \tag{A3}$$

where $\mu = \cos\theta$, and $I'(y) = dI(y)/dy$ is the usual derivative.

Equation (A2) thus demonstrates by construction that, given a complete knowledge of the intensity distribution, one can fully recover the radioisotope source distribution. Thus, measurement of the geoneutrino angular distribution truly does carry the promise of tomographic imaging of the earth's interior. In addition, with $q(x)$ in hand, one can completely determine the radiogenic heat production of the earth, both globally and as a function of depth.

Furthermore, Equation (A2) has the properties one would expect on physical grounds. The density at $r = R_{\oplus}x$ depends only on the intensity derivative for the region $\sin\theta \geq x$, i.e., angles along or exterior to the tangent angle. Thus, inferring the outer density structure requires only knowledge of the peripheral intensity. On the other hand, to recover the inner density structure requires both peripheral and central intensities. This is indeed sensible if one thinks of the angular distribution roughly as a linear combination of intensities along the line of sight: outer angles have only a few "terms" in the sum, while inner angles contain all "terms."

One consequence of this result is that the peripheral intensity constrains the central intensity, by setting a lower limit on it. If we consider $I(\sigma)$ only for $\sigma > \sigma_0$, we can infer a lower limit $q_{\min}$ to the density distribution at $x < \sigma_0$, namely

$$q(x) > q_{\min}(x) = -\frac{1}{\pi R_{\oplus}} \int_{\sigma_0}^{1} d\sigma \frac{I'(\sigma)}{\sqrt{\sigma^2 - x^2}} \tag{A4}$$

This example illustrates that even with an incomplete or low-resolution determination of the intensity pattern, one can draw powerful physical conclusions.

References

Albarede, F. and van der Hilst, R.: 2002, *Phil. Trans. R. Soc. Lond. A* **360**, 2569.

Apollonio, M. et al. [CHOOZ Collaboration]: 2000, *Phys. Rev. D* **61**, 012001 [arXiv:hep-ex/9906011].

Araki, T. et al.: 2005a, *Nature* **436**, 499.

Araki, T. et al. [KamLAND Collaboration]: 2005b, *Phys. Rev. Lett.* **94**, 081801 [arXiv:hep-ex/0406035].

Barger, V., Whisnant, K., Pakvasa, S. and Phillips, R.: 1994, in J. N. Bahcall et al. (eds.), *Solar Neutrinos-The First 30 years*, Addison-Wesley, Reading, p. 300.

Binney, J. and Tremaine, S.: 1987. *Galactic Dynamics*, Princeton University Press, Princeton.

Bracewell, R. N.: 1986. *The Fourier Transform and its Applications*, McGraw-Hill, New York.

Chandrasekhar, S.: 1950, *Radiative Transfer*, Oxford University Press, Oxford, Chap. 1.

Dahlen, F. A.: 2004, *Geophys. J. Int.* **157**, 315.

Dziewonski, A. M. and Anderson, D. L.: 1981, *Phys. Earth Plan. Int.* **25**, 297 http://solid_Earth.ou.edu/prem.html.

Eder, G.: 1966, *Nucl. Phys.* **78**, 657.

Eguchi, K. et al. [KamLAND Collaboration]: 2003, *Phys. Rev. Lett.* **90**, 021802 [arXiv:hep-ex/0212021].

Fiorentini, G., Lasserre, T., Lissia, M., Ricci, B. and Schonert, S.: 2003a, *Phys. Lett. B* **558**, 15 [arXiv:hep-ph/0301042].

Fiorentini, G., Mantovani, F. and Ricci, B.: 2003b, *Phys. Lett. B* **557**, 139 [arXiv:nucl-ex/0212008].

Gessmann, C. K. and Wood, B. J.: 2002, *Earth Planet Sci. Lett.* **200**, 63.

Herndon, J. M.: 2003, *PNAS* **100**(6), 3047.

Hochmuth, K. A. et al.: 2005, Astropart. Phys. [arXiv:hep-ph/0509136] (in press).

Hofmeister, A. M. and Criss, R. E.: 2005, *Tectonophysics* **395**(3–4), 159.

Krauss, L. M., Glashow, S. L. and Schramm, D. N.: 1984, *Nature* **310**, 191.

Lee, K. K. M. and Jeanloz, R.: 2003, *Geophys. Res. Lett.* **30**, 2212.

Mantovani, F., Carmignani, L., Fiorentini, G. and Lissia, M.: 2004, *Phys. Rev. D* **69**, 013001 [arXiv:hep-ph/0309013].
McDonough, W.: 2002, http://mahi.ucsd.edu/cathy/SEDI2002/ABST/SEDI1-2.html.
McDonough, W. F. and Sun, S.-s.: 1995, *Chem. Geol.* **120**, 223.
McKenzie, D. and Richter, F.: 1981, *J. Geophys. Res.* **86**, 11667.
Nunokawa, H., Teves, W. J. C. and Zukanovich Funchal, R.: 2003, *JHEP* **0311**, 020 [arXiv:hep-ph/0308175].
Pollack, H. N., Hurter, S. J. and Johnson, J. R.: 1993, *Rev. Geophys.* **31**, 267.
Rama Murthy, V., van Westrenen, W. and Fei, Y.: 2003, *Nature* **423**, 163.
Rothschild, C. G., Chen, M. C. and Calaprice, F. P.: 1998, *Geophys. Res. Lett.* **25**, 1083 [arXiv:nucl-ex/9710001].
Stein, C.: 1995, in T. J. Ahrens (ed.), *Global Earth Physics: A Handbook of Physical Constants*, AGU Reference Shelf 1, American Geophysical Union, Washington, p. 144.
Van Schmus, W. R.: 1995, in T. J. Ahrens (ed.), *Global Earth Physics: A Handbook of Physical Constants*, AGU Reference Shelf 1, American Geophysical Union, Washington, p. 283.
Vogel, P. and Beacom, J. F.: 1999, *Phys. Rev. D* **60**, 053003 [arXiv:hep-ph/9903554].
Wasserburg, G. J., MacDonald, G. J. F., Hoyle, F. and Fowler, W. A.: 1964, *Science* **143**, 465.

Earth, Moon, and Planets (2006) 99:183–192
DOI 10.1007/s11038-006-9130-6

On the Possibility of Directional Analysis for Geo-neutrinos

MIKHAIL BATYGOV
University of Tennessee, Knoxville, TN 37996, USA
(E-mail: msb@positron.phys.utk.edu)

(Received 5 June 2006; Accepted 15 October 2006)

Abstract. The possibility of terrestrial antineutrino directionality studies is considered for future unloaded liquid scintillator detectors. Monte-Carlo simulations suggest that the measurable displacement between prompt and delayed antineutrino signals makes such studies possible. However, it is estimated that on the order of 1000 terrestrial antineutrino events are required to test the simplest models, demanding detectors of 100 kt size to collect sufficient data in a reasonable period of time.

Keywords: Directionality, geo-reactor, KamLAND, simulation, terrestrial antineutrinos

1. Introduction

In 2005, KamLAND collaboration reported the first result on the observation of terrestrial antineutrinos and their flux estimation (Araki et al., 2005). This result is of prominent importance for geophysics. With significantly higher statistics and smaller background levels in future experiments it will be possible to check different geological models. Nonetheless, the measurement of the incident antineutrino direction distribution would have even more geological and geophysical implications.

First of all, according to commonly accepted geological models, (e.g., McDonough and Sun, 1995), the Earth's continental crust is considered to have relatively high concentrations of long-living radioactive isotopes (Taylor and McLennan, 1985). Hence it is expected to produce more geoneutrinos than the oceanic crust. If a KamLAND-like experiment is located on the border between the oceanic and continental crusts, there might be a measurable horizontal anisotropy in the antineutrino flux, provided the detector is directionality sensitive. Second, since the mantle and especially the core are believed to be depleted in uranium and thorium, which are the primary sources of detectable terrestrial antineutrinos, the flux from the center of the Earth should be substantially smaller than that from near-horizontal directions. Third, the highly debated geo-reactor hypothesis (Herndon, 1996; Hollenbach and Herndon, 2001) can be finally tested, based on the antineutrino directionality study (Raghavan, 2002). Due to the differences in the antineutrino spectra, the geo-reactor hypothesis test is practically independent from the measurement of the radioactive isotope distribution inside the Earth, as long as the detector is sensitive to the antineutrino energy.

Besides geoneutrino studies, the directionality measurement can provide an additional tool to suppress background and to distinguish between different sources of antineutrinos, thus being useful for nearly all antineutrino detection experiments. So the benefits of directionality measurements are quite obvious. The goal of this paper is, using our experience from KamLAND, to explore the feasibility of such measurements.

2. Detection Method in KamLAND

This study was performed for liquid scintillator (LS) antineutrino experiments of KamLAND type and was originally intended to find out whether directionality measurement was possible with KamLAND. It was realized that such a measurement requires detectors much larger than KamLAND. The main conclusions of this paper are relevant for future, large antineutrino detectors using unloaded organic LS.

Water-Cherenkov detectors have a different detection mechanism and are not considered here due to their high energy threshold (3.5–5 MeV) not suitable for most terrestrial antineutrino studies.

In LS antineutrino experiments, electron antineutrinos are detected through the inverse beta decay reaction:

$$\bar{\nu}_e + p \rightarrow e^+ + n \tag{1}$$

The main feature of this detection mechanism is that each antineutrino event gives two signals, a prompt positron and a delayed neutron capture, well correlated in space and time. This coincidence suppresses background tremendously. However, the reaction has an energy threshold of about 1.8 MeV and terrestrial antineutrinos with energies below that cannot be detected (e.g. $\bar{\nu}_e$ from ^{40}K decay are undetectable through this mechanism).

Another property of the inverse beta decay is essential for the directionality study in any scintillator-based experiment. The direction from the prompt signal coordinate to the delayed one is correlated with the direction of the incident antineutrino, although rather weakly. Due to the weakness of this correlation, the detection of antineutrino directions on the event-by-event basis is impossible. However, given enough events, it may still be feasible to extract the antineutrino direction distribution function statistically, by unfolding the angular correlation. This property has been successfully used in the CHOOZ experiment for the first practical directionality study with a LS antineutrino detector (Apollonio et al., 2000).

Unlike KamLAND, CHOOZ used gadolinium-loaded LS, resulting in a more timely, energetic, and isotropic delayed signal, so the simulations performed for that experiment are not directly applicable to experiments using unloaded LS. Thus the feasibility of a similar analysis with such detectors, especially in the context of lower-energy terrestrial antineutrino studies, is the subject of a quantitative study presented below. Before going on, it would be instructive to cover the physics of event detection in more detail.

When an antineutrino interacting with a proton releases a positron and a neutron in reaction (1), the residual energy is divided between the two products. Owing to the laws of kinematics, the positron as the much lighter particle gets most of the available energy.

The prompt signal is generated from the energy deposited by the positron through ionization and Cherenkov radiation, followed by the positron annihilation. The ionization generates scintillation directly, while the UV part of Cherenkov radiation can be absorbed and re-emitted in longer wavelengths by the scintillator. The positron annihilation with an electron releases two 0.511 MeV gammas. The gammas lose energy mostly to Compton scattering on electrons; the electrons cause ionization and hence scintillation. All these processes take place on a time scale well within the time resolution of the detector and are observed as a single event.

The kinetic energy of the neutron generated in the inverse beta decay is below 100 keV and quickly diminishes through elastic scattering on nuclei of the LS, primarily protons. These protons, being charged particles, do produce ionization as well, contributing a very small addition to the prompt signal. After thermalization, the neutron experiences many elastic collisions before being captured on a nucleus. In KamLAND LS, the most likely neutron capturer is a proton, although a neutron does have a small probability to be captured on a ^{12}C nucleus as well. The capture on a proton yields a 2.2 MeV gamma which, similarly to the annihilation gammas, loses energy through Compton scattering on electrons. The latter lose energy to ionization producing scintillation light. This signal forms the delayed event. Since the average neutron lifetime before its capture on a proton is around 200 μs, KamLAND electronics easily distinguish between the two events.

3. Directionality Detection Principle

The possibility for directionality analysis comes from the kinematics of inverse beta-decay. The relativistic 4-vector momentum conservation equations:

$$p^{\mu}_{\bar{\nu}_e} + p^{\mu}_{p} = p^{\mu}_{e^+} + p^{\mu}_{n}, \tag{2}$$

where $\mu = 0, 1, 2, 3$ can be satisfied for any positron scattering angle, from 0° to 180°. Given the initial $\bar{\nu}_e$ energy and this angle, the kinetic energies of both neutron and positron, as well as the neutron scattering angle, are determined from (2).

The relationship between the positron and the neutron scattering angles is shown in Figure 1. Note that the component of neutron momentum in the direction of the incident $\bar{\nu}_e$ is always positive. In other words, neutrons cannot "backscatter" in the inverse-beta reaction. This property is the key to the directionality measurement.

The distribution of positron angles can be calculated (Vogel, 1999). It is almost uniform with a slight preference for "backscattering". This study uses the approximation:

$$\frac{d\sigma}{d\cos\xi} \propto 1 - 0.102\beta\cos\xi, \tag{3}$$

where σ is the cross section of the reaction, β is the positron velocity in terms of the speed of light and ξ is the scattering angle.

Starting with this distribution, a Monte-Carlo simulation is used to generate and track positrons and neutrons according to given $\bar{\nu}_e$ energies. The simulation yields spatial distributions of energy depositions for prompt and

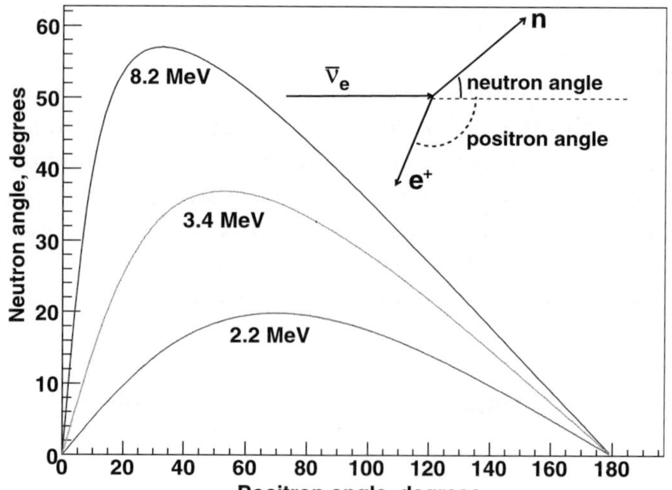

Figure 1. Neutron direction as a function of positron direction for $\bar{\nu}_e$ energies (bottom to top) 2.2, 3.4, 8.2 MeV.

delayed events. Due to the limited number of photoelectrons obtainable in existing liquid scintillators from events with energies around 1–2 MeV, a reliable reconstruction of the energy deposition spatial distributions is practically impossible. Instead, some reference point for each event is estimated by means of a maximum likelihood approach. Normally, it is associated with the three-dimensional center of light emission and is referred to as the "reconstructed vertex".

Two statistical uncertainties arise here. First of all, the center of light emission is not the same as the origin of the event that caused it. This is especially true for the delayed signal in which the gamma can sometimes travel more than one meter from the point of the neutron capture that generated it. Second, vertex reconstruction tools can not be perfect and always introduce additional smearing.

4. Simulation Procedure

The study employed GEANT4-07-01 Monte-Carlo simulation package which is freely available from CERN. To track the neutrons, the G4NDL3.7 neutron scattering cross section database was used. The positron scattering angle was simulated according to (3). As was mentioned above, once this value is fixed, the angle of the neutron and the initial energies of both particles are known. From this point, the positron and the neutron are tracked separately with GEANT4. The coordinates of detailed energy deposition events are stored and later averaged to provide the estimations for reconstructed vertices.

Two technical details are essential here. First, the position of the inverse beta reaction is unknown, so the absolute displacement of either prompt or delayed signals relative to its origin is useless. What can be measured is the prompt-to-delayed displacement vector and this has to be simulated for any comparison with the experiment. Second, the aforementioned additional smearing by the vertex reconstruction should be included in the simulation.

Upon fairly general assumptions, the vertex reconstruction resolution is inversely proportional to the square root of "Visible Energy" (essentially the number of detectable photons generated in the event): $\sigma = k/\sqrt{E_{\text{vis}}}$, which has been confirmed in KamLAND. The coefficient "k" characterizes the quality of vertex reconstruction and depends on many factors, including the detector design, the scintillator composition, the quality of photomultiplier tubes and electronics, the size of the detector and the accuracy of the reconstruction algorithms.

Although long living isotopes emit almost no terrestrial antineutrinos with energies above 3.5 MeV, this work includes simulations for up to 8.2 MeV to cover most of the hypothetical geo-reactor spectrum as well. Simulation runs with different $\bar{\nu}_e$ energies and different vertex reconstruction resolutions study the significance of vertex finding quality for directionality analysis.

5. Simulation Results

A simulation of absolute displacements for low-energy $\bar{\nu}_e$ signals, without vertex reconstruction smearing is shown in Figure 2. The prompt signal does not show appreciable systematic deviation from the origin. This is due to the fact that the distribution (3) is almost flat, while the free path of positrons in scintillator before annihilation is quite short. On the other hand, the delayed signal exhibits a clear bias in the direction of the incident antineutrino as expected. So does the vector between the prompt and delayed signal. Therefore, the directionality analysis in an unloaded scintillator is, in principle, possible. When all antineutrinos come from a single source direction, a displacement of about 1.9–2.0 cm is expected. This is close to the 1.7 cm obtained in a similar simulation for the CHOOZ experiment (Apollonio et al., 2000). The difference might be attributable to different scintillator composition.

The simulation of observable displacements smeared with vertex resolution equal to 12 $\text{cm}/\sqrt{E_{\text{vis}}[\text{MeV}]}$ in each projection (x, y, z) is presented in Figure 3. The displacement does not depend substantially on the antineutrino energy. Its spread decreases slightly with increasing energy due to the better absolute resolution of the prompt signal but the effect is quite small.

6. Prospects for Model Checking

The simulation generated 240,000 events to achieve a statistically significant result. In a real experiment, this many may not be available. When the statistical error of the average displacement becomes comparable to or exceeds the displacement itself, no conclusion about the direction of incident antineutrinos can be made. In general, the question about the sufficient number of candidates is quite complicated and model dependent. Distinguishing between two geological models with similar directionality patterns would require more statistics.

Perhaps the easiest and the least demanding case here is a monodirectional $\bar{\nu}_e$ flux, e.g. the geo-reactor hypothesis test. The numbers of events necessary to test the hypothesis at different confidence levels (CL) are estimated according to the generated vertex distributions and presented in Table I. The table shows the number of $\bar{\nu}_e$ events necessary to test the geo-reactor hypothesis by means of directionality alone. Of course, from the practical

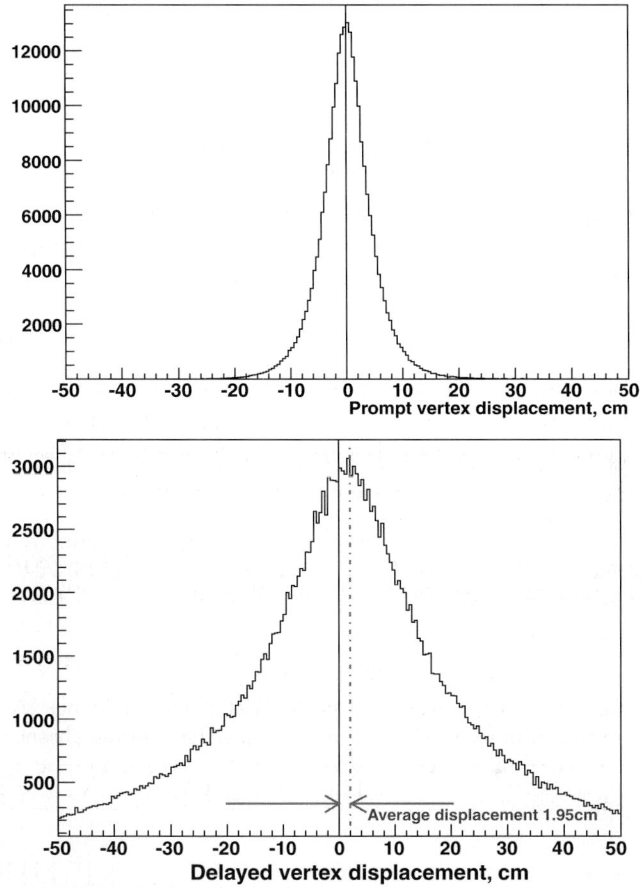

Figure 2. Displacements of the prompt (top) and delayed (bottom) signals from the origin for $E_{\bar{\nu}_e} = 2.6\,\text{MeV}$. The delayed signal is displaced by 1.95 cm in the direction of the incident $\bar{\nu}_e$.

point of view, this hypothesis can be tested by rate and spectrum shape analysis, which would probably require fewer events.

The "Perfect vertex" represents the "raw" signal displacement with no reconstruction smearing, just to show the absolute sensitivity limit of such a study. In real experiments finite smearing is unavoidable but, as was mentioned above, can vary substantially. KamLAND achieves spatial resolution of 12 cm/$\sqrt{E_{\text{vis}}[\text{MeV}]}$, so this can be considered a realistic reference for future experiments. A resolution of 30 cm/$\sqrt{E_{\text{vis}}[\text{MeV}]}$ is presented in the table to illustrate the significance of reconstruction resolution. If the resolution is compromised to that level by the detector design or by other factors, then almost 2/3 of all statistics can be effectively lost for directionality purposes.

As for the absolute event numbers required for the directionality studies, they are quite high. For instance, during four years of operation, KamLAND

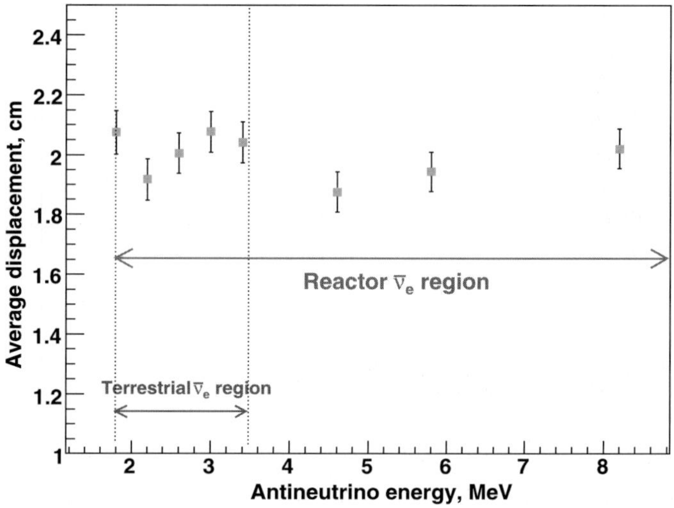

Figure 3. Prompt-to-delayed vertex displacement for reconstruction resolution 12 cm/$\sqrt{E_{vis}[\mathrm{MeV}]}$ as a function of $\bar{\nu}_e$ energy. Error bars correspond to RMS/$\sqrt{240,000}$ to account for the simulation accuracy.

TABLE I

Necessary event counts to test geo-reactor hypothesis at given confidence levels (CL) for different vertex reconstruction resolutions *by directionality only*, in the absence of background

CL	Perfect vertex	12 cm/$\sqrt{E_{vis}[\mathrm{MeV}]}$	30 cm/$\sqrt{E_{vis}[\mathrm{MeV}]}$
0.683	157	236	432
0.900	424	639	1171
0.950	600	905	1659
0.990	1036	1563	2863

has collected less than 1000 $\bar{\nu}_e$ candidates and more than 90% of them are reactor antineutrinos from nuclear power plants. Moreover, the values in Table I are obtained for an ideal, no-background scenario.

In KamLAND, the biggest background source for terrestrial antineutrino studies is the reactor antineutrinos. Most other sources (including uncorrelated accidental coincidences and the α-n correlated events) are caused by the presence of radioactive isotopes in the detector. Currently, they are smaller than but comparable to the reactor background. However, KamLAND expects to reduce them by several orders of magnitude through the detector purification. When this is completed those background sources will become negligible. On the other hand, the reactor $\bar{\nu}_e$ background is unavoidable for KamLAND geoneutrino studies and the only way to minimize its impact in the future is to build new detectors much further from working reactors.

The practical effect of the remaining background on the directionality measurement is similar to that of a finite vertex resolution: still more events are necessary for the same confidence level. Table II provides the quantitative summary of such an effect. Testing more sophisticated geological models and, in particular, the attempts to unfold the angular distribution of anti-neutrino directions would also require more events than can be found in Tables I and II.

7. Conclusion and Discussion

Monte-Carlo simulations taking into account the kinematics of inverse beta decay and the properties of neutron scattering predict a statistically measurable delayed signal displacement in the direction of the incident anti-neutrino. This makes directionality studies with unloaded LS detectors possible. However, the expected effect is small compared to the natural spread of the reconstructed signals. Obtaining a statistically significant result requires thousands of $\bar{\nu}_e$ events (Tables I, II) even for the easiest case (i.e. single antineutrino source). For KamLAND-sized detectors this implies at least a century of run time, which is impractical.

It is mainly the spread of the delayed signal from neutron capture that causes the need for high statistics. The single gamma resulting from the neutron capture on hydrogen and sometimes traveling more than one meter in a random direction greatly weakens the correlation between the antineutrino direction and the reconstructed vertex displacement.

There seem to be two ways to improve the situation. First, a bigger detector located further from working reactors may provide enough statistics in reasonable time. Thus, a 100-kt unloaded LS detector should collect

TABLE II

Necessary event counts to test geo-reactor hypothesis at given confidence levels (CL) for vertex resolutions of 12 cm/$\sqrt{E_{\text{vis}}[\text{MeV}]}$ in the presence of background, from the background-free case to the case when background exceeds the geo-reactor signal by a factor of four

CL	Background to signal ratio						
	0.0	0.1	0.25	0.5	1.0	2.0	4.0
0.683	236	258	290	350	465	678	938
0.900	639	698	785	947	1260	1836	2542
0.950	905	988	1113	1342	1786	2602	3602
0.990	1563	1706	1921	2317	3083	4491	6219

enough events for simple terrestrial antineutrino directionality studies in several years, provided the vertex reconstruction accuracy is not worse than that in KamLAND and the background is low enough. Second, it may be possible to minimize the spatial spread of the delayed event with specially chosen LS loading. Preliminary estimations suggest that loading the scintillator with ^{6}Li to yield ^{3}H $+\alpha$ delayed signal instead of a single gamma should offer about two to four times better statistical efficiency compared to unloaded scintillators. High vertex reconstruction quality is especially important in this case. These two approaches are not mutually exclusive and can be combined in future experiments to provide the best directionality detection characteristics.

The simulations show that background substantially decreases the statistical significance of data that can be collected through a given exposure. To be an efficient geoneutrino instrument, a detector must be designed with special care to avoid radioactive contamination and placed as far as possible from working reactors. This is true for terrestrial antineutrino studies in general but is really critical for the directionality analysis.

While practical directionality measurement is a very demanding task and is impossible with existing detectors, including KamLAND, the scientific significance of such research makes it worth constructing detectors better suited for this purpose.

Acknowledgements

The author would like to thank Yuri Kamyshkov and Yuri Efremenko (University of Tennessee), Charley Rasco (ORNL) and Glenn Horton-Smith (Kansas State University) for useful discussions and suggestions.

References

Apollonio, M. et al.: 2000, *Phys. Rev. D* **61**, 012001.

Araki, T. et al.: 2005, *Nature* **436**, 499.

Herndon, J. M.: 1996, *Proc. Nat. Acad. Sci.* **93**, 646.

Hollenbach, D. F. and Herndon, J. M.: 2001, *Proc. Nat. Acad. Sci.* **98**, 11085.

McDonough, W. F. and Sun, S.-s.: 1995, *Chem. Geol.* **120**, 223.

Raghavan, R. S.: 2002, Detecting a nuclear fission reactor at the center of the earth, arXiv:hep-ex/0208038.

Taylor, S. R. and McLennan, S. M.: 1985. *The Continental Crust its Composition and Evolution*, Blackwell, Oxford.

Vogel, P. and Beacom, J. F.: 1999, *Phys. Rev. D* **60**, 053003.

Earth, Moon, and Planets (2006) 99:193–206
DOI 10.1007/s11038-006-9104-8

Towards Earth AntineutRino TomograpHy (EARTH)

R. J. DE MEIJER
Stichting EARTH, de Weehorst, 9321 XS2, Peize, The Netherlands
Department of Physics, University of the Western Cape, Private Bag X17, Belleville 7535, South Africa
(E-mail: demeijer@geoneutrino.nl)

F. D. SMIT
iThemba LABS, PO Box 722, Somerset West, 7129, South Africa

F. D. BROOKS, R. W. FEARICK
Department of Physics, University of Cape Town, Private Bag, Rondebosch, 7701, South Africa

H. J. WÖRTCHE
Rijksuniversiteit Groningen, 9747AA, Groningen, The Netherlands

F. MANTOVANI
Centro di GeoTecnologie, Siena University, Via Vetri Vecchi 34, I-52027, San Giovanni Valdarno, Arezzo, Italy

(Received 28 February 2006; Accepted 28 June 2006)

Abstract. The programme Earth AntineutRino TomograpHy (EARTH) proposes to build ten underground facilities each hosting a telescope. Each telescope consists of many detector modules, to map the radiogenic heat sources deep in the interior of the Earth by utilising direction sensitive geoneutrino detection. Recent hypotheses target the core-mantle boundary (CMB) as a major source of natural radionuclides and therefore of radiogenic heat. A typical scale of the processes that take place at the CMB is about 200 km. To observe these processes from the surface requires an angular resolution of about 3°. EARTH aims at creating a high-resolution 3D-map of the radiogenic heat sources in the Earth's interior. It will thereby contribute to a better understanding of a number of geophysical phenomena observed at the Earth's surface. This condition requires a completely different approach from the monolithic detector systems as e.g. KamLAND. This paper presents, for such telescopes, the boundary conditions set by physics, the estimated count rates, and the first initial results from Monte-Carlo simulations and laboratory experiments. The Monte-Carlo simulations indicate that the large volume telescope should consist of detector modules each comprising a very large number of detector units, with a cross section of roughly a few square centimetres. The signature of an antineutrino event will be a double pulse event. One pulse arises from the slowing down of the emitted positron, the other from the neutron capture. In laboratory experiments small sized, ^{10}B-loaded liquid scintillation detectors were investigated as candidates for direction sensitive, low-energy antineutrino detection.

Keywords: antineutrino, antineutrino detector, core-mantle boundary, direction sensitive antineutrino detection, EARTH, geoneutrino, inverse beta-decay, 10B-loaded liquid scintillator, 3D-mapping of the Earth's interior, Monte-Carlo simulations, radiogenic heat sources, undergound antineutrino telescope

1. Introduction

1.1. How does the Earth's Interior Work?

In its special issue Science in July 2005 listed this question as one of the 25 most prominent questions for the next 25 years (Kerr, 2005). In October 2005 the Scientific American produced a special edition on "Our Ever Changing Earth". This indicates a revitalisation of widespread interest in the interior of our planet. At first glance this sounds surprising. It seems that in contrast to the successful exploration of our solar system and parts of the Universe, we have a very limited knowledge of the interior of our planet. The deepest that has been drilled into the Earth was ~13 km deep, a mere 0.1% of the Earth diameter and corresponding to the cruising altitude of jetliners. With present techniques, to "descending" deeper is prevented by a rapid increase in temperature and pressure.

Since the beginning of the 20th century information on the deeper parts has been derived from the speed, reflection and refraction of seismic waves, the moment of inertia and the precession motion of the planet, and the physical, chemical and mineralogical information obtained from meteorites and xenolithes. Our present knowledge is often schematically in spherical symmetric models having a crust floating on a viscous mantle, subdivided into a number of concentric shells and encompassing a partially liquid core (Oldham, 1906; Gutenberg, 1914). Only in the crust and the upper mantle usually some structure is indicated.

In the last decades of the 20th century through developments in seismic tomography it has been revealed that parts of the crust are being subducted and have reached the deeper parts of the mantle. The previous view that the convective flow is stratified at a depth of about 670 km and an unmixed or pristine lower mantle is preserved is no longer tenable (van der Hilst and Karason, 1999; Zhao, 2004).

Boyet and Carlson (2005) present a new view on the Earth's interior, which is based on the differences in the isotopic abundance of ^{142}Nd found in meteorites and mantle-derived terrestrial samples. One of the new features is that the layer at the core-mantle boundary (CMB) is a reservoir enriched in radiogenic heat producing sources, resulting from a distillation of the earlier magma ocean and subducted crust. This layer is likely to be the origin of the deep volcanic plumes that manifest themselves at the Earth's surface as ocean islands (e.g. Hawaii, Iceland, Galapagos and Curaçao).

Wilson (2005) quoting Tolstikhin and Hoffmann (2005) speculates that this 'hidden' reservoir is composed out of the ancient primordial crust formed from the solidifying magma ocean. Regardless of its precise dimensions and location, the hidden reservoir is thought to contain over 40% of Earth's K, Th and U, the main heat producing elements. If it resides on the core-mantle boundary, the layer would form a blanket of heat, consistent with the temperature jump of 1000–2000 K within a few hundred kilometres as proposed by Lay et al. (1998).

Presently, little detail is known regarding the fate of subducting slabs. It is clear that large earthquakes occur at the slab/continent interface, and within the slab down to about 670 km depth. But just how far the slab penetrates into the mantle and how the rheology of lithospheric materials behaves once they reach the lower mantle are currently matters of active debate. These questions are fundamental issues in Earth Sciences since they relate to the nature of mantle convection as well as how the Earth evolved and cools off: or, in other words, how the Earth's Interior "works"?

Processes in the deeper Earth manifest themselves at the surface. The convection in the liquid core produces the geomagnetic field, while the convection in the mantle leads to drift of ocean plates and continents as well as volcanic plumes forming ocean islands. These processes are driven by heat flow. The location, type and size of the heat sources are still a topic of debate. We know the heat flow at the Earth's surface from measurements at about 25,000 locations. These measurements have lead to a rather detailed heat-flow map (Pollack et al., 1993). The map shows a large variation (factor 20) in heat flow at the surface with maxima at the mid-oceanic ridges on the southern hemisphere. Integrating the mapped yields, gives a total heat flow of about 45 TW, which is equivalent to the heat production of 15,000 power plants of 1000 MW_e each, with an efficiency of 33%.

1.2. Geoneutrinos

According to Buffett (2003), 6–12 TW is produced in the crust and is of radiogenic origin (decay of natural radionuclides). Radiogenic processes are considered to be predominantly responsible for also the remaining part of the heat flow. In addition to heat the radiogenic processes in the Earth produce antineutrinos and neutrinos: they have been named geoneutrinos (Araki et al., 2005).

The heat produced in nuclear decay is directly related (Fiorentini et al., 2003) to the flux of antineutrinos, as is illustrated in Table I. Detection of low-energy antineutrinos produced in the U and Th decay processes has been demonstrated by liquid scintillator detectors in Kamioka, Japan and in Chooz and Bugey-3 in France. These set-ups primarily address the fundamental aspects of antineutrinos such as their flavour changes and the related mixing angles. The principle of these detectors is based on the capture of an electron–antineutrino, $\overline{\nu_e}$, by a proton of the scintillator material, producing a positron and a neutron. In a simplified picture the positron carries the energy information and the neutron is emitted preferentially in the same direction as

TABLE I
Maximum electron-(anti)neutrino energy and heat production in natural decay processes

Decay	E_{max} (MeV)	Heat (W/kg)
$^{238}U \rightarrow ^{206}Pb + 8^4He + 6e + 6\overline{\nu_e}$	3.26	0.95×10^{-4}
$^{232}Th \rightarrow ^{208}Pb + 6^4He + 4e + 4\overline{\nu_e}$	2.25	0.27×10^{-4}
$^{40}K \rightarrow ^{40}Ca + e + \overline{\nu_e}$ (88.8%)	1.31	0.36×10^{-8}
$^{40}K + e \rightarrow ^{40}Ar + \nu_e$ (11.2%)	1.51	

the incoming antineutrino (Beacom and Vogel, 1996). The neutron travels only a few centimetres before it is captured. In these detectors the neutron capture is detected by the emitted capture γ-ray. The delayed coincident detection between positron emission and neutron capture characterises an antineutrino detection signature.

The KamLAND collaboration published the first official results on the detection of geoneutrinos (Araki et al., 2005) in July 2005. Their results were obtained with a 1 kiloton, monolithic detector, filled with liquid scintillator and housed in an underground mine near Kamioka, Japan. The detector was originally designed for the detection of fundamental properties of antineutrinos and for this reason is located in the vicinity of nuclear power plants. The geoneutrinos therefore are superimposed on a bell-shaped continuum ranging up to about 8 MeV in the antineutrino spectrum caused by antineutrinos resulting from fission processes in the power reactors. The data presented by Araki et al. (2005) correspond to a measuring period of about 750 days. They have been analysed after making extensive corrections for antineutrinos from the power plants and spurious events due to cosmic-ray induced reactions and due to the $^{13}C(\alpha, n)^{16}O$ reaction introduced in particular by the decay of ^{210}Po, which mimic geoneutrino events.

2. Proposed Geoneutrino Telescope

The need for high-resolution antineutrino tomography to map the radiogenic heat sources in the Earth's interior has set the goals for the Earth AntineutRino TomograpHy (EARTH) programme, initially presented in 2004 (de Meijer et al., 2004a, b). Waveform studies of seismic waves by closed spaced seismometers record differences in waveforms, which can best be explained by heterogeneities occurring over lateral distances as small as a few tens of kilometres. Seismic-wave reflections have revealed that the layer thickness varies between non-detectable up to 300 km (Lay et al., 1998; Jeanloz and Lay, 2005). To resolve structures of 150–300 km diameter in the CMB requires an angular resolution of about 3–4°. This goal is to be realised by a set of ten telescopes distributed worldwide, each with a resolution of about

10–15° achieved by using direction sensitive detector modules. This goal sets a number of boundary conditions that more or less dictates our starting point and the initial direction of our technological research. We are fully aware that our task is ambitious and not straightforward and cannot be achieved with state-of-the-art technology. Hence it requires a step-wise approach with manageable tasks, clear deliverables and go/no–go decisions. It seems feasible, but requires considerable technological development, with therefore, in all likelihood quite a number of spin-offs.

The results of KamLAND confirm the feasibility of geoneutrino detection by large volume detectors, but with the present monolithic detectors, no location of the geoneutrino sources can be made. As indicated above, the antineutrino capture contains information on the direction of the incoming antineutrino and the challenge becomes how to utilise this information. As we will demonstrate in this proposal for the localisation of the radiogenic heat sources, directional sensitive detectors are to be developed. These detectors will be placed in a modular detector set-up to form a telescope with a detector mass of four times that of KamLAND.

To check the feasibility and the degree of directional discrimination we place our first telescope, TeleLENS (Telescope for Low-Energy Neutrino based Sciences), on the island of Curaçao, The Netherlands Antilles, situated at about 12° N; 69° W. Using the crustal reference model as used by Mantovani et al., (2004) and assuming 20 TW homogeneously distributed in the mantle as well as a localised hypothetic source of 5 TW in the CMB at 30° S; 69° W, we estimate an expected signal of 24 TNU[1] from the continental and oceanic crusts, 17 TNU from the mantle and 6 TNU from the hypothetical localised source. Since the conversion to the number of detected geoneutrinos depends on the actual volume of the detectors, the detector material and the detection efficiency, it is hard to produce reliable numbers for TeleLENS at this stage of its development.

The choice for Curaçao comes from its large distance to operating nuclear power stations in Florida. The ratio between the fluxes of electron–antineutrinos from the power reactors and geoneutrinos at Curaçao is 0.1. This value is similar to the ratio for Hawaii and two orders of magnitude smaller than for Kamioka, Japan. According to geological information obtained from surface studies (Beets, 1972; Klaver, 1987) the western part of the island contains a large and deep body of limestone on top of basalt. Plans for an analysis of seismic data and pilot drilling are presently being discussed, partly in the framework of the International Continental Scientific Drilling

[1] TNU stands for Terrestrial Neutrino Unit and corresponds to one electron–antineutrino event per year and per 10^{32} proton.

Programme (ICDP). A first exploratory drill to 70 m depth revealed a temperature drop with depth similar to that in the ocean waters.

Direction sensitive antineutrino detectors have not yet successfully been demonstrated and therefore the design, construction and test of these detectors will be one of the objectives of the first phase of the EARTH programme. At this time, our prime emphasis is on the development of the detector units and modules as well as their associated electronics and read-out systems. This development should lead to a Proof of Principle test of our direction sensitive detectors, planned to be carried out at the nuclear power plant of Koeberg, ~25 km north of Cape Town. The outcome of this test is the first go/no–go decision point. We therefore refrain from further speculations on the subsequent trajectories or the details of TeleLENS.

2.1. Technological Approach

Antineutrino detection is traditionally based on the inverse β-decay:

$$\bar{\nu}_e + p \rightarrow n + e^+$$

in which an electron–antineutrino, $\overline{\nu_e}$, is captured by a proton producing a positron and a neutron. The reaction has a Q-value of -1.8 MeV, hence in a scintillator, geoneutrinos produced by ^{40}K are not detected (see Table I). As mentioned above, in a simplified picture the positron carries the energy information, and the neutron is emitted preferentially in the same direction as the incoming antineutrino. The neutron is detected via charged particles or photons emitted directly after it is captured by a nucleus in the scintillator material. The neutron travels a few centimetres in a few microseconds before it is captured. The delayed coincident detection between the positron and the neutron defines an electron–antineutrino detection signature.

Traditionally the neutron capture takes place on a H or a Gd nucleus within the scintillator. Prompt γ-rays resulting from neutron capture are detected. The γ-ray emission is isotropic and the mean free path of the γ-rays is considerably larger than the few centimetres the neutron travels. Hence, the direction information carried by the neutron is lost. In our proposal ^{10}B is used as a neutron catcher. Capture of a thermal or epithermal neutron on ^{10}B leads to disintegration into two charged particles (α and ^{7}Li nucleus) which are then brought to rest in the scintillator within a few microns from the point of capture. In addition, the velocity dependence (v_n) of the neutron capture cross section ($1/v_n$) of ^{10}B leads to an earlier capture of the neutron. This narrows the time window of the event and ensures that the neutron does not deviate too much from its original direction. Our Monte-Carlo simulations (see below) show a reduction in the number of collisions (between positron emission and neutron capture) by a factor of two in scintillators containing 5% (by weight) ^{10}B compared with those containing no boron.

Wang et al. (1999) have discussed the feasibility of using liquid boron-loaded scintillators (BLS) for the detection of antineutrinos. Based on their work together with earlier studies of boron-loaded scintillators we consider that this detection medium is not suitable for use in large scale monolithic detectors such as KamLAND but could nevertheless be useful in a large detector system consisting of a large number of relatively small (< 1 l) detector units. Two important factors that prohibit the use of BLS in a large monolithic antineutrino detector are the following: (a) only a liquid BLS could be considered for use in such a large detector and the liquid BLS presently available are all highly hygroscopic and would thus be extremely difficult to handle and to contain in large volumes; and (b) even though the kinetic energy released to the charged products of the neutron capture reaction in BLS is > 2.3 MeV the light output resulting from this energy is very small, equivalent to that produced by an electron of energy about 60 keV, due to the well-known ionisation density quenching characteristics of organic scintillators. In a very large scintillator the light attenuation due to the long travel distance through the scintillator to the photomultiplier tubes can be expected to reduce the weak neutron capture signal to a level at which it cannot be distinguished from photomultiplier noise and low energy background.

These problems can be avoided in a modular system if the design of a basic single detector unit is very simple and limited to a maximum volume of about 1 l. In such a system there is first the possibility of using a plastic BLS which is rugged and chemically stable, unlike the liquid BLS. However, even if a liquid BLS is used, handling and containment of the liquid should not be a problem when the volume is small. In addition, the good light collection properties that can be achieved using the proposed size of module should avoid problems in the detection of the small amplitude of the neutron capture signal in BLS.

2.2. Simulations

The influence of ^{10}B on the direction information carried by the neutron has been investigated by simulating the capture reaction of an antineutrino by a proton according to the kinematics as described by Beacom and Vogel (1996). Figure 1 clearly shows the effect of ^{10}B on the longitudinal and transverse distribution of the position where the neutron is captured. It clearly shows that the transverse distribution is much narrower and the longitudinal distribution is also considerably more focussed. These effects are

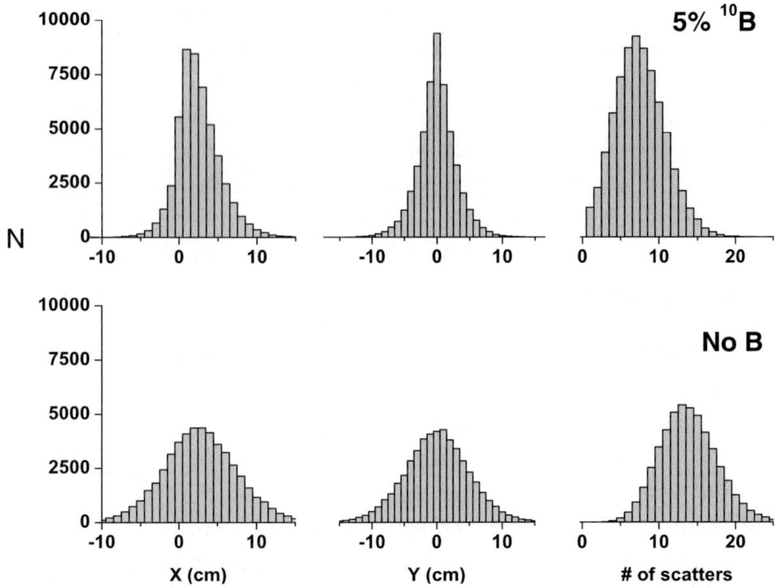

Figure 1. The effect of ^{10}B loading on the neutron capture location. At the top a detector loaded with 5% ^{10}B; at the bottom a detector without B. The results represent 50,000 antineutrino-capture simulations of reactor antineutrinos coming in along the negative x-axis and captured at $(x,y) = (0,0)$ in a large volume detector.

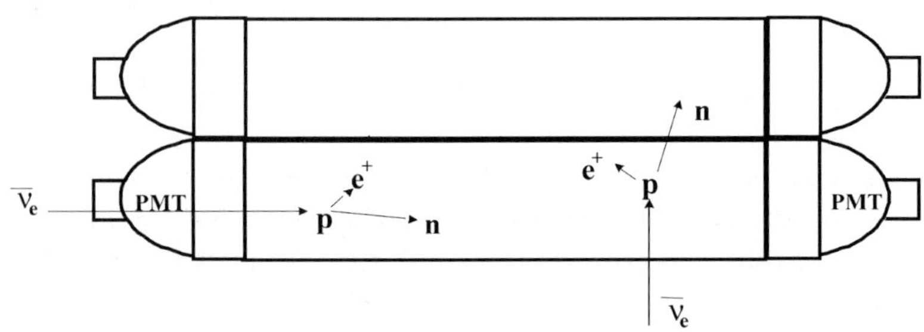

Figure 2. Schematic view of two detector units in an antineutrino module.

clearly related to the fact that the neutron, on average, has half the number of collisions before it is captured.

These results have been used to simulate, in a single detector unit, the sensitivity of the detection probability for neutron detection to the angle of incidence of the antineutrino. For simplicity the unit is assumed to be very long relative to its cross section (two units are schematically presented in Figure 2). Figure 3 shows our simulated neutron detection probability as function of incident antineutrino angle relative to the detector axis for various detector cross sections. It clearly shows that direction sensitivity can

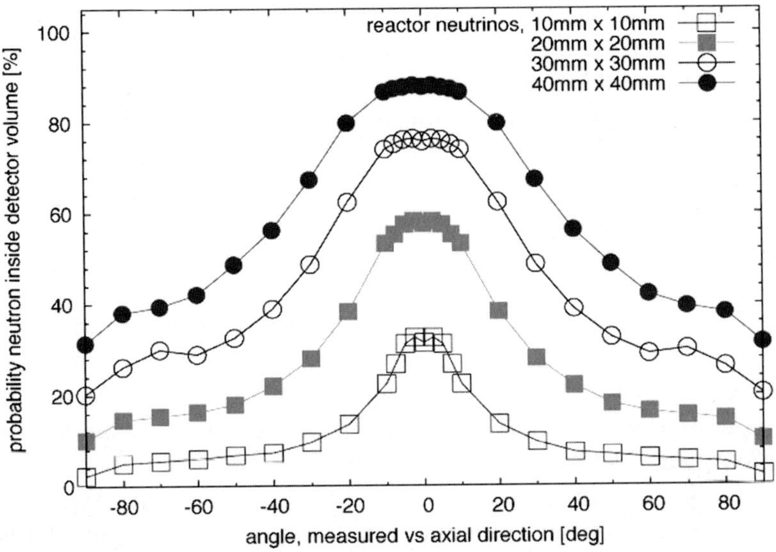

Figure 3. Neutron-detection probability as function of incident angle and detector cross section.

only be obtained for small physical cross section, modular detector systems. It also shows that the efficiency of a single-unit detector is directly proportional to its physical cross section, indicating that without appropriate measures events will be discarded. With the low reaction cross section for antineutrino capture this will be unacceptable. To solve this problem detectors are stacked in modules as e.g. in Figures 2 and 4. Neutrons arising from radially entering antineutrinos would be lost from a single detector module, but can be recorded in an adjacent detector.

The simulations confirm the fact that the neutron travels only a few centimetres, which dictates that for direction sensitivity the physical cross section of the detectors should be restricted to a few centimetres or less. Consequently, to obtain a large volume implies that a very large number of units is imperative.

2.3. Experimental Tests

To test the feasibility of our approach we have started to carry out some experiments at iThemba LABS, South Africa. In these experiments the detection mechanism is mimicked by using neutrons from a ^{252}Cf spontaneous fission source, and 3.8 cm diameter, 2.5 cm long, sealed glass cells filled with NE311A liquid scintillator containing 1 or 5% ^{10}B by mass. The neutrons elastically scatter off protons and produce a recoil-proton scintillation that simulates the positron emitted in antineutrino capture. A double-pulse event results if the neutron is moderated after multiple scatterings and eventually captured by a ^{10}B nucleus in the BLS. The test experiments were carried out using a 8-bit, digital sampling oscilloscope to digitise the photomultiplier-output pulse shapes and a desktop PC to read out, record and analyse the digital output information.

Figure 5 shows examples of two types of a double pulse event recorded in the experiments. Figure 5a shows a typical true double-pulse event. The initial pulse can be recognised as due to a recoil proton from the fact that it displays a distinct low-amplitude tail (slow scintillation component) that continues for 200–300 ns after the start of the pulse. The second pulse, due to neutron capture, stands out clearly above the background noise. Figure 5b shows an example of a spurious double-pulse event that can occur very easily in this type of detector and therefore needs to be well understood and carefully avoided. The event shown in this figure was obtained using a ^{60}Co gamma source. Similar results can be obtained using any type of source, including both neutron and gamma. The initial pulse in Figure 5b is attributed to a recoil electron associated with Compton scattering in the BLS. The "tail" of this pulse is small in comparison with that of the initial pulse in Figure 5a and the two pulses can easily be distinguished as due to "electron" and "proton" respectively by means of a pulse-shape discrimination algorithm operating on the digital output data. The second pulse in Figure 5b is attributed to "after-pulsing" associated with ion and/or optical feedback effects inside the photomultiplier tube of the detector. The after-pulse occurs at a characteristic time after the initial pulse (about 480 ns in the test experiments), depending on the operating conditions. It can be suppressed or controlled by careful selection of the high voltage applied to the photomultiplier and selection of the photomultiplier itself.

Figure 6 shows results from test measurements made under conditions in which after-pulsing was suppressed. The frequency of double-pulse events produced by neutrons from the ^{252}Cf source was measured as a function of the time delay T between the two pulses. The plot shows the number $N(T)$ of events for which this delay exceeds T as a function of T. From simple considerations this distribution is expected to drop off exponentially with a decay time T_0 that depends on the concentration of ^{10}B in the liquid scintillator, the detector geometry and perhaps other factors as well. Monte-Carlo simulations are in progress to determine T_0 for comparison with the experimental measurements.

2.4. Background Reduction

One of the challenges for monolithic detectors is to reduce background; for modular detectors a similar challenge will be faced. In comparison to the monolithic detectors the modular detector is expected to have advantages in

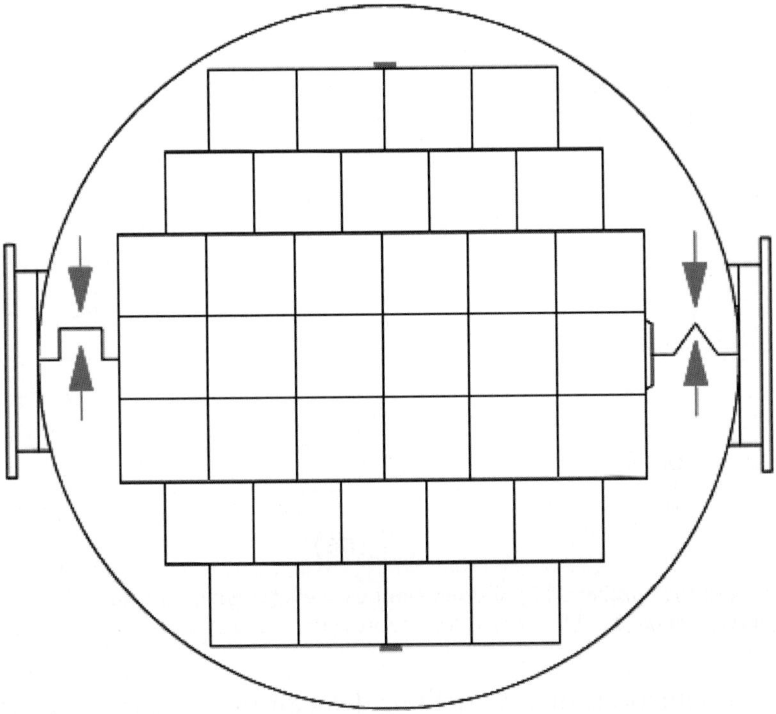

Figure 4. Schematic representation of a multi-unit antineutrino module.

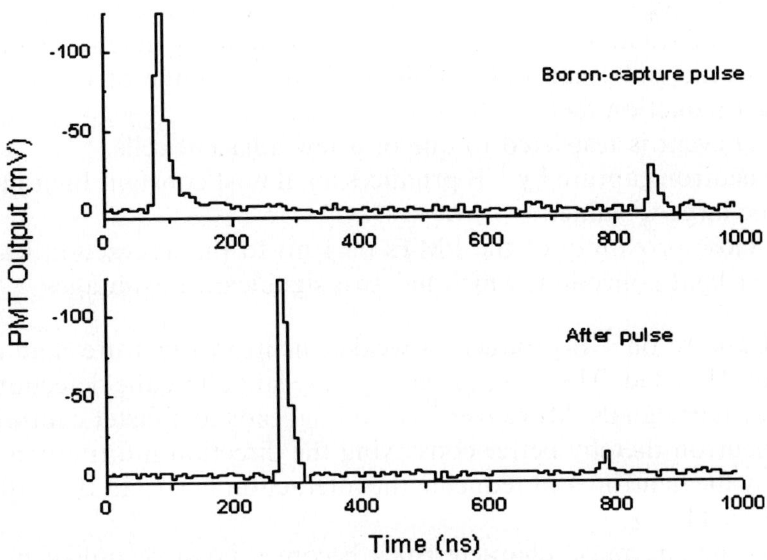

Figure 5. An example of two types of a double pulse event. Top: a recoil proton and a boron-capture pulse and bottom: a γ-ray pulse and an after pulse.

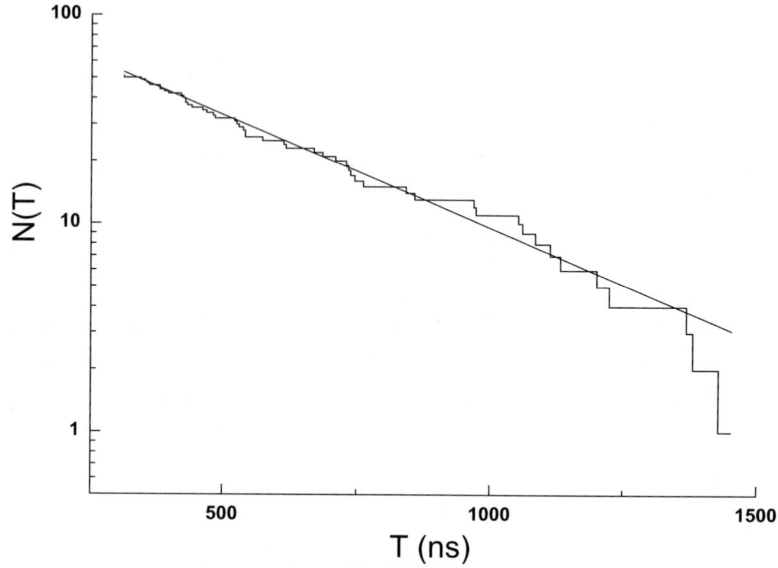

Figure 6. Plot of the number $N(T)$ of double pulse events for which the time t between the two pulses is larger than T. The fitted line represents the function 116 exp (T/T_0) with T_0 = 400 ns.

background reduction. In a monolithic detector every light producing event is detected by every PMT, if the signal is above its threshold. For a modular detector system the following factors contribute to background reduction:

1. The light pulse is only detected in a single cell, which comprises only a very small fraction of the total volume. For the coincidence requirement the second pulse can only originate from a volume of one cell. (Estimated reduction factor: 10^{-6}–10^{-7}.)
2. A real event is restricted to one or a few adjacent cells.
3. The neutron capture by ^{10}B produces an almost constant light pulse due to the large Q-value.
4. The close proximity of the PMTs (< 1 m) to the interaction leads to a higher light collection, which has two significant advantages

- Capture on ^{10}B produces a weaker neutron signature than capture on H or Gd. The close proximity may still allow the detection of the weaker signals. Moreover ^{10}B loading leads to a faster capture of the neutron thereby better conserving the direction information carried by the neutron and reducing the interaction in space (10^{-1}–10^{-2}) and time (10^{-6}).
- Using the pulse characteristics becomes possible unlike in monolithic detectors. Quantification of these background-reduction factors will be part of the Proof of Principle test.

2.5. Proof of Principle

The results obtained thus far give us confidence to proceed to the next stage on the route to the Proof of Principle test. In the next step we will first investigate the properties of boron-loaded plastic detectors as well as the use of natural boron-loaded liquid scintillators. These detectors will then be exposed to the high antineutrino flux at one of the Koeberg reactors (0.92 GW_e), located ~25 km north of Cape Town. Based on the estimates of Bernstein et al. (2002) we expect about 2 to 3 events per day per kilogram detector material. Initially we will mainly be interested in detection of double pulses and analysing the scintillator properties. After optimising the detectors and their electronics, we will construct a number of test detectors and investigate their individual direction sensitive detection efficiency.

3. Conclusions

A 3D image of the radiogenic heat sources in the Earth's interior with a spatial resolution of about 150 km at the Core-Mantle-Boundary (CMB) will certainly revolutionise the understanding of how the Earth works and may lead to better knowledge on a number of phenomena observed at the Earth surface. In addition to seismic tomography, antineutrino tomography seems to be the only additional method to reach this goal. To obtain antineutrino tomography with a spatial resolution comparable or better than seismic tomography requires direction sensitive antineutrino detection. The existing large monolithic antineutrino detector set-ups will not be able to provide sufficient resolution.

The proposed detector system in this paper is a consequence of the goal to eventually map the radiogenic heat sources with high resolution by antineutrino tomography (e.g. located at the CMB with a size of about 200 km). It starts by exploiting the direction information contained in the kinematics of the antineutrino capture by a proton. Based on simulations of the neutron tracking we conclude that the detectors should have a physical cross section of the order a few centimetre squared. In this paper we have demonstrated that from the physics point of view such detectors provide sufficient direction sensitivity and sufficient background reduction. In this paper we have not yet addressed the technical question of how a very large number of detectors may be equipped and read out. The presentation of Daniel Ferenc (submitted) at this conference is an indication that with time these technical challenges can be resolved. We are fully aware that this is an indication of a solution to only one of the many technical developments that need to take place. On the one hand there is no guarantee of success but on the other hand we see no other obvious solution.

References

Araki, T. et al.: 2005, *Nature* **436**, 499–503.

Beacom, J. F. and Vogel, P.: 1996, *Phys. Rev. D* **60**, 033077.

Beets, D. J.: 1972, Lithology and stratigraphy of the cretaceous and Danian succession of Curaçao. Natuurwetenschappelijke Studiekring voor Suriname en de Nederlandse Antillen. no. 70.

Bernstein, A. et al.: 2002, *J. Appl. Phys.* **91**(7), 4672–4676.

Boyet, M. and Carlson, R. W.: 2005, *Science* **309**, 576.

Buffett, B. A.: 2003, *Science* **299**, 1675–1677.

Ferenc D.: *Earth Moon and Planets*, (submitted).

Fiorentini, G., Mantovani, F., and Ricci, B.: 2003, *Phys. Lett. B.* **557**, 139–146.

Gutenberg, B.: 1914, *Göttinger Nachrichten* **166**, 218.

van der Hilst, R. D. and Karason, H.: 1999, *Science* **283**, 1885–1888.

Jeanloz, R. and Lay, T.: 2005, *Sci. Am. Special* **15**(2), 36–43.

Kerr, R. A.: 2005, *Science* **309**, 87.

Klaver, G. Th.: 1987, The Curaçao lava formation. An ophiolitic analogue of the anomalous thick layer 2B of the mid-cretaceous oceanic plateaus in the western pacific and central Caribbean. *GUA papers of geology*, series 1, no. 27, 1987.

Lay, T., Williams, Q., and Garnero, E.: 1998, *Nature* **392**, 461–468.

Mantovani, F. et al.: 2004, *Phys. Rev. D* **69**, 013001.

de Meijer, R. J., van der Graaf, E. R., and Jungmann, K. P.: 2004, *Nucl. Phys. News* **14**(2), 20–25.

de Meijer, R. J., van der Graaf, E. R., and Jungmann, K. P.: 2004, *Rad. Phys. Chem* **71**, 769–774.

Oldham, R. D.: 1906, *Quat. J. Geol. Soc. London* **62**, 456–475.

Pollack, H. N., Hurter, S. J., and Johnson, J. R.: 1993, *Rev. Geophys* **31**(3), 267–280.

Tolstikhin, I. and Hoffmann, A. W.: 2005, *Phys. Earth. Planet. Inter.* **148**, 108.

Wang, S. C. et al.: 1999, *Nucl. Instr. Meth.* **432**, 111–121.

Wilson, M.: 2005, *Physics Today* 19–21.

Zhao, D.: 2004, *Phys Earth Planet Inter* **146**, 3–34.

Earth, Moon, and Planets (2006) 99:207–220
DOI 10.1007/s11038-006-9106-6

Geoneutrinos in Borexino

MARCO G. GIAMMARCHI and LINO MIRAMONTI
Dipartimento di Fisica dell'Università di Milano and Infn, Via Celoria 16, 20133, Milano, Italy
(E-mail: giammarchi@mi.infn.it)

(Received 11 March 2006; Accepted 28 June 2006)

Abstract. This paper describes the Borexino detector and the high-radiopurity studies and tests that are integral part of the Borexino technology and development. The application of Borexino to the detection and studies of geoneutrinos is discussed.

Keywords: geoneutrinos, low background, scintillation detector

1. The Gran Sasso National Laboratory

The Gran Sasso National Laboratory (Laboratori Nazionali del Gran Sasso - LNGS), home of the Borexino experiment, is the world's largest underground laboratory. It is located in the center of Italy in the highway tunnel between Teramo and L'Aquila under the "Monte Aquila" (Gran Sasso mountain). The laboratory, located at 42°27′ N, 13°34′ E, is financed and operated by the Italian National Institute for Nuclear Physics (INFN). Its total underground volume is about 180,000 m^3 with an area greater than 13500 m^2. It is composed of three main experimental halls (20 m high, 18 m wide and 100 m long). The overburden rock is on the average about 1400 m, equivalent to 3700 m of water. The muon flux is reduced by about 6 orders of magnitude to a value of approximately 1.1 muons per square meter per hour, whereas the neutron flux is of the order of 3×10^{-6} neutrons per square centimeter per second with energies greater than 2.5 MeV.

The rock of the Gran Sasso mountain has a density of 2.71 ± 0.05 g cm^{-3}, and consists mainly of $CaCO_3$ and $MgCO_3$ (Catalanou et al., 1986). The primordial radionuclide content of the rock of Hall C is 0.66 ± 0.14 ppm for ^{238}U, 0.066 ± 0.025 ppm for the ^{232}Th and 160 ppm for K (Wulandari et al., 2004). The radioactive content of the concrete employed as experimental hall liner is 1.05 ± 0.12 ppm for ^{238}U and 0.656 ± 0.028 ppm for the ^{232}Th (Bellini et al., 1991).

The LNGS hosts about 15 experiments of astroparticle physics such as neutrino research, double beta decay physics, dark matter studies and nuclear

Talk given by M.G. Giammarchi

astrophysics. Interdisciplinary studies (biology, geology) are also conducted in the LNGS underground location.

The Borexino detector is located in one of the big underground experimental halls, hall C.

2. The Borexino Detector

Borexino is a real time experiment whose main goal is to study the low energy (sub-MeV) solar neutrinos, and in particular the 862 keV ^{7}Be solar neutrino line, through the neutrino-electron elastic scattering reaction. The maximum energy of the recoiling electron is 664 keV and the experimental design threshold is set at 250 keV (Alimonti, 2002).

Borexino is an unsegmented scintillation detector featuring 300 tonnes of well shielded liquid ultra-pure scintillator viewed by 2200 photomultipliers (PMT). The detector core is a transparent spherical vessel (Nylon Sphere, 100 μm thick), 8.5 m of diameter, filled with 300 tonnes of liquid scintillator and surrounded by 1,000 tonnes of high-purity buffer liquid. The scintillator mixture is PC (Pseudocumene) and PPO (1.5 g l^{-1}) as a fluor, while the buffer liquid will be PC alone (with the addition of DMP as light quencher). The photomultipliers are supported by a Stainless Steel Sphere, which also separates the inner part of the detector from the external shielding, provided by 2400 tonnes of pure water (water buffer), see Figure 1.

An additional containment vessel (Nylon film radon barrier) is interposed between the Nylon Sphere and the photomultipliers, with the goal of reducing radon diffusion towards the internal part of the detector.

The outer water shield is instrumented with 200 outward-pointing PMT's serving as a veto for penetrating muons, the only significant remaining cosmic ray background at the Gran Sasso depth.

The innermost 2200 photomultipliers are divided into a set of 1800 PMT's equipped with light cones (so that they see light only from the Nylon Sphere region) and a set of 400 PMT's without light cones, sensitive to light originated in the whole Stainless Steel Sphere volume. This design greatly increases the capability of the system to identify muons crossing the PC buffer (and not the scintillator).

The Borexino design is based on the concept of a graded shield of progressively lower intrinsic radioactivity as one approaches the sensitive volume of the detector; this culminates in the use of 200 tonnes of the low background scintillator to shield the 100 tonnes innermost Fiducial Volume. In these conditions, the ultimate background will be dominated by the intrinsic contamination of the scintillator, while all backgrounds from the construction materials and external shieldings will be negligible. For instance, the external 2.614 MeV γ line gives about 1010 external gammas per day

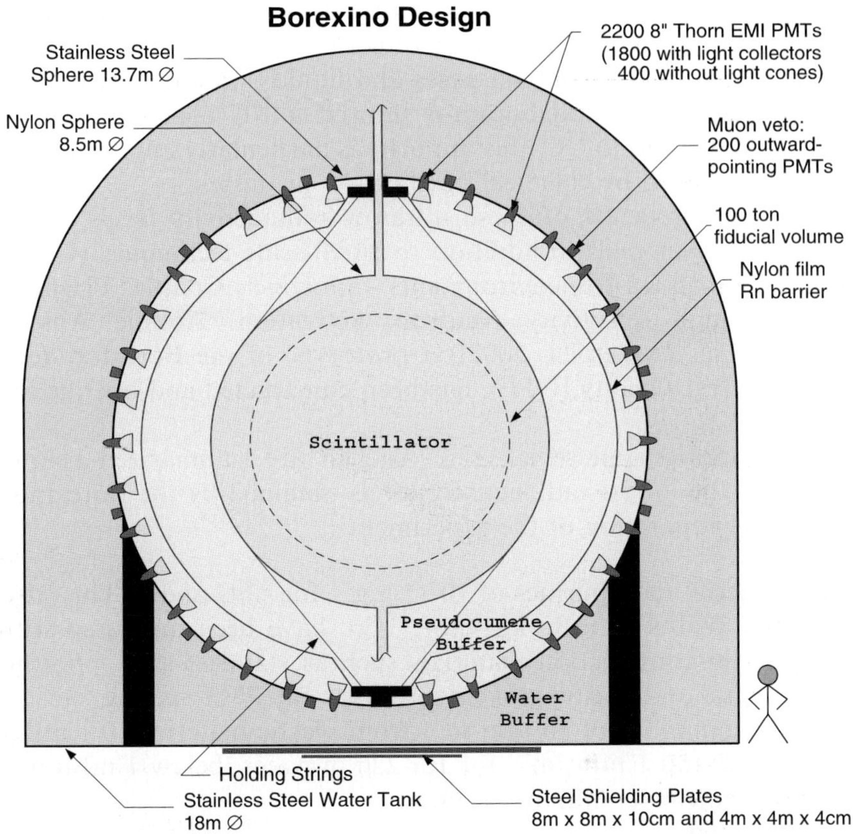

Figure 1. Schematic of the Borexino detector (see text).

impinging on the shielding, to be reduced to less than 1 count per day in the fiducial volume due to the ~5 m of liquid.

Borexino also features several external plants and purification systems conceived to purify the experimental fluids (water, nitrogen and scintillator) used by the experiment.

The main problem of a real time experiment with such a low energy threshold is the natural radioactivity which is present in any environment and in any material. For these reasons an intense R&D program has been carried out in the last ten years to develop methods for selecting low radioactivity materials and/or purify them. An effort in this field has to be complemented by a comparably thorough research concerning detection and measurement of very low radioactivity levels. In this context four purification methods have been developed: distillation, water extraction, stripping with ultrapure N_2, solid gel column (Si gel, Al gel) adsorption.

Significant results have been achieved by the Collaboration as for example: $10^{-16} - 10^{-17}$ (g of contaminants/g of material) for ^{232}Th and ^{238}U family and a few μBq m^{-3} of Rn-222 in gases and liquids. In addition the organic solvent selected by the collaboration showed a ^{14}C concentration clearly below 10^{-17} in its ratio to ^{12}C; this impurity is particularly important because it cannot be removed by chemical purification processes.

For the measurements of these ultralow radioactivity levels, dedicated methods were developed. In addition to small-scale techniques (Ge underground detectors in Rn-free environments, Inductively Coupled Plasma Mass Spectometer, high sensitivity Neutron Activation, Atomic Absorption Spectroscopy etc...(Arpesella, 2002)) a prototype of the Borexino detector, the Counting Test Facility (CTF), has been constructed and operated in the Hall C of LNGS.

The radiopurities and sensitivities reached are summarized below and correspond to the lowest radioactivity levels obtained by the Borexino Collaboration, in preparation of the experiment:

- Bulk material radiopurities of 10^{-10} g g^{-1} for ^{238}U and ^{232}Th, ~10^{-5} for ^{nat}K, few tenths of mBq kg^{-1} for ^{60}Co, have been measured with Ge detectors in construction materials such as stainless steel, photomultipliers, metal and plastic gaskets, products for PMT sealing, etc...
- Radon emanations of 10 μBq m^{-2} from plastic materials, 0.1 mBq m^{-3} for Rn-222 and 1 mBq m^{-3} for Ra-226 in water, below 1 mBq m^{-3} for the N_2 used for scintillator stripping.
- Radiopurity levels of a few times 10^{-15} g g^{-1} ^{238}U, ^{232}Th and ^{40}K have been reached with ICMPS in measuring the Borexino and CTF shielding water.
- Sensitivities of few ppt for ^{238}U and ^{232}Th concentrations have been obtained in the Nylon Sphere material measurements.
- The radiopurity of the scintillator itself was measured to be at the level of few 10^{-16} g g^{-1} for ^{238}U, ^{232}Th and ~10^{-18} for ^{14}C/^{12}C in the Counting Test Facility.
- Bulk radiopurity levels of $10^{-13} - 10^{-14}$ g g^{-1} for Au, Ba, Ce, Co, Cr, Cs, Ga, Hg, In, Mo, Rb; less than few 10^{-15} g g^{-1} for Cd, Sb, Ta, W; $10^{-16} - 10^{-17}$ g g^{-1} for La, Lu, Re, Sc, Th; less than 1×10^{-17} g g^{-1} for U, have been reached by means of Neutron Activation followed by $\beta - \gamma$ delayed coincidence analysis applied to the scintillator.
- Kr and Ar contamination in nitrogen at 0.005 ppm (for Ar) and 0.06 ppt (for Kr) were obtained and measured with noble gas mass spectrometry.

These results represent a milestone in the development of the Borexino detector and technique. Several of these concepts were incorporated in the construction of the high purity systems for the treatment of the most critical liquid, the scintillator of the experiment.

3. The Counting Test Facility

The CTF description and its performance have been published elsewhere (Bellini, 1996; Alimonti et al.,1998a, b). In this section we simply review the main features of this detector.

The CTF consists of an external cylindrical water tank ($\oslash 11 \times 10$ m; $\simeq$1,000 t of water) serving as passive shielding for 4.8 m^3 of liquid scintillator contained in an inner spherical vessel (Inner Vessel) of 2.1 m in diameter and observed by 100 PMT's. An additional nylon barrier against radon convection and a muon veto system were installed in 1999. Figure 2 shows a picture of the CTF detector.

The radio-purity level of the water is $\simeq 10^{-14}$ g g^{-1} (U, Th), $\simeq 10^{-10}$ g g^{-1} (^{nat}K) and < 5 μBq l^{-1} for ^{222}Rn (Bellini, 1996, Alimonti et al., 1998b, Balata, 1996].

The organic liquid scintillator has the same composition as in Borexino. The yield of emitted photons is $\simeq 10^4$ per MeV of energy deposited and the fluorescence peak emission is located at 365 nm. The principal scintillator decay time is $\simeq$3.5 ns in a small volume, while for large volume (because of absorbtion and re-emission) this value is 4.5–5.0 ns. The attenuation length is larger than 5 m above 380 nm [Alimonti et al., 2000].

Figure 2. Internal view of the Counting Test Facility (see text).

The purification of the scintillator is performed by recirculation from the Inner Vessel through a radon stripping tower, a water extraction unit, a Si-Gel column extraction unit, and a vacuum distillation unit. The ^{232}Th and ^{238}U contaminations in the CTF liquid scintillator were found to be less than (2–5) $\times$ 10^{-16} g g^{-1}.

The Inner Vessel for the liquid scintillator containment is made of nylon with a thickness of 500 μm, with excellent optical clarity at 350-500 nm. The collection of scintillation light is ensured by 100 PMT's mounted to a 7 m diameter support structure inside the CTF tank.

The photomultiplier tubes are 8 inches (Thorn EMI 9351, the same as for Borexino) made of low radioactivity Schott 8246 glass and characterized by high quantum efficiency (26% at 420 nm), limited transit time spread (σ = 1 ns), good pulse height resolution for single photoelectron pulses (Peak/Valley = 2.5), low dark noise rate (0.5 kHz), low after pulse probability (2.5%), and a gain of 10^7.

The PMT's are equipped with light concentrators 57 cm long and with 50 cm diameter aperture. The PMT system provides an overall 20% optical coverage for events taking place inside the Inner Vessel. The number of photoelectrons per MeV measured experimentally is (300 $\pm$ 30) MeV^{-1} on average.

The total background rate in the 250-800 keV energy range is about 0.3 counts year^{-1} keV^{-1} kg^{-1} and appears to be dominated by external background from radon in the shielding water ($\approx$30 mBq m^{-3} in the region surrounding the Inner Vessel). The internal background was measured to be less than 0.01 counts year^{-1} keV^{-1} kg^{-1}. The total count rate in the higher 1.0/2.6 MeV energy range was found to be of 10 counts day^{-1} ton^{-1}.

4. The Counting Test Facility Related Publications

Data collected with the Counting Test Facility have contributed significantly to the best limits on quantities such as neutrino magnetic moment, electron lifetime, nucleon decays in invisible channels, violation of the Pauli exclusion principle, production of heavy-neutrinos in the Sun.

Concerning the study of the stability of the electron, the CTF data have been analyzed to search for the 256 keV line of the gamma emitted in the decay channel e $\rightarrow$ γ ν. Since we have found no signal, we established a limit on the electron lifetime of $\tau \geq 4 \times 10^{26}$ (90% C.L.); this is still the best world limit for the electron decay in this channel (Back et al., 2002).

CTF data analysis has allowed the study of the neutrino magnetic moment, obtaining the limit of $\mu_\nu \leq 0.5 \times 10^{-10}$ μ_B, still a very competitive result (Back et al., 2003a).

We have also investigated the possibility of heavy neutrinos ($M \geq m_e$) emitted in the ^8B reaction in the sun. Heavy neutrinos would decay to light neutrinos via the reaction $\nu_H \rightarrow \nu_L + e^+ + e^-$. The analysis of the CTF energy spectrum has allowed to significantly enlarge the excluded region of the parameter space with respect to previous experiments (Back et al., 2003b).

The stability of nucleons bounded in nuclei has been studied in the Counting Test Facility searching for decays of single nucleon or pair of nucleons into invisible channels. The limits are comparable to or improve the previously set world limits (Back et al., 2003c). Furthermore a search was made for non-Paulian transitions of nucleons from nuclear $1P$ shell to a filled $1S_{1/2}$ shell obtaining the best limit on the Pauli exclusion principle (Back et al., 2004).

Other studies have concerned the search for anti-neutrinos coming from the sun (Balata et al., 2006a) and the cosmogenic ^{11}C underground production (Balata et al., 2006b).

5. Geoneutrinos Detection

One possible application of a high mass, well-shielded scintillator detector such as Borexino is the search for geoneutrinos, a new and very interesting subject which we will discuss in the remainder of this paper.

The conceptual foundations of Earth science rest on a variety of observables as well as interior characteristics. One important parameter is the internally produced heat which is currently measured to be in the ~60 mW m^{-2} range (or 30 TW when integrated on the planet surface).

Part of this energy flow is due to the presence of radioactive elements in the Earth interior, mainly naturally occurring uranium and thorium chain elements and potassium. Models of the Earth disperse about 50% of the total U, Th in the crust while leaving the remaining half to the mantle. Roughly speaking, the 35 km thick continental crust contains a few ppm of U, Th while the much thinner (~6 km) oceanic crust has a typical concentration of ~0.1 ppm.

Our goal is to measure this radiogenic heat by detecting neutrinos emitted during the decays of the radioactive chains. For a given structure of naturally occurring radioactive families, a measurement of antineutrino flux can be related to the U, Th family content of the Earth.

In the case of the ^{238}U family (the ^{235}U leftover can be neglected) one can globally represent the full decay chain as:

$$^{238}\mathrm{U} \rightarrow {}^{206}\mathrm{Pb} + 8\,\alpha + 6\,\mathrm{e}^- + 6\,\overline{\nu}_e + 51.7\,\mathrm{MeV} \tag{1}$$

and therefore the detected number of antineutrinos is related to the number of times this reaction has taken place.

The ^{232}Th family presents a similar case:

$$^{232}\mathrm{Th} \rightarrow {}^{208}\,\mathrm{Pb} + 6\,\alpha + 4\,\mathrm{e}^- + 4\,\overline{\nu}_\mathrm{e} + 42.8\,\mathrm{MeV} \qquad (2)$$

where again a definite number of antineutrinos is emitted and a well defined energy is released.

Finally, antineutrinos are also emitted in the K terminations:

$$^{40}\mathrm{K} \rightarrow {}^{40}\,\mathrm{Ca} + \mathrm{e}^- + \overline{\nu}_\mathrm{e} + 1.32\,\mathrm{MeV} \qquad (3)$$

The $\overline{\nu}$ energy spectra produced by these three sources are plotted in Figure 3.

Our goal will be to detect antineutrinos emitted by these sources in order to measure the rates of occurrence of the above reactions.

5.1. Principle of geoneutrino detection

While low energy neutrino detection offers formidable experimental challenges due to backgrounds as explained above, antineutrinos were discovered in a reaction that naturally affords a nice way to cope with the unwanted background events. The proposed detection reaction:

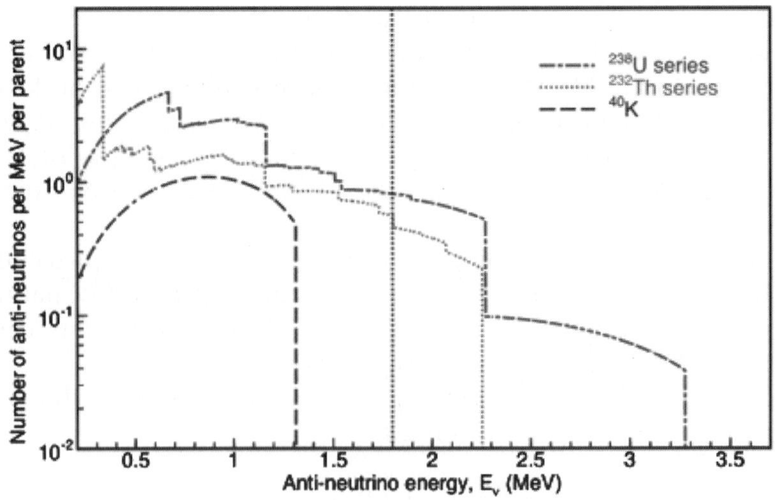

Figure 3. Antineutrino spectra from the three main Earth sources. The line indicates the threshold for the Reines-Cowan detection reaction (see text).

$$\bar{\nu}_e + p \rightarrow n + e^+ \tag{4}$$

is the inverse beta decay (or *Reines–Cowan*) reaction and generates a positron and a neutron in the final state. The positron gets quickly absorbed in ordinary matter, coupling with an electron and generating two gammas ($e^+e^- \rightarrow \gamma\,\gamma$) with a total energy release of 1.02 MeV.

The neutron, on the other hand, gets slowed down in the material and finally thermalizes to be absorbed as shown in Figure 4. The lifetime of this process in the Borexino scintillator is about 200μs and at the end the neutron is absorbed by a free proton in the scintillator:

$$H(n,\gamma)D \tag{5}$$

with the emission of a 2.2 MeV energy gamma.

From the detection viewpoint, this cascade of processes allows a very favorable tag of energies and time.

First of all, the energy of the positron and neutron electromagnetic cascades are above ~1 MeV and 2.2 MeV, respectively. Secondly, the time delay between the two events is short, very difficult to mimic by a couple of accidental background events. Finally, these two events are also subjected to a mild spatial condition of coincidence.

In summary, the tagging of an antineutrino event will be an $E > 1$ MeV e/γ event followed (in a narrow time window of, say, 0.5 ms) by a 2.2 MeV γ event. The two events must lie within the typical (1 m) neutron diffusion length.

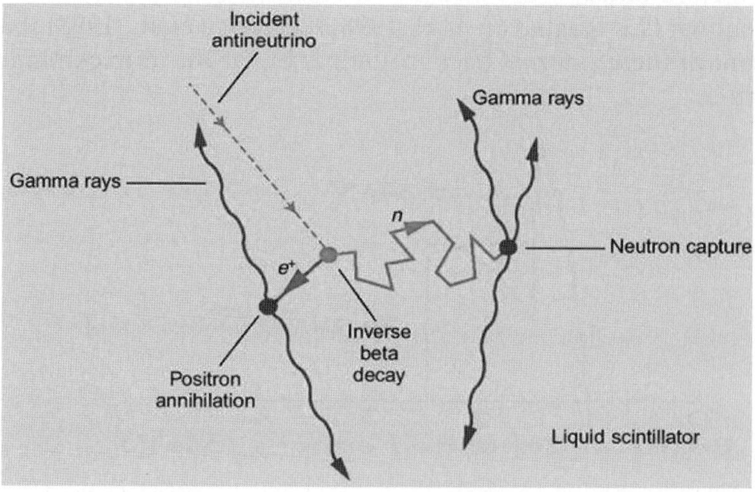

Figure 4. Antineutrino detection principle in the inverse-beta reaction.

The $\overline{\nu}_e + p \rightarrow n + e^+$ detection reaction has a threshold of 1.8 MeV which is determined by the mass difference between neutron plus positron and the initial state proton. Therefore this reaction has the drawback of not being sensitive to the detection of K antineutrinos (see Figure 3).

The detector will reconstruct the antineutrino energy based on the observed positron kinetic energy. The visible energy of the event has to take into account also the 1.02 MeV energy due to annihilation:

$$E(\text{vis}) = \text{K}(e^+) + 1.02\,\text{MeV} \tag{6}$$

which is plotted in Figure 5 for the case of the U and Th chain.

In turn, the energy of the antineutrino is shifted with respect to $\text{K}(e^+)$ by the Q-value of the reaction:

$$E(\overline{\nu}) = \text{K}(e^+) - 1.81\,\text{MeV} \tag{7}$$

So, the final relation between the visible energy and the $\overline{\nu}$ energy is

$$E(\text{vis}) = E(\overline{\nu}) - 0.78\,\text{MeV} \tag{8}$$

The observed kinetic energy spectrum E(vis) will begin at 1.02 MeV (the case when the positron kinetic energy is zero).

Apart from the internal (radioactive) contamination of the scintillator, the background to the antineutrino signal can come in principle from a variety of sources, including atmospheric shower particle decays (π, μ, K), relic of past supernovas, non-standard $\nu \rightarrow \overline{\nu}$ oscillations in the Sun, muon-induced neutron production and $\overline{\nu}_e$ from nuclear reactors.

It can be shown (see Raghavan et al.,1998; Calaprice et al., 1998) that the most significant of these external backgrounds is by far the term coming from nuclear reactors.

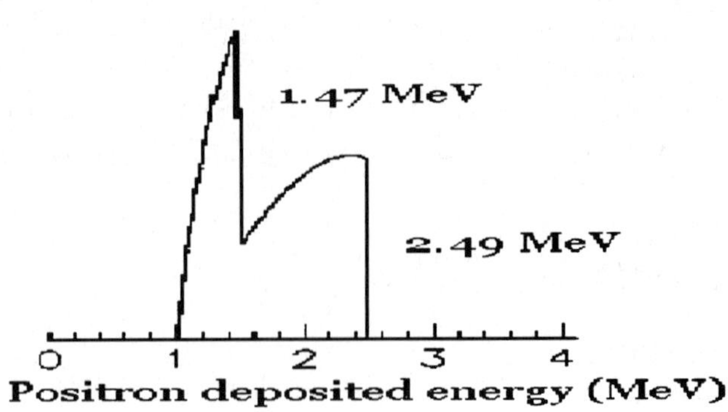

Figure 5. Energy deposited by the positron for antineutrinos belonging to the U and Th chains.

5.2. Background from nuclear reactors

Nuclear power reactors produce energy by fission of heavy nuclei. Since a super-heavy nucleus has a ~30% excess of neutrons over protons, this excess will be transferred to the lighter fission products which therefore will be beta-instable and produce antineutrinos during their de-excitation.

In order to produce background signal for the inverse-beta decay detection reaction the antineutrinos must have more than 1.8 MeV kinetic energy. The fissile nuclides featuring fission fragments of such energy are ^{235}U, ^{239}Pu, ^{238}U and ^{241}Pu.

The number of fissions generated is typically calculated from the reactor thermal power and the specific fission energy release. In addition, some mild dependence over time is introduced by the initial composition and time evolution of nuclear fuel at a specific reactor.

However, in spite of this parameter variability, antineutrino reactor spectra are all very similar, producing a background that extends from our detection threshold and up to about 9 MeV.

For the case of Borexino, the relevant reactors are the ones situated in Europe, which are relatively far away (700–800 km) from the Gran Sasso location (Figure 6). Neutrino oscillation effects will then be washed off in

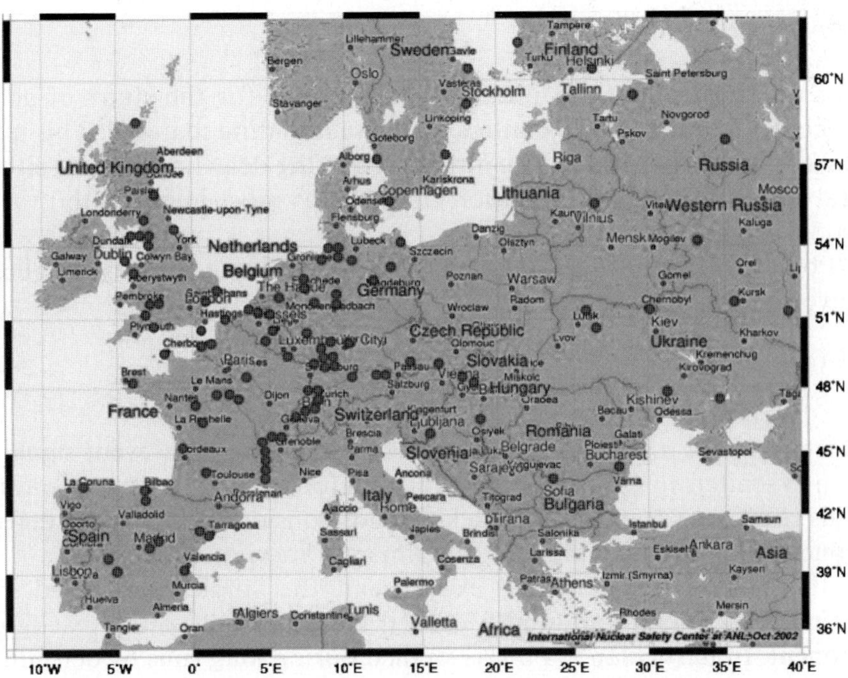

Figure 6. Location of nuclear reactors in Europe.

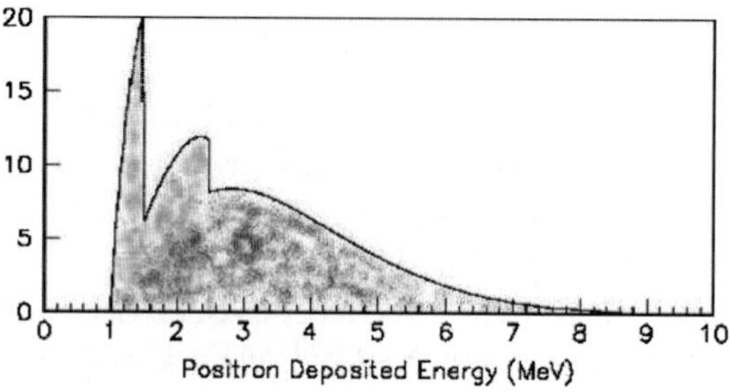

Figure 7. Predicted spectrum for antineutrino detection in Borexino. The signal (U, Th chains) and background (in arbitrary units) are relatively normalized to 6 events year^{-1} (signal, yellow) and is 20 events year^{-1} (nuclear reactors, red).

phase giving only an overall reduction factor to about 60% of the original reactor flux.

Figure 7 shows both the reactor background spectrum and the U, Th signal to be searched for. The normalization of the spectra is made accordingly to the expected number of events from European reactors (~20 events year^{-1}) and the estimated geoneutrino signal in Borexino (6 events year^{-1}). More on this later on.

5.3. Background from ^{210}Pb

One of the most important internal backgrounds for the study of geoneutrinos comes from the ^{210}Pb content of the scintillating material. This nuclide has a 22.3 yr half-life and can be introduced in the detecting volume either by natural bulk (^{226}Ra) contamination or (perhaps most importantly) through ^{222}Rn diffusion.

^{210}Pb has an α-emitting daughter (^{210}Po) which has a 138 day half-life and can give rise to the reaction

$$^{13}\mathrm{C}(\alpha, n)^{16}\mathrm{O}$$

This originates a chain of events closely mimicking the $\bar{\nu}$ event signal with a first n–p scattering release (or the α release) followed by neutron capture. In fact this reaction was one of the backgrounds considered in the first detection of geoneutrinos by the Kamland experiment (Araki et al., 2005).

In the case of the Borexino experiment, the studies conducted in the Counting Test Facility allowed a careful measurement of this background component. In particular ^{210}Pb was studied during long runs by detecting the ^{210}Po alpha and by studying the total single rate spectra (see Figure 8 for an example of such fits.)

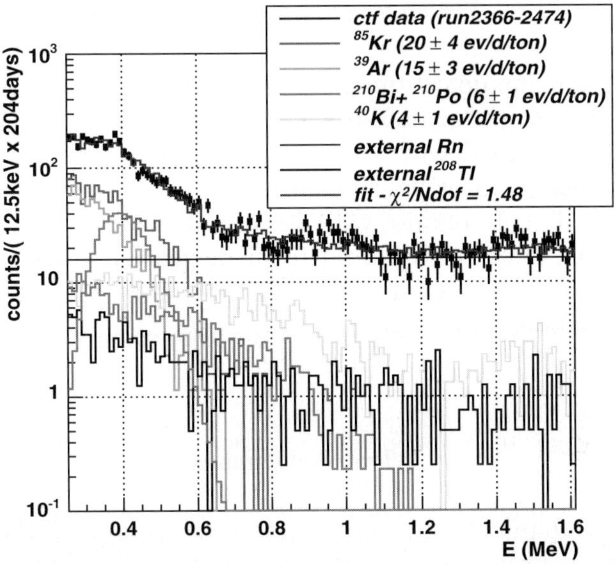

Figure 8. Single counts spectra showing different contributions in the Counting Test Facility.

These studies showed that the specific ^{210}Pb background in Borexino is limited to ~20μBq ton^{-1}, a value much lower than the one quoted in (Araki et al., 2005). Therefore, this background will be negligible in Borexino.

6. Antineutrino Signal and Detection Sensitivity

Having demonstrated that the dominant background in Borexino will be due to European reactors, we now proceed to the evaluation of the terrestrial antineutrino signal.

In order to do this, it is necessary to take into account the detailed distribution of U, Th in the Earth's crust, particularly within 500 km distance from the detector (a region that generates about 1/2 of the total signal).

For this evaluation we have exploited the Earth model given in Mantovani et al., (2004) which gives U, Th distributions and detailed evaluations for specific detector locations. This evaluation, together with the high efficiency (assumed 1) of the detection, leads to the prediction of ~6 events year^{-1} (oscillation effects included) detected in Borexino, as shown in Figure 7.

7. Conclusion

Terrestrial antineutrino detection requires shielded low background detectors of high mass. Borexino, located at Gran Sasso Laboratory, will tackle this fascinating subject armed with its low background capability and its relatively long distance from nuclear reactors.

We estimate that Borexino will detect ~6 events year^{-1} coming from geoneutrinos with some background from nuclear reactors (Figure 7). The statistical significance of such a signal is predicted to be better than 30% in 5 years of data taking.

Acknowledgements

We would like to thank Stephen Dye, John Learned and all the organizers of this conference for providing us with a stimulating scientific atmosphere in a wonderful setting.

References

Alimonti, G. et al.: 1998a, *Nucl. Instrum. Meth. A* **406**, 411.
Alimonti, G. et al.: 1998b, *Astroparticle Phys.* **8**, 141.
Alimonti, G. et al.: 2000, *Nucl. Instrum. Meth. A* 440.
Alimonti, G. et al.: 2002, *Astroparticle Phys.* **16**, 205.
Araki, T. et al.: 2005, *Nature* **436**, 499.
Arpesella, C. et al.: 2002, *Astroparticle Physics* **18**, 1.
Back, H. O. et al.: 2002, *Phys. Lett. B* **525**, 29.
Back, H. O. et al.: 2003a, *Physics Letters B* **563**, 35.
Back, H. O. et al.: 2003b, *JETP Lett.* **78**, 261.
Back, H. O. et al.: 2003c, *Phys. Lett. B* **563**, 23.
Back, H. O. et al.: 2004, *Eur. Phys. J. C* **37**, 421.
Balata, M. et al.: 1996, *Nucl. Instrum. Meth. A* **370**, 605.
Balata, M. et al.: 2006a, hep-ex/0602027.
Balata, M. et al.: 2006b, hep-ex/0601035.
Bellini, G.: 1991, *Borexino Proposal*, 184.
Bellini, G.: 1996, (for the Borexino Collaboration), Nucl. Phys. B (Proc. Suppl.) **48**, 363.
Calaprice, F. et al.: 1998, *Geophys. Res. Lett.* **25**, 1083.
Catalano, P. G. et al.: 1986, *Men. Soc. Geol. It.* **35**, 647.
Mantovani, F. et al.: 2004, *Phys. Rev. D* **69**, 013001.
Raghavan, R. S. et al.: 1998, *Phys. Rev. Lett.* **80**, 635.
Wulandari, H. et al.: 2004, *Astroparticle Phys.* **22**, 313–322.

Earth, Moon, and Planets (2006) 99:221–228
DOI 10.1007/s11038-006-9116-4

Geo-neutrinos in SNO+

M. C. CHEN
Department of Physics, Queen's University, Kingston, ON, K7L 3N6, Canada
(E-mail: mchen@post.queensu.ca)

(Received 16 April 2006; Accepted 25 July 2006)

Abstract. There are plans to fill the Sudbury Neutrino Observatory with liquid scintillator after measurements with heavy water are completed. The new experiment, known as SNO+, would make an excellent detector for geo-neutrinos. SNO+ would be located amidst a thick and uniform region of continental crust, away from nuclear power reactors. As a result, the geo-neutrino signal to reactor background ratio in SNO+ will exceed that from previous measurements. Geo-neutrino measurements by SNO+ will shed light on the amount of uranium and thorium radioactivity in the crust, as well as deeper inside the Earth. Spectral information from SNO+ geo-neutrino detection will provide the first direct measurement of the U/Th ratio.

Keywords: Continental crust, geo-neutrinos, geochemistry, neutrino geophysics, radioactivity

1. Introduction

A follow-up experiment to the Sudbury Neutrino Observatory is being developed. The experiment is called SNO+ and it consists of filling SNO with liquid scintillator after the physics program with heavy water is completed and the heavy water is removed in 2007. SNO+ would be a large, low background, liquid scintillator detector in the deepest underground site for neutrino physics. The physics capabilities of this detector include the precision study of the pep solar neutrinos, confirmation of reactor neutrino oscillations at a longer baseline, and, of particular interest in this article, the study of geo-neutrinos.

Transforming SNO into a liquid scintillator detector would boost the light yield in the detector by a factor of ~100 or greater, compared to Čerenkov light. The new detector would have a large number of protons in the CH_2-based liquid scintillator instead of the deuterons currently in SNO's heavy water. A liquid scintillator is thus an ideal detector for electron antineutrinos.

Using the inverse beta decay reaction, electron antineutrinos ($\bar{\nu}_e$) interacting with protons in a liquid scintillator produce positrons and neutrons

$$\bar{\nu}_e + p \rightarrow e^+ + n.$$

Scintillation light is emitted when the positron is produced in this reaction. The neutron in the event can be detected in delayed coincidence, when it captures about 200 μs later on hydrogen in the scintillator producing a 2.2 MeV gamma. This coincidence signal is distinctive and correlated backgrounds are few and limited, especially for a detector that is deep underground. The depth of the SNO (or SNO+) detector is 6 km water equivalent of overburden and this basically eliminates correlated backgrounds from muon-induced fast neutrons, or from cosmogenically produced β-n decaying isotopes.

The visible energy produced by the positron in these events is equal to the kinetic energy of the positron plus 1.022 MeV, coming from the annihilation of the positron. Thus, $\bar{\nu}_e$ signals produce at least 1 MeV of visible energy (even for positrons with zero kinetic energy), easily observable in a large scintillator detector. A liquid scintillator with suitably low background can thus detect $\bar{\nu}_e$ interactions all the way down to the reaction threshold of E_ν = 1.804 MeV. These are the reasons why a large, low background liquid scintillator detector is well suited for detecting geo-neutrinos from uranium and thorium, for which the signal is $\bar{\nu}_e$ with energy up to 3.3 MeV and an event rate of the order of tens of events per kiloton of scintillator per year.

KamLAND is a neutrino detector with 1 kiloton of liquid scintillator. KamLAND reported the first detection of geo-neutrinos in 2005 (Araki et al., 2005). Ways in which to improve upon KamLAND's first observation include:

- increased statistics, which KamLAND is acquiring
- reducing the correlated background from $^{13}C(\alpha, n)$, which KamLAND is also working on by purifying their liquid scintillator
- ultimately, KamLAND's geo-neutrino measurements will be limited by their large reactor neutrino event rate, so another geo-neutrino experiment with a smaller reactor background is required.

SNO+ would also be a 1 kiloton liquid scintillator detector. The province of Ontario has three nuclear power generating stations with a total of 14 operating reactor cores. The total power output of these reactors is smaller than what's in Japan and distances from Sudbury to the nuclear generating stations are longer. SNO+ thus offers the possibility to improve upon KamLAND's geo-neutrino studies by having a larger signal to reactor background. Since the SNO+ experiment is a retrofit of an existing detector, it is an imminently realizable experiment for relatively little cost. SNO+ and KamLAND geo-neutrino signal rates and reactor background rates will be compared below.

2. The SNO+ detector

The SNO+ detector will be located where SNO is currently, at 46.475° N, 81.201° W in INCO's Creighton Mine, near Sudbury, Ontario, Canada. The center of the detector is 2039 m below the surface. The surface is 309 m above sea level. Figure 1 is an illustration of the detector configuration for SNO. The only change for SNO+ is that rather than heavy water (D_2O) inside the acrylic vessel, there would be liquid scintillator. PSUP in the figure refers to the photomultiplier tube (PMT) support structure, which holds 9438 PMTs viewing the inner volume. The 20-cm diameter Hamamatsu R1408 PMTs are mounted on the PSUP with light collecting reflectors and provide 54% effective photocathode cover. Further technical details on the existing SNO detector can be found in (Boger et al., 2000).

The liquid scintillator to be used in SNO+ needs to be compatible with acrylic. The SNO+ project initiated the development of liquid scintillator using linear alkylbenzene (LAB) as a solvent. Undiluted LAB appears to be

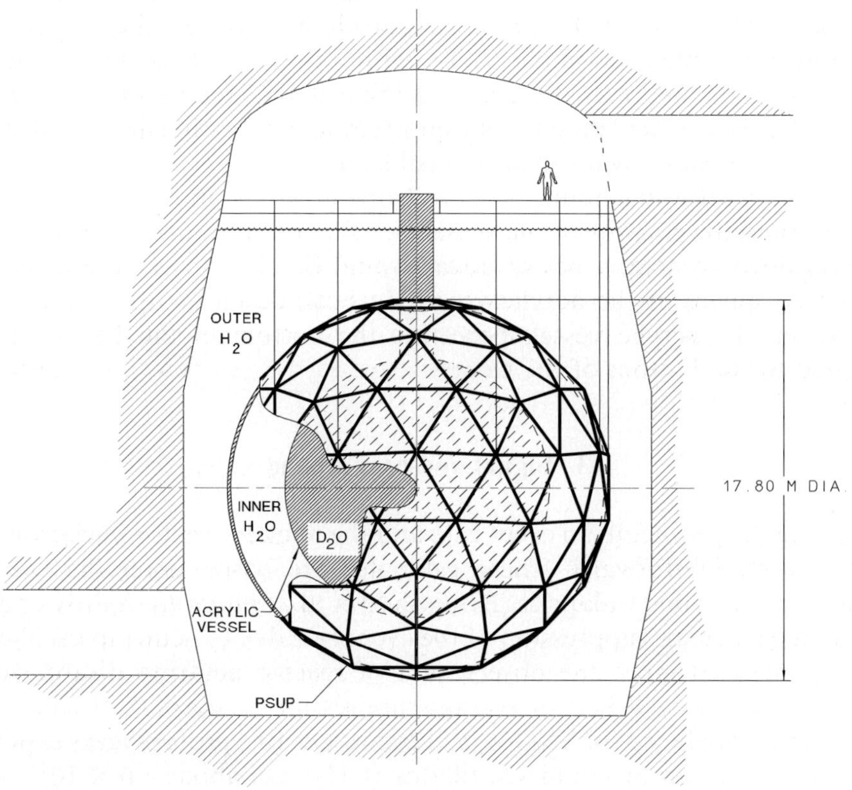

Figure 1. A drawing of the SNO detector showing the geodesic PMT support structure and acrylic vessel. SNO+ will maintain the same general configuration.

compatible with acrylic. LAB is very transparent; we have measured attenuation lengths greater than 20 m at 420 nm. LAB has high light yield. In comparison with KamLAND scintillator which is 20% pseudocumene diluted in dodecane, we have measured the light yield of undiluted LAB as being 50–75% greater than the KamLAND scintillator, depending on the concentration of fluor used. LAB is also a relatively inexpensive substance (it is much cheaper than most high quality mineral oils). Our study of the long-term properties of this scintillator are ongoing. Details of our development of what appears to be a very suitable scintillator for SNO+ and for other neutrino experiments will appear in a technical publication, in preparation.

Monte Carlo simulations (using essentially the same code that is being used in SNO simulations and data analysis) of the amount of light collected in SNO+ estimate 1200 photoelectrons per MeV of energy deposited, which is exceedingly good for a large liquid scintillator detector. This high light yield should not be surprising. KamLAND detects 300 photoelectrons/MeV for 22% photocathode coverage. SNO has 54% coverage and undiluted LAB has 50–75% greater light output.

The density of LAB is 0.86 g/cm^3. Filling the SNO+ acrylic vessel with LAB, and keeping water (H_2O) on the outside would place the acrylic vessel in a buoyant configuration. In SNO, with heavy water on the inside, the acrylic vessel is supported by ropes holding it up, due to the 10% higher density of heavy water. Preliminary engineering studies have concluded that supporting a 14% buoyant load is feasible. Two designs are being explored. One would involve machining reverse "rope grooves" in the existing acrylic plates around the equator of the vessel (see Figure 2). In the other option for the hold down system, a net or collar would be placed over the top of the upper hemisphere of the acrylic vessel. In both designs, ropes would hang down from the acrylic vessel, penetrate the bottom of the PSUP and be anchored to the bottom of the cavity.

3. Signal and background

Estimating the geo-neutrino event rate and the nuclear reactor background in SNO+ is straightforward. These estimates are based upon the work of Rothschild, Chen and Calaprice (Rothschild et al., 1998) with slightly updated values. In particular, suppression of the event rate due to neutrino oscillations must be included given the observation of reactor neutrino disappearance (Eguchi et al., 2003). When an average suppression factor of 0.59 is included, the event rate predicted for SNO+ is 51 events per 10^{32} proton-years exposure. Note that a kiloton of liquid scintillator (CH_2) has about 8.6×10^{31} target protons. This rate in SNO+ is for a Bulk Silicate Earth-based "reference" model, a concept first introduced in the Rothschild, Chen and Calaprice paper.

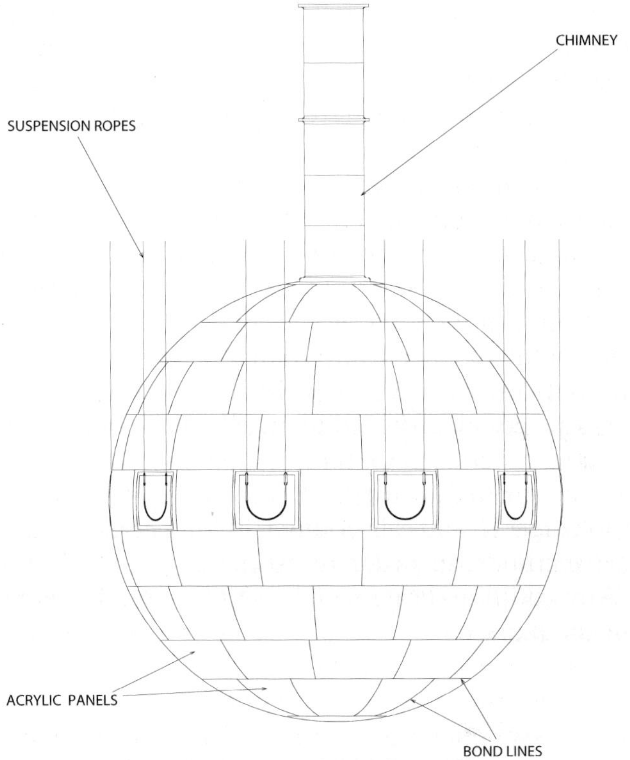

Figure 2. The existing SNO acrylic vessel, showing the ropes that hold up the acrylic vessel. SNO+ requires engineering a system to hold down a buoyant acrylic vessel filled with $\rho = 0.86$ g/cm^3 liquid scintillator.

The rate in SNO+ can be compared with an event rate of 38 events per 10^{32} proton-years in KamLAND, as derived from values given in Araki et al. (2005) and Enomoto et al. (2005). The SNO+ detector would be located in a region surrounded by thick continental crust. This results in a greater geo-neutrino signal rate in SNO+. Oceanic crust is thin and relatively depleted in uranium and thorium, in comparison. The consequence is a somewhat lower geo-neutrino signal rate in KamLAND, which is in the vicinity of both continental crust and oceanic crust rock.

The geological region around SNO+ is known as the Canadian Shield. It is an old mass of continental crust rock that is fairly uniform and well understood. Because of the mining activity occurring around Sudbury, the local geology has been extensively studied. Because of this knowledge, one might be able to calculate the local contribution to the geo-neutrino rate reasonably accurately. Hence, a SNO+ geo-neutrino measurement could reveal the uranium and thorium content of the deep Earth – one would be subtracting from the total SNO+ signal a known, local contribution, to get at the amount of radioactivity deeper in the Earth (e.g. in the mantle).

Figure 3 shows the fractional contribution to the geo-neutrino signal in SNO+ at different distances of integration. One sees that the total signal (integrating over the whole Earth) is derived 80% from continental crust radioactivity and 20% from radioactivity in the mantle. This is a calculation that employs a crust map (Laske et al., 2001) and reference model distributions of radioactivity in the Earth.

One concludes from Figure 3 that SNO+ will test our understanding of the amount of radioactivity in the continental crust. This is the most accessible layer of the Earth, thought to be well characterized. Testing our understanding of the continental crust using geo-neutrinos serves to confirm basic and fundamental geochemical paradigms and assumptions.

One also finds from Figure 3 that 70% of the 80% continental crust contribution to the geo-neutrino rate in SNO+ comes from the closest 1200 km of continental crust rock (approximately). Since most of the signal rate comes from continental crust that is relatively near, one would aim to determine the average U and Th content of this continental crust rock by other geological methods, in order to estimate its contribution to the geo-neutrino rate. Any additional rate seen in SNO+ could then be attributed to radioactivity in the mantle.

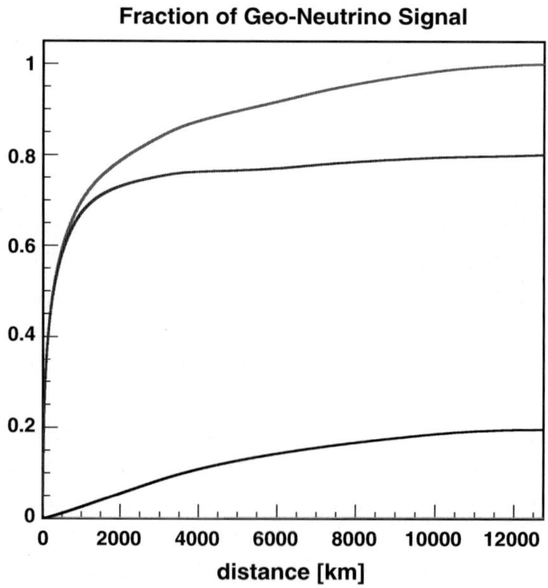

Figure 3. The fractional contribution to the geo-neutrino signal in SNO+ at different distances of integration. The upper curve is the total; below it, is the contribution from continental crust; the lower curve is the contribution from the mantle. There is also a curve for the contribution from oceanic crust which is barely visible on this plot and negligible.

Nuclear reactor backgrounds is present at a lower level in SNO+ compared to KamLAND. For six operating cores at the Bruce generating station (the closest reactors to Sudbury), the reactor neutrino rate in SNO+ is 179 events per 10^{32} proton-years (oscillations included). Only 49 of those events are in the same spectral region as the geo-neutrinos. The signal-to-background in SNO+ is 51/49 or about 1:1. In contrast, KamLAND has 572 events of reactor background per 10^{32} proton-years (oscillations included and corrected for detection efficiency). In the spectral region of interest for geo-neutrinos are 165 reactor background events per 10^{32} proton-years, a signal-to-reactor background ratio of about 1:4 in KamLAND.

Figure 4 shows the expected spectrum of geo-neutrino events compared with the reactor neutrino background in SNO+. As expected, the signal-to-background is favorable and the geo-neutrino events can be easily identified. An important feature for SNO+ should be the ability to separately determine the uranium and thorium contribution to the geo-neutrino signal. The higher energy spectral feature or lobe in the geo-neutrino signal spectrum is due purely to uranium. The lower energy lobe includes geo-neutrinos produced by thorium in addition to uranium. Clearly measuring the relative rates in those two lobes enables the average U/Th ratio in the Earth to be determined. Spectral features in the reactor background arise from neutrino oscillations.

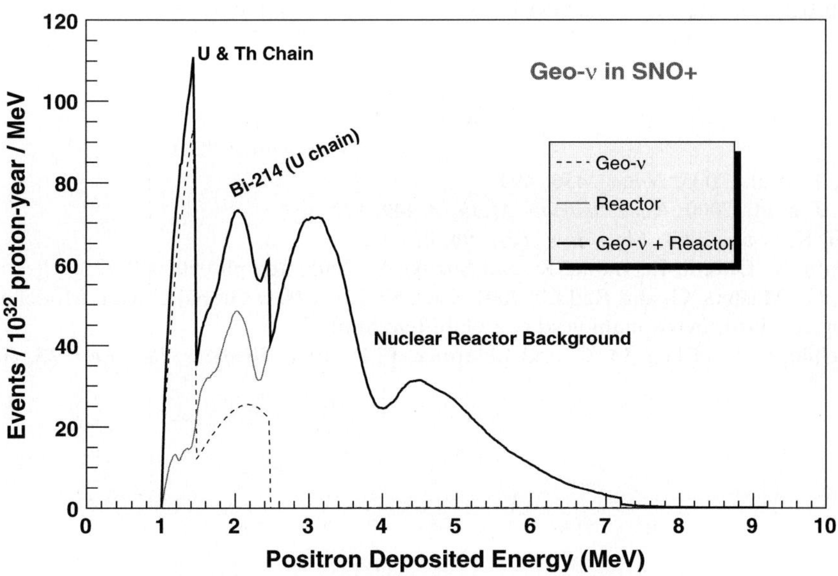

Figure 4. The energy spectrum of the geo-neutrino flux at SNO+. The background from reactor neutrinos is shown. The vertical axis is in events per MeV for an exposure of 10^{32} proton-years. Energy resolution was not included in this plot. Oscillations were included in calculating the geo-neutrino signal and the reactor background rates and spectra.

The significant reactor neutrino background in KamLAND prevents a clear separation of the two spectral lobes in the geo-neutrino signal. Projections by KamLAND suggest that this may still be difficult even with improved statistics (Enomoto et al., 2005). SNO+ could thus provide the first experimental determination, using geo-neutrinos, of the average U/Th ratio in the Earth (SNO+ measurement dominated by the U and Th composition of the surrounding continental crust).

4. Summary

With diverse physics goals and an operational start date within the next 2–3 years, SNO+ represents an opportunity to extend the science output from SNO for relatively little cost, by continuing to use much of the existing detector and investment in infrastructure and capability. One of the key science objectives for SNO+ will be the study of geo-neutrinos. Improved signal-to-background, confirmation of our understanding of the geochemistry and composition of the continental crust, probing the uranium and thorium radioactivity in the deep Earth by subtracting a knowable local contribution, and the ability to separately determine the amounts of uranium and thorium in the Earth are the main features of SNO+ geo-neutrinos. This experiment should be a very good follow-up to the first studies of geo-neutrinos by Kam-LAND.

References

Araki, T. et al.: 2005, *Nature* **436**, 499.
Boger, J. et al.: 2000, *Nucl. Instrum. Meth. A* **449**, 172.
Eguchi, K. et al.: 2003, *Phys. Rev. Lett.* **90**, 021802.
Enomoto, S., Ohtani, E., Inoue, K. and Suzuki A.: 2005, hep-ph/0508049 v2.
Laske, G., Masters, G. and Reif C.: 2001, CRUST 2.0 A New Global Crustal Model at 2 × 2 Degrees, http://www.mahi.ucsd.edu/Gabi/rem.html.
Rothschild, C. G., Chen, M. C. and Calaprice, F. P.: 1998, *Geophys. Res. Lett.* **25**, 1083.

Earth, Moon, and Planets (2006) 99:229–240
DOI 10.1007/s11038-006-9112-8

A Geoneutrino Experiment at Homestake

N. TOLICH, Y. -D. CHAN, C. A. CURRAT, B. K. FUJIKAWA, R. HENNING, K. T. LESKO and A. W. P. POON
Institute for Nuclear and Particle Astrophysics and Nuclear Science Division, Lawrence Berkeley National Laboratory, Berkeley, CA 94720, USA
(E-mail: nrtolich@lbl.gov)

M. P. DECOWSKI
Physics Department, University of California, Berkeley, CA 94720, USA

J. WANG
Earth Sciences Division, Lawrence Berkeley National Laboratory, Berkeley, CA 94720, USA

K. TOLICH
Department of Physics, Stanford University, Stanford, CA 94305, USA

(Received 31 May 2006; Accepted 11 July 2006)

Abstract. A significant fraction of the 44TW of heat dissipation from the Earth's interior is believed to originate from the decays of terrestrial uranium and thorium. The only estimates of this radiogenic heat, which is the driving force for mantle convection, come from Earth models based on meteorites, and have large systematic errors. The detection of electron antineutrinos produced by these uranium and thorium decays would allow a more direct measure of the total uranium and thorium content, and hence radiogenic heat production in the Earth. We discuss the prospect of building an electron antineutrino detector approximately 700 m^3 in size in the Homestake mine at the 4850' level. This would allow us to make a measurement of the total uranium and thorium content with a statistical error less than the systematic error from our current knowledge of neutrino oscillation parameters. It would also allow us to test the hypothesis of a naturally occurring nuclear reactor at the center of the Earth.

Keywords: Geoneutrino, Electron antineutrino, Georeactor, Homestake mine

1. Introduction

Thanks in part to Ray Davis' pioneering neutrino experiment (Davis et al., 1968) located in the Homestake mine (44.35° N, 103.75° W), more is now known about the interior workings of the Sun than the Earth. The KamLAND collaboration has recently investigated electron antineutrinos originating from the interior of the Earth (Araki et al., 2005a); however, the sensitivity achieved was limited by a large background from surrounding nuclear power reactors. A similar experiment located deep underground to reduce cosmic-ray backgrounds, and away from nuclear power plants, could reach a sensitivity that would allow constraints to be placed on our current knowledge of the Earth's interior.

The idea of using electron antineutrinos, $\bar{\nu}_e$'s, to study processes inside the Earth was first suggested by Eder (1966) and Marx (1969). ^{238}U, ^{232}Th, and ^{40}K decays within the Earth are believed to be responsible for the majority of the current radiogenic heat production, which is the driving force for Earth mantle convection, the process which causes plate tectonics and earthquakes. These decays also produce $\bar{\nu}_e$'s, the vast majority of which reach the Earth's surface since neutrinos hardly interact with matter, allowing a direct measurement of the total Earth radiogenic heat production by these isotopes.

The regional composition of the Earth is determined by a number of different methods. The deepest hole ever dug penetrates 12 km of the crust (Kremenetsky and Ovhinnikov, 1986), allowing direct sampling from only a small fraction of the Earth. Lava flows bring xenoliths, foreign crystals in igneous rock, from the upper mantle to the surface. The regional composition of the Earth can also be modeled by comparing physical properties determined from seismic data to laboratory measurements. Our current knowledge suggests that the crust and mantle are composed mainly of silica, with the crust enriched in U, Th, and K. The core is composed mainly of Fe but includes a small fraction of lighter elements. Table I shows the estimated concentration of U, Th, and K in the different Earth regions.

Models of Earth composition based on the solar abundance data (McDonough and Sun, 1995) establish the composition of the undifferentiated mantle in the early formation stage of the Earth, referred to as "Bulk Silicate Earth" (BSE). Table I includes the estimated concentration of U, Th, and K in the BSE model. The ratio of Th/U by weight, between 3.7 and 4.1

TABLE I

Estimated total mass of the major Earth regions (Schubert et al., 2001), and the estimated concentration of U, Th, and K in each region. It is assumed that there is no U, Th, or K in the Earth's core. The concentration of radiogenic elements in the mantle is obtained by subtracting the isotope mass in the crust from the Bulk Silicate Earth (BSE) model

Region	Total mass	Concentration		
	[10^{21} kg]	U[ppb]	Th[ppb]	K[ppm]
Oceanic crust (Taylor and McLennan, 1985)	6	100	220	1250
Continental crust (Rudnick and Fountain, 1995)	19	1400	5600	15600
Mantle	3985	13.6	53.0	165
BSE (McDonough and Sun, 1995)	4010	20.3	79.5	240

(Rocholl and Jochum, 1993), is known better than the total abundance of each element.

The rate of radiogenic heat released from U, Th, and K decays are 98.1 μW kg^{-1}, 26.4 μW kg^{-1}, and 0.0035 μW kg^{-1} (Schubert et al., 2001), respectively. Table II summarizes the total radiogenic heat production rate of these elements in the Earth regions based on the masses and concentrations given in Table I. For comparison, the rate of mantle heating due to lunar tides is a negligible ~0.12 TW (Zschau 1978).

The radiogenic heat production within the Earth can be compared to the measured heat dissipation rate at the surface. Based on the rock conductivity and temperature gradient in bore holes measured at 20,201 sites, the estimated heat dissipation rate from oceanic and continental crust, respectively, is 31.2 $\pm$ 0.7 TW and 13.0 $\pm$ 0.3 TW, resulting in a total of 44.2 $\pm$ 1.0 TW (Pollack et al., 1993). In this study the majority of the heat is lost through the oceanic crust, despite the fact that the continental crust contains the majority of the radiogenic heat producing elements. A recent re-evalutaion of the same data (Hofmeister and Criss, 2005) suggests that the heat dissipation rate in the oceanic crust is significantly less, resulting in a total heat dissipation rate of 31.0 $\pm$ 1.0TW. The measured heat flow per unit area at the Earth's surface surrounding the Homestake mine (International Heat Flow commission, 2005) is consistent with the continental crust average, which suggests that increased local uranium concentration is not significant.

The Urey ratio, the ratio between mantle heat dissipation and production, indicates what fraction of the current heat flow is due to primordial heat. Subtracting the continental crust heat production rate of 6.5 TW, the mantle is dissipating heat at a rate of 37.7 TW and, assuming the BSE model, generating heat at a rate of 13.2 TW, giving a Urey ratio of ~0.35. It is widely believed that the mantle convects although the exact nature of that convection is still unclear. Models of mantle convection give Urey ratios greater than ~0.69 (Spohn and Schubert, 1982; Jackson and Pollack, 1984; Richter, 1984), which is inconsistent with the value obtained from heat considerations. A direct measurement of the terrestrial radiogenic heat production rate would help our understanding of this apparent inconsistency.

TABLE II
Radiogenic heat production rate in different Earth regions

Region	U [TW]	Th [TW]	K [TW]	Total [TW]
Oceanic crust	0.06	0.03	0.03	0.12
Continental crust	2.61	2.81	1.04	6.46
Mantle	5.32	5.57	2.30	13.19
BSE	7.99	8.42	3.37	19.78

2. Geoneutrino signal

A $\bar{\nu}_e$ is produced whenever a nucleus β^- decays. The ^{238}U and ^{232}Th decay chains (Firestone, 2005) both contain at least four β^- decays, and ^{40}K β^- decays with a branching fraction of 89.28%. These β^- decays result in the well established $\bar{\nu}_e$ energy distributions for ^{238}U, ^{232}Th, and ^{40}K shown in Figure 1. Because $\bar{\nu}_e$'s have such a small cross-section for interaction with matter, the majority of these $\bar{\nu}_e$'s produced within the Earth reach the surface. However, due to a phenomenon usually referred to as "neutrino oscillation", the $\bar{\nu}_e$ may change into a $\bar{\nu}_\mu$ or $\bar{\nu}_\tau$. The probability of the $\bar{\nu}_e$ being found in the same state as a function of distance traveled, L, can be approximated as,

$$P(E_\nu, L) = 1 - \sin^2 2\theta_{12} \sin^2\left(\frac{1.27\Delta m_{12}^2[\text{eV}^2]L[\text{m}]}{E_\nu[\text{MeV}]}\right), \tag{1}$$

where $\Delta m_{12}^2 = 7.9^{+0.6}_{-0.5} \times 10^{-5}\,\text{eV}^2$, and $\sin^2 2\theta_{12} = 0.816^{+0.073}_{-0.070}$ (Araki et al., 2005b). This assumes two "flavor" oscillation and neglects "matter effects" both of which are less than 5% corrections (Araki et al.,2005a).

The most common method (Boehm et al., 2001; Appollonio et al., 2003; Araki et al.,2005a) for detecting $\bar{\nu}_e$'s is neutron inverse β decay,

$$\bar{\nu}_e + \text{p} \longrightarrow \text{e}^+ + \text{n}. \tag{2}$$

The detection of both the positron, e^+, and neutron, n, separated by a small distance and time, greatly reduces the number of backgrounds. Due to the reaction threshold, the minimum $\bar{\nu}_e$ energy detectable by this method is 1.8 MeV, which has the disadvantage that the ^{40}K $\bar{\nu}_e$'s cannot be detected since they have a maximum energy of 1.3 MeV. To zeroth order in $1/M$,

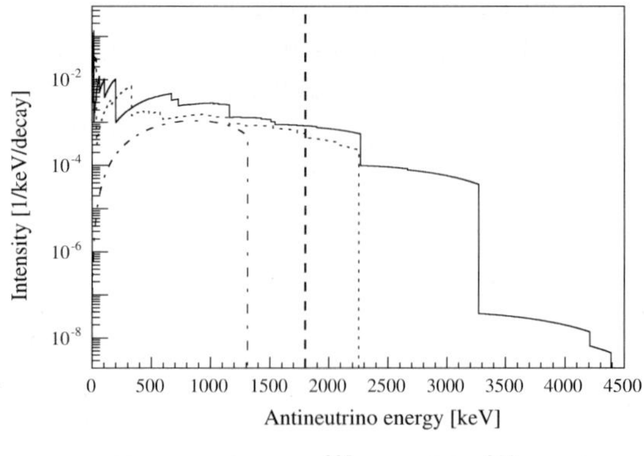

Figure 1. The $\bar{\nu}_e$ energy distributions for the ^{238}U (solid), ^{232}Th (dash), and ^{40}K (dot-dash) decay chains. The vertical line represents the $\bar{\nu}_e$ detection threshold for neutron inverse β decay. Only the ^{238}U and ^{232}Th chains are measurable with neutron inverse β decay.

where M is the nucleon mass, the total positron energy, $W^{(0)}{}_e$, is related to the total antineutrino energy, W_ν, by

$$W_e^{(0)} = W_\nu - m_n + m_p, \tag{3}$$

where m_n and m_p are the neutron and proton masses, respectively. Therefore, the $\bar{\nu}_e$ energy can be estimated from a measurement of the positron kinetic energy. This allows spectral separation of the $\bar{\nu}_e$'s from ^{238}U and ^{232}Th decays.

The geoneutrino observation rate depends on the decay rate of ^{238}U and ^{232}Th, the resulting $\bar{\nu}_e$ energy distribution, the detection cross-section, the neutrino oscillation parameters, and the distribution of the ^{238}U and ^{232}Th in the Earth. Based on a detailed simulation (Mantovani et al., 2004), including seismic models of crustal thickness, the number of neutron inverse β decays at Homestake due to terrestrial ^{238}U and ^{232}Th is estimated to be 54 per 10^{32} target protons per year, assuming $\sin^2 2\theta_{12} = 0.816$. The lines labeled 1, 2, and 3 in Figure 2 show the cumulative geoneutrino fluxes as a function of distance from detectors located over continental crust of varying thickness and with varying contributions from neighboring oceanic crust. A detector located in the Homestake mine could expect ~50% of the geoneutrino flux originating within ~500 km of the detector.

With ~50% of the geoneutrino flux originating within ~500 km of the detector it is important to remove the effects of local geology to obtain a global measurement of the total ^{238}U and ^{232}Th concentration. The estimated error in the signal from local geology for the recent KamLAND geoneutrino measurement is 16% (Enomoto, 2005). The nearest known uranium reserve (Energy Information, 2004) is located ~100 km from the Homestake mine at the boundary of Wyoming and South Dakota. To place an upper limit on the impact of local concentrations of uranium and thorium, we assume that the Earth's total reasonably assured uranium reserves of 3600 kton uranium (World Nuclear Association, 2006) were located 100 km from the proposed detector. This would contribute less than 0.03% to the expected global signal. A possible heat flow measurement in the Homestake mine and uranium and thorium concentrations obtained from the Homestake mine core samples (South Dakota Geological Survey, 2006) could be used to reduce the systematic uncertainties associated with geoneutrinos originating from within ~10 km of the detector.

It has been suggested that a large amount of uranium may be located in the core of the Earth (Herndon, 2003) forming a natural nuclear reactor. This could produce up to 6 TW of heat, powering the Earth's dynamo. It would also lead to ^{3}He production which could explain the observed anomaly in the ^{3}He/^{4}He ratio for gases from the Earth's mantle. Excluding neutrino oscillation, a natural reactor at the Earth's core would produce a very similar

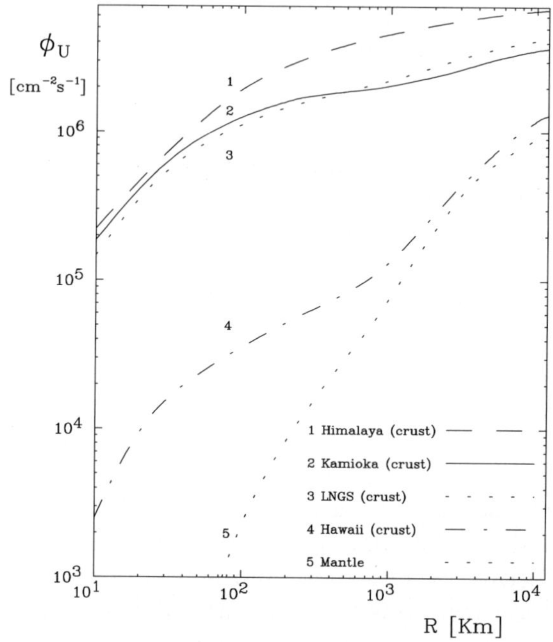

Figure 2. Cumulative geoneutrino flux as a function of distance to the source (Mantovani et al., 2004). The Himalaya curve is for a detector located over thick continental crust. The Kamioka curve is for a detector located at the boundary of continental and oceanic crust. The LNGS curve is for a detector located over continental crust; this probably best represents a detector located at the Homestake mine. The Hawaii curve is for a detector located over oceanic crust; this most closely matches the flux from the mantle, since there is no high local uranium and thorium concentrations.

energy spectrum of detected $\bar{\nu}_e$'s to that from commercial nuclear power reactors, which is peaked at ~4 MeV and extends up to ~9 MeV, see Figure 4. In order to accurately test this hypothesis it is necessary to have a very low commercial nuclear reactor background. The number of $\bar{\nu}_e$'s detected by neutron inverse β decay near the surface of the Earth is expected to be approximately 40 per 10^{32} target protons per year due to a 6 TW nuclear reactor at the Earth's core, assuming $\sin^2 2\theta_{12} = 0.816$.

3. Backgrounds

The detection of correlated signals in neutron inverse β decay significantly enhances the detectability of the geoneutrinos. Nevertheless, other events contribute backgrounds to the measurement. Backgrounds can typically be subdivided into three main categories: natural radioactivity, cosmic-rays and associated spallation products, and other $\bar{\nu}_e$ sources. The most significant backgrounds in the recent KamLAND geoneutrino measurement (Araki et al., 2005a) were $\bar{\nu}_e$'s from nearby nuclear power reactors and $^{13}C(\alpha, n)$ reactions where the α is primarily from ^{210}Pb decay.

3.1. $\bar{\nu}_e$ SOURCES

Figure 3 shows that the Homestake mine is located more than 750 km away from any major nuclear power reactor. Based on the rated maximum thermal power, and excluding neutrino oscillation, the expected rate of $\bar{\nu}_e$'s from nuclear reactors is calculated to be 64 per 10^{32} target proton yr. Since the $\bar{\nu}_e$'s typically travel distances greater than 1000 km, the $\bar{\nu}_e$ survival probability due to neutrino oscillation can be approximated by $P(E_\nu, L) \approx 1 - 0.5 \sin^2 2\theta_{12}$, which equals 0.592 assuming $\sin^2 2\theta_{12} = 0.816$. Therefore, the expected rate in the geoneutrino region, below 3.4 MeV, and including neutrino oscillation, is only 11 per 10^{32} target proton yr, which is ~7% of the expected rate at KamLAND (Araki 2005a).

Figure 4 shows the expected spectra for geoneutrinos, commercial nuclear reactors, and a natural nuclear reactor at the Earth's core. The commercial reactor background is insignificant for the geoneutrino measurement. It is also small enough to allow an ultimate sensitivity of 1.3 TW at 99% CL, limited by the systematic uncertainty in the commercial reactor flux, for a nuclear reactor at the Earth's core. This assumes that the commercial reactor background can be obtained to 10% accuracy, which should be possible based on the published electrical power and an averaged core nuclear cycle. If the isotopic fission rates of the reactors can be obtained the reactor background could be determined to ~2% accuracy (Boehm et al., 2001; Appollonio et al., 2003; Eguchi et al., 2003).

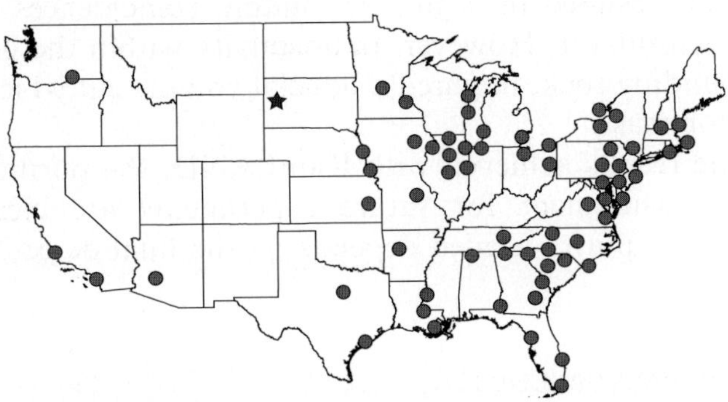

Figure 3. Location of nuclear power reactors (circles) in the USA, modified from the reference International Nuclear Safety Center (2005). The closest nuclear power reactor to Homestake (star) is ~750 km away.

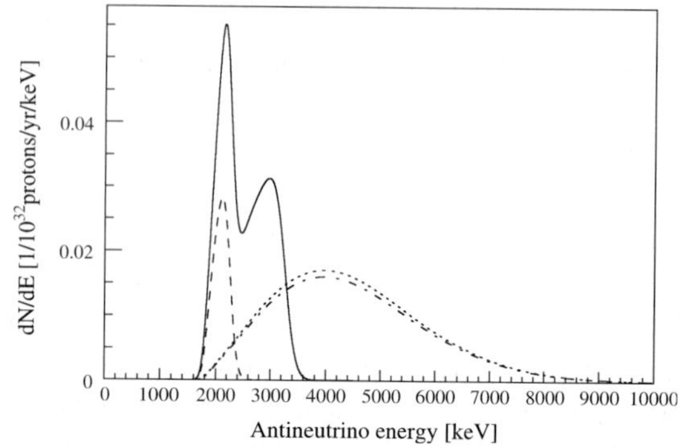

Figure 4. The expected $\bar{\nu}_e$ energy distribution with a detector energy resolution of $6\%/\sqrt{E[\mathrm{MeV}]}$ for the ^{238}U (solid) and ^{232}Th (dash) geoneutrinos, expected commercial reactor background (dot-dash), and the expected spectrum from a 6 TW natural reactor at the Earth's core (dot).

3.2. Radioactive backgrounds

The largest radioactive background in the recent KamLAND measurement was due to the reaction ^{13}C(α, n)17O. The neutron produces two events, one as it loses energy, and the second when it captures on a proton. This mimics the $\bar{\nu}_e$ events. The α in this reaction is a product of ^{210}Pb decay, which is itself a product from the decay of radon (Rn) gas present in the detector during construction. There is a plan to purify the KamLAND detector, reducing this background by a factor of one million, making this background negligible in the liquid scintillator.

The next most significant radioactive background is due to random correlations caused by radioactivity in the detector, mostly from U, Th, K, and Rn decays. The KamLAND experiment achieved U, Th, and ^{40}K concentrations in the scintillator of 6×10^{-16} g/g, 2×10^{-16} g/g, and 2×10^{-16} g/g, respectively. This resulted in negligible random coincidences due to radioactivity in the scintillator. However, radioactivity within the detector enclosure and surrounding rock, required a fiducial volume cut which reduced the effective detector mass.

Based on the results achieved with KamLAND, the purities required to perform this measurement for future experiments are clearly possible. However, the exact purity needed depends on the final detector design, discussed in Section 4.

3.3. Cosmic-ray backgrounds

Cosmic-ray muons produce energetic neutrons and radioactive isotopes which can mimic the neutron inverse β decay signature. The effect of energetic neutrons and cosmogenic radioactivity is reduced by vetoing the detector after a muon passes through. There is a small residual background due to muon vetoing inefficiency and backgrounds caused by muons that pass through the rock surrounding the detector without detection in the muon veto.

The recent KamLAND result (Araki et al., 2005a) had a negligible background due to energetic neutrons and a background due to cosmogenic radioactivity of 0.6 per 10^{32} target proton yr. Because of the greater rock overburden at the 4850' level of the Homestake mine, the energetic neutron and cosmogenic backgrounds are expected to be ~20 times less than those at the KamLAND site (Mei and Hime, 2006). The exact cosmic-ray background rates will depend on the detector material, layout, and veto efficiency, although it is expected that in almost any final design this will be negligible.

4. The Detector

In past experiments, both liquid scintillator and water Cherenkov detectors have been used to observe the positron and neutron produced in neutron inverse β decay. Both techniques detect the photons emitted as charged particles move through the detector. The neutron is detected via the γ-ray emitted from its capture by a nucleus in the detecting material.

Water Cherenkov detectors produce a cone of light which allows the direction of the charged particle to be determined, and therefore allowing the direction of the $\bar{\nu}_e$ to be inferred. The Cherenkov photon yield is generally much less than the scintillation light yield, and consequently water Cherenkov detectors have poor sensitivity to events with energy less than ~4 MeV. A typical liquid scintillator detector easily observes $\bar{\nu}_e$'s at the neutron inverse β decay threshold. However, liquid scintillator detectors do not have very good directional information.

4.1. Requirement

The following measurements and tests should be performed by the proposed detector: measure the total geoneutrino rate, measure the ratio of ^{238}U to ^{232}Th in the Earth's interior, and test the hypothesis of a natural nuclear reactor at the Earth's core. Excluding the ^{238}U and ^{232}Th distribution in the Earth, the largest error in determining the expected geoneutrino rate is due to the uncertainty in the neutrino oscillation parameter $\sin^2 2\theta_{12}$ which is known to ~6%. It is unlikely that the accuracy of this parameter will be determined to better than a few percent within the next decade. Therefore, it does not make sense to plan on measuring the geoneutrino rate to much better than

6%. Assuming a 10% error in the commercial nuclear reactor background, and only using the $\bar{\nu}_e$ spectrum below 3.4 MeV, to measure the total geoneutrino rate to ~10% the required exposure is estimated to be $\sim 2.3 \times 10^{32}$ target proton yr. This does not include systematic errors, but these should be constrained to better than 10%.

In determining the ratio of ^{238}U to ^{232}Th many errors cancel, therefore it should be possible to obtain a measurement to better than 10% uncertainty. The Th/U ratio is currently estimated from meteorites to be between 3.7 and 4.1. To do this measurement, the $\bar{\nu}_e$ energy spectrum could be split into two regions, one between 2.5 MeV and 3.5 MeV, which contains only ^{238}U $\bar{\nu}_e$'s, and the other between 1.5 MeV and 2.5 MeV. An exposure of $\sim 20 \times 10^{32}$ target proton yr is required to measure the ratio to 10% accuracy. The uncertainty could be slightly reduced by a full spectral shape analysis.

Assuming a 10% error in the commercial nuclear reactor background, and only using the $\bar{\nu}_e$ spectrum above 4 MeV, the required exposure to observe a 6 TW georeactor at 3 sigma above zero is estimated to be $\sim 0.8 \times 10^{32}$ target proton yr.

For a measurement of the total geoneutrino rate and an observation of a hypothetical georeactor, we need an exposure of about 2×10^{32} target proton yr. This could be achieved in approximately four years assuming a similar fiducial volume, 700 m^3, and target proton density to KamLAND (Eguchi, 2003). A detector much larger than this is not required, since this detector will already reach the sensitivity imposed by the uncertainty in the neutrino oscillation parameters. However, an accurate measurement of the ^{238}U to ^{232}Th ratio would require a larger detector, or a longer exposure time.

4.2. Detector design

There are two main types of large scintillator $\bar{\nu}_e$ detectors: monolithic, such as the 1 kton KamLAND detector (Eguchi, 2003); and segmented, such as the 11 ton Palo Verde detector (Boehm, 2001). The advantage of a monolithic detector is reduced random coincidence backgrounds due to reduced support material, which is typically harder to purify than the scintillator. However, it would not be possible to build a KamLAND shaped detector at Homestake mine without further excavation since the detector is spherically symmetric. The advantage of a segmented detector is it could be constructed in sections above ground and transported below for assembly. Depending on the segment size, it could also be placed in one of the larger existing cavities.

In the KamLAND detector, the neutron produced in the neutron inverse β decay is captured by a proton with a mean capture time of ~200 μs producing a 2.2 MeV γ-ray. Gadolinium (Gd) was added to the Palo Verde detector scintillator in order to reduce backgrounds. Neutron capture by Gd produces γ-rays with a total energy of ~8 MeV, which is a higher energy than that produced by most radioactive backgrounds, greatly reducing the accidental background. Additionally, the Gd neutron capture cross-section is high, resulting in a mean neutron capture time of only ~27 μs in the Palo Verde detector, which would further decrease the accidentals due to the shorter correlation time window.

5. Conclusion

A measurement of geoneutrinos is an important step in constraining our understanding of the Earth's uranium and thorium distributions. The heat from the decay of these isotopes is the driving force for plate tectonics and earthquakes, and this is the only technique that allows us to directly observe these decays occurring at the inner depths of the Earth. The KamLAND experiment (Araki et al., 2005a) has recently shown the viability of such a measurement; however, it was limited by backgrounds from nearby nuclear power plants. A similar experiment at the Homestake mine does not have the same problem with nearby nuclear power plants, and other backgrounds should be small or negligible, but depend on the final detector design. It is envisioned that a detector could be located in an existing cavity in the Homestake mine, such as the one used by the Davis experiment (Davis et al., 1968).

References

Apollonio, M. et al.: 2003, *Eur. Phys. J. C* **27**, 331.

Araki, T. et al.: 2005a, *Nature* **436**, 499.

Araki T, et al.: 2005b, *Phys. Rev. Lett.* **94**, 081,801, arXiv:hep-ex/040635.

Boehm, F. et al.: 2001, *Phys. Rev. D* **64**, 112,001.

Davis, R., Harmer, D. S. and Hoffman, K. C.: 1968, *Phys. Rev. Lett.* **20**, 1205.

Eder, G.: 1966, *Nucl. Phys.* **78**, 657.

Eguchi, K. et al.: 2003, *Phys. Rev. Lett.* **90**, 021,802.

Energy Information Administration: 2004, URL http://www.eia.doe.gov/cneaf/nuclear/page/reserves/ures.html.

Enomoto, S.: 2005, Neutrino Geophysics and Observation of Geo-Neutrinos at KamLAND. PhD thesis, Tohoku University, URL http://www.awa.tohoku.ac.jp/KamLAND/publications/Sanshiro_thesis.pdf.

Firestone, R. B.: 2005, URL http://ie.lbl.gov/ensdf/.

Herndon, J. M.: 2003, *Proc. Nat. Acad. Sci.* **100**, 3047.

Hofmeister, A. and Criss, R.: 2005, *Tectonophysics* **395**, 159.

International Heat Flow Commission: 2005, URL http://www.heatflow.und.edu/.

International Nuclear Safety Center: 2005, URL http://www.insc.anl.gov/pwrmaps/index.php.

Jackson, M. J. and Pollack, H. N.: 1984, *J. Geophys. Res.* **89**, 10,103.
Kremenetsky, A. A. and Ovhinnikov, L. N.: 1986, *Precambrian. Res.* **33**, 11.
Mantovani, F., Carmignani, L., Fiorentini, G. and Lissia, M.: 2004, *Phys. Rev. D* **69**, 013,001.
Marx, G.: 1969, *Czech J. Phys. B* **19**, 1471.
McDonough, W. F. and Sun, S.: 1995, *Chem. Geol.* **120**, 223.
Mei, D. M. and Hime, A.: 2006, Muon-induced background study for underground laboratories. *Phys. Rev. D* **73**:053,004, arXiv:astro-ph/0512125.
Pollack, H. N., Hurter, S. J. and Johnson, J. R.: 1993, *Rev. Geophys.* **31**, 267.
Richter, F. M.: 1984, *Earth Planet Sci. Lett.* **68**, 471.
Rocholl, A. and Jochum, K. P.: 1993, *Earth Planet Sci. Lett.* **117**, 265.
Rudnick, R. L. and Fountain, D. M.: 1995, *Rev. Geophys.* **33**, 267.
Schubert, G., Turcotte, D. L. and Olson, P.: 2001. *Mantle Convection in the Earth and Planets*, Cambridge University Press, Cambridge.
South Dakota Geological Survey: 2006, URL http://www.sdgs.usd.edu/Homestake.html.
Spohn, T. and Schubert, G.: 1982, *J. Geophys. Res.* **87**, 4682.
Taylor, S. R. and McLennan, S. M.: 1985. *The Continental Crust: Its Composition and Evolution: An Examination of the Geochemical Record Preserved in Sedimentary Rocks*, Blackwell Scientific Publications, Oxford.
World Nuclear Association: 2006, URL http://www.world-nuclear.org/info/inf48.htm.
Zschau, J.: 1978, in Brosche, P. and Sundermann, J. (eds.), *Space Geodesy and Geodynamics*, Springer-Verlag, Berlin, pp. 315.

Earth, Moon, and Planets (2006) 99:241–252
DOI 10.1007/s11038-006-9129-z

Earth Radioactivity Measurements with a Deep Ocean Anti-neutrino Observatory

S. T. DYE*, E. GUILLIAN, J. G. LEARNED, J. MARICIC, S. MATSUNO, S. PAKVASA, G. VARNER and M. WILCOX

Department of Physics and Astronomy, University of Hawaii, 2505 Correa Road, Honolulu, Hawaii 96822, USA
(E-mail: sdye@hpu.edu)

**College of Natural Sciences, Hawaii Pacific University, 45-045 Kamehameha Highway, Kaneohe, Hawaii 96744, USA*

(Received 22 September 2006; Accepted 17 October 2006)

Abstract. We consider the detector size, location, depth, background, and radio-purity required of a mid-Pacific deep-ocean instrument to accomplish the twin goals of making a definitive measurement of the electron anti-neutrino flux due to uranium and thorium decays from Earth's mantle and core, and of testing the hypothesis for a natural nuclear reactor at the core of Earth. We take the experience with the KamLAND detector in Japan as our baseline for sensitivity and background estimates. We conclude that an instrument adequate to accomplish these tasks should have an exposure of at least 10 kilotonne-years (kT-y), should be placed at least at 4 km depth, may be located close to the Hawaiian Islands (no significant background from them), and should aim for KamLAND radio-purity levels, except for radon where it should be improved by a factor of at least 100. With an exposure of 10 kT-y we should achieve a 25% measurement of the flux of U/Th neutrinos from the mantle plus core. Exposure at multiple ocean locations for testing lateral heterogeneity is possible.

Keywords: Anti-neutrino, mantle, uranium, thorium, geo-neutrino, geo-reactor

1. Introduction

This report furnishes recommendations for the size and sensitivity needed by a deep ocean anti-neutrino detector near Hawaii (Hawaii Anti-Neutrino Observatory – Hanohano) to perform geophysics measurements of Earth radioactivity. The design and experience of the KamLAND project in Japan provides an excellent guide to detector size and location needed to approach two geophysics goals. This detector would be more than an order of magnitude larger than KamLAND with appropriate modifications and adaptations for operation in the deep ocean. Event rates quoted for this detector are based on an exposure of 10 kilotonne-years (kT-y) of KamLAND scintillating oil, which contains some 8.5×10^{31} free protons per kilotonne.

1.1. Measurement of Geo-neutrinos

The first geophysics goal is measurement, not merely detection, of the electron anti-neutrino flux from the mantle and core of Earth due to uranium and thorium (U/Th) decays. Because the concentrations of U/Th are much higher in the continental crust than in the oceanic crust and mantle, locations far from the continental crust, like Hawaii, are well suited to this measurement. Although geologists usually predict the core to be free of U/Th, the flux measurement described herein is not sensitive to electron anti-neutrino direction and therefore does not differentiate between the mantle and core. In subsequent descriptions of measurements of electron anti-neutrino flux from U/Th decays mantle refers to mantle plus core.

Whereas the concentrations of U/Th in the outermost Earth's crust can be sampled directly, measurement of the "geo-neutrino" flux provides the only viable method for determining these concentrations in the mantle. These values are known poorly at present and speculated upon by geologists. Concentrations are typically inferred from the U/Th concentrations in meteorites plus assumptions about Earth accretion and differentiation. Note that geologists generally quote such numbers without error ranges, simply because there are too many unknowns: U/Th concentrations and distributions are generally acknowledged to be informed guesses. Hence the geological community welcomes information carried by geo-neutrinos from the otherwise inaccessible inner Earth. This was made evident by the reception of the first KamLAND results on the measurement of geo-neutrinos (mostly from the local crust in Japan), published as a cover article in the 28 July 2005 issue of Nature (Araki et al., 2005a). This paper heralding the first positive detection of Earth's total radioactivity marks a beginning for neutrino geophysics, long a tantalizing goal (Eder, 1966; Avilez et al., 1981; Krauss et al., 1984). The article reported 28 geo-neutrino events above background from an exposure of about 1 kT-y. Seven of these events can be attributed to the mantle (Enomoto 2005). While a start, the report does not add much new information about Earth's composition or radiogenic heat.

Several groups have made calculations of geo-neutrino fluxes (Raghavan et al., 1998; Rothschild et al., 1998; Fiorentini et al., 2004; Enomoto et al., 2005), and there are two PhD dissertations from KamLAND which contain significant modeling (Tolich, 2005; Enomoto, 2005).

1.2. Search for Hypothetical Geo-reactor

The second goal is a definitive search for a hypothetical nuclear reactor at Earth's core. This theory (Herndon, 1996; Hollenbach and Herndon, 2001) has not met wide acceptance by the geological community, who have generally preferred the idea that much of the U/Th rose from the molten, early inner Earth as slag, rather than sank to the core as elemental metal. Yet, many geologists say that there really is no evidence against the hypothesis since the conditions at Earth's formation are little known. Moreover, there are peculiarities in the isotopic content of Earth, and most particularly the observed high ratio of $^3He/^4He$ coming out of oceanic volcanic hot spots (such as Hawaii and Iceland), which a natural reactor could explain (^3He would come from tritium decay, made abundantly in reactors).

As discussed elsewhere (Raghavan, 2002), this hypothetical energy source in the range of 1–10 terawatts of thermal (TW_t) power could be the enigmatic power source driving the deep Earth plumes, and hence ultimately responsible for the motion of landmasses (plate tectonics) as well as Earth's magnetic field (geo-dynamo). The neutrino flux from this putative geo-reactor is very hard to measure at locations anywhere near electrical power reactors, especially in places such as Japan, Europe and North America (Raghavan, 2002; Domogatski et al., 2004).

At KamLAND the geo-reactor would present a flux of only a few percent of that due to power reactors around Japan. This is very hard to distinguish from the power reactor flux because the energy spectrum at the source of natural or man-made reactors is essentially indistinguishable. However, since reactors at distances of a few hundred kilometers are not so far as to have all neutrino oscillations effects (see next section) washed out, a study has been made seeking an unchanging geo-reactor signal added to the power reactor flux with a time varying spectrum (Maricic, 2005). This is sufficient for claiming an upper limit on such a power source (< 20 TW_t). Although tantalizing, the study lacks the sensitivity required to detect and measure the power of the geo-reactor if it exists.

1.3. Neutrino Oscillations

In the following discussion, all processes that involve neutrino production and subsequent detection assume that neutrino oscillations occur, as is now established. The oscillation parameters employed are the best-fit values from global fits to all solar and reactor neutrino experiments as of this time (Araki et al., 2005b). Since the baseline of neutrino propagation considered in this context (thousands of kilometers) is much larger than the oscillation lengths (< 100 km) for the energy scale under consideration, the neutrinos can be considered to a good approximation to be fully mixed. The effect of oscillations can be accounted for by reducing the event rate by a factor of

$$P(\nu_e \rightarrow \nu_e) = 1 - {}^1\!/\!_2\{\cos^4(\theta_{13})\sin^2(2\theta_{12}) + \sin^2(2\theta_{13})\} \approx 0.6 \tag{1}$$

compared to the rate without oscillations, using current experimental values for the mixing angles.

1.4. Electron Anti-neutrino Detection and Analysis Windows

Electron anti-neutrinos are observed by the detection of positrons and neutrons produced by inverse neutron decays in scintillating liquid by the standard technique. The positron produces a prompt signal boosted by positron-electron annihilation with a visible energy in the detector ~0.8 MeV less than the electron anti-neutrino energy. A delayed signal at 2.2 MeV of visible energy from the formation of deuterium tags the neutron capture.

The threshold energy for inverse neutron decay is 1.8 MeV. This sets the lower bound of the geo-neutrino analysis at 1.7 MeV, adjusted for detector energy resolution. Geo-neutrino energies extend up to 3.4 MeV, which sets the upper bound for the analysis. The window for the geo-reactor analysis is 3.4 MeV–9.3 MeV.

2. Geo-neutrino Detection Sensitivity

Hanohano's location in the middle of the Pacific Ocean makes it sensitive primarily to geo-neutrinos originating from Earth's mantle and core. The nominally expected event rate of geo-neutrinos of mantle origin based upon the Bulk Silicate Earth (BSE) model (Fiorentini et al., 2004) is 79 events per 10 kT-y. This is more than 2.5 times larger than the 31 events per 10 kT-y for geo-neutrinos from the oceanic and continental crusts. Because of the uncertainty of the modeling we do not assign an error estimate to the event rate from the mantle.

A geo-neutrino flux dominated by the mantle at sites near Hawaii is noted by several authors (Rothschild et al., 1998; Mantovani et al., 2004; Enomoto et al., 2005; Pakvasa, 2005). The situation is reversed at a continental location with the same flux from the mantle but about eight times the flux from the crust. In this analysis the signal is the geo-neutrino event rate from the mantle. Geo-neutrinos from the crust are considered part of the background. The conclusion is that the flux of geo-neutrinos from the mantle is extremely difficult to measure at continental locations, which yield mainly a flux from the crust.

2.1. Geo-neutrino Background

Expected background for geo-neutrinos (based upon KamLAND experience) includes

- ^{9}Li produced by cosmic rays traversing the detector,
- Fast neutrons from cosmic rays passing near the detector,
- α decay of ^{210}Po followed by $^{13}C(\alpha,n)^{16}O$ in the scintillating oil,
- Accidental or random coincidences,
- Anti-neutrinos from commercial nuclear reactors, and
- Anti-neutrinos from a geo-reactor if it exists.

The lithium background is due to cosmic ray muons traversing the detector, decreasing with increasing depth. At the depth of KamLAND, equivalent to ~2.1 km of water (Mei, 2006), it is a major nuisance. By 4 km depth the Li background is almost negligible. The low lithium background level in KamLAND is achieved at the cost of applying tight cuts around reconstructed muon tracks, which results in the removal of a significant amount of good data. These cuts, moreover, introduce systematic errors that obscure the signal. For these reasons, the most favorable strategy is to go as deep as possible so that the cosmic ray background rate is so low that the application of cuts to remove ^{9}Li events becomes unnecessary. Greater depth alleviates a multiplicity of background problems, including entering fast neutrons and significant dead time around muon transits. Background contribution due to fast neutrons is negligible in KamLAND and is ignored in this analysis. Since fast neutrons occur at the edge of the fiducial (inner, software-defined) volume they are efficiently removed in the data analysis.

We determine that overburden-dependent background is reduced to a comfortable level by a depth of 4 km. Fortunately the abyssal plane of the ocean is 4–5 km in depth. A potential site 34 km west of the Big Island of Hawaii (19.72N, 156.32W) at about 4.5 km depth meets our requirements.

The polonium background, due to alphas which interact with ^{13}C, stems mostly from radon contamination at KamLAND. The mine levels of radon are 40 times those in the free air outside. In the course of experimental preparation the KamLAND scintillating oil was possibly exposed to mine air. While the radon itself decays in a matter of months, further decay products lead to the initially unrecognized polonium background, which can simulate inverse beta decay. The background level used for this analysis assumes a concentration of ^{210}Po 1/100 of that at KamLAND, which is conservatively high (Suzuki, 2006).

We define "accidental" background to be due to random coincidences. Radioactivity of the detector itself contributes. Some of this radioactivity comes from the periphery of the detector, the balloon and supporting ropes in KamLAND. Thus this scales with detector outer surface area not volume. For present purposes we take the conservative assumption that the rate per unit volume will be the same as in KamLAND, but we can doubtless do better than indicated.

The calculation of the contribution from distant commercial power reactors can be carried out to about 2% precision, and should be well known. A detector near Hawaii would record about 12 events per 10 kT-y exposure. Locations in the southern ocean and near Australia realize contributions lower by about a factor of two, whereas contributions at continental locations in the northern hemisphere are typically higher by at least an order of magnitude (Rothschild et al., 1998). Clearly, Hawaii has a very low rate compared to other possible locations making it extremely suitable for measuring geo-neutrinos from the mantle. Our estimates include the small contribution due to long-lived reactor products. The contribution from nuclear powered ships and submarines warrants consideration (Detwiler et al., 2002). Submarine power plants are in the range of 100 MW_t, as compared to 2 GW_t for power reactors. Ships cruise at a small fraction of maximal power generally, and are shut down in port at Pearl Harbor. Although not included in our background estimates, these could be accounted for with cooperation of the military (since we only need to know flux at the detector and not power or range of the ships).

In arriving at detector parameters for geo-neutrino measurement, we assume that the geo-reactor power is zero. Were it to exist, then we would know the power quite well from the measurements above 3.4 MeV of neutrino energy. An Earth-centered geo-reactor would contribute 19 events per TW_t per 10 kT-y to the geo-neutrino measurement.

The final "background" is the contribution to the U/Th neutrino flux from the oceanic and distant continental crusts. An uncertainty of about 20% is assigned. This is consistent with geological models (McDonough and Sun, 1995) and the detailed studies for the KamLAND site (Enomoto, 2005).

We have made a preliminary estimate of the additional flux of neutrinos due to the proximity of the Hawaiian Islands. When we assume that the Big Island can be modeled as a cone of mid-ocean ridge basalt of height 10 km and radius 100 km, and that our detector is located a mere 10 km from the effective source center, the contribution would amount to only 5% of the mantle flux. When a site is finally chosen we will have to do a more careful calculation, but for now we can safely conclude that proximity to the Hawaiian Islands will not affect the experimental goal of measuring the mantle flux.

2.2. Geo-neutrino Signal Analysis

Table I presents the numbers of events expected for the geo-neutrino signal analysis. The total geo-neutrino "background" rate at 4-km depth (without geo-reactor but including neutrinos from commercial reactors and the crusts) is 96 ± 7 per 10 kT-y, compared to a mantle signal rate of 79 per 10 kT-y. The background subtracted mantle geo-neutrino signal would be 79 ± 20, a 25% measurement on the total rate alone (not using spectrum).

In order to confirm the above conclusion, we performed simulations where the combined energy spectrum of the signal and background are varied randomly and a multi-component fit is done for the number of signal events. The results improve, particularly depending upon how well we are able to constrain the background components.

2.3. Advantage of Oceanic Site

The advantage of an oceanic site for measuring the geo-neutrino signal from the mantle and thereby Earth radioactivity is demonstrated by the following example. We consider equal 10 kT-y exposures for several potential geo-neutrino detectors including Borexino (Giammarchi and Miramonti, 2006), SNO+ (Chen, 2006), and Hanohano. Assuming lateral homogeneity in the mantle each detector would record a signal of 79 events. Background event numbers are estimated from the depth and geographic location of each detector and an assumed level of ^{210}Po radio-purity 100 times better than

TABLE I

The numbers of events expected for the mantle geo-neutrino analysis for energies between 1.7 and 3.4 MeV

	Events (10 kT-y)$^{-1}$		
	SNO+	Borexino	Hanohano
^{9}Li	0 ± 0	3 ± 1	3 ± 1
^{210}Po	8 ± 2	8 ± 2	8 ± 2
Accidental	42 ± 1	42 ± 1	42 ± 1
Reactor	528 ± 21	295 ± 12	12 ± 1
Crust Geo-νs	368 ± 74	279 ± 56	31 ± 6
Background	946 ± 77	627 ± 57	96 ± 7
Mantle	79	79	79
Total ($N \pm \sqrt{N}$)	1025 ± 32	706 ± 27	175 ± 13
Expected Signal	79 ± 109	79 ± 84	79 ± 20

We assume the ^{210}Po background to be 100 times purer than the level for the scintillating oil at KamLAND and the reactor background to be known to 4%.

reported by KamLAND (Araki et al., 2005a, b). Table I presents the numbers of events from each source along with their uncertainties expected in the detectors considered.

A measurement of the mantle signal M requires subtracting the non-mantle background B from the total N. The uncertainty is $\delta M = \delta N + \delta B$. This is shown graphically in Figure 1. Hanohano is capable of measuring the mantle U/Th neutrino flux to 25% in 1 year. After 4 years of operation, the 20% systematic uncertainty in background from the crust would begin to dominate, ultimately limiting the measurement at the 8% level. This same uncertainty severely limits the capability of detectors at continental sites for measuring the mantle flux. For example neither of the other detectors considered would make a positive detection from the same exposure. We note that it would take SNO+ and Borexino 15 and 50 years, respectively, to achieve this exposure. This result demonstrates the advantage of an oceanic site over continental sites for measuring the U/Th neutrino flux from Earth's mantle.

3. Geo-reactor Neutrino Detection Sensitivity

The energy spectrum of anti-neutrinos produced in a nuclear fission reactor extends from 0 MeV up to about 10 MeV, much wider than the 0 to 3.4 MeV

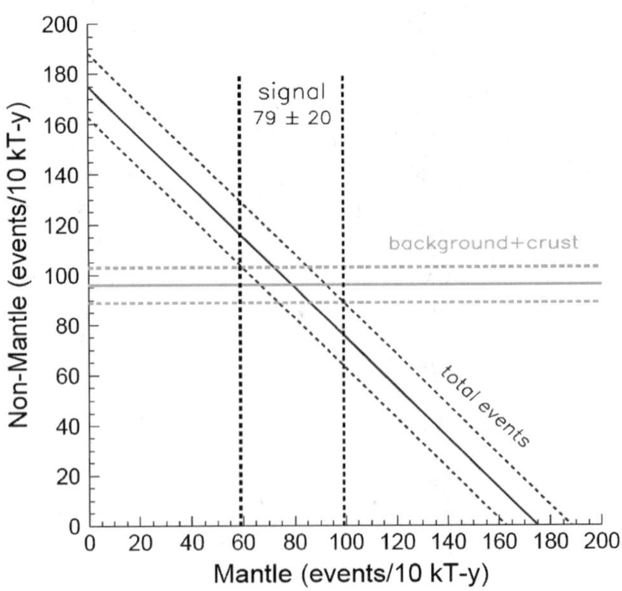

Figure 1. The background-subtracted mantle signal measured by Hanohano with a 10 kT-y exposure.

for geo-neutrinos. In the energy region above the threshold energy for anti-neutrino interaction with target 1.8–3.4 MeV, geo-neutrinos are a background to the geo-reactor. Therefore a lower energy threshold of 3.4 MeV is applied to the geo-reactor search to completely remove this background. We set an upper bound of 9.3 MeV for convenience (the probability that a geo-reactor neutrino has greater energy is negligible). In this energy range (3.4–9.3 MeV) the geo-reactor neutrino event rate is 38 events/TW_t/10 kT-y.

3.1. Geo-reactor Neutrino Background

The background to geo-reactor neutrino detection comes from the same sources as for geo-neutrino detection, but is lower at higher energies with the exception of the contribution due to lithium, which is perhaps a factor of two higher (Tolich, 2005). The event rates differ because of the different energy window and different analysis cuts. We summarize them in Table II. At a depth of 4 km the total background rate is 30 per 10 kT-y, compared to the signal for a 1 TW_t geo-reactor of 38 per 10 kT-y. The signal detection significance for a 1 TW_t geo-reactor from a 10 kT-y exposure is $S/\sqrt{S+B} = 4.6$ sigma. If the geo-reactor exists at the high power end of predicted range, 10 TW_t, the detection significance would increase to 19 sigma. Assuming we have pinned down the background (via fitting the spectrum and other means), the geo-reactor power can be resolved to 5–22% going from the upper to lower expected power levels (10–1 TW_t), limited by statistical fluctuations, not systematic uncertainty.

As with geo-neutrino detection, we performed simulation studies of Hanohano's sensitivity to the geo-reactor assuming various signal and background levels and an initial 10 kT-y exposure. Using the knowledge of expected spectrum for both signal and background will add to our confidence

TABLE II

Background sources and corresponding rates expected for mid-ocean geo-reactor neutrino detection

Geo-reactor Background	Rate $(10\text{ kT-y})^{-1}$
^{9}Li (4 km)	4 ± 1
^{210}Po	1 ± 1
Accidentals	1 ± 0
Commercial reactors	24 ± 1
Total background	30 ± 2
Geo-reactor signal	38/TW_t

We assume the ^{210}Po background to be 100 times purer than the level for the scintillating oil at KamLAND and the reactor background to be known to 4%.

in detecting any geo-reactor signal, as well as improving background estimate precision. We confirm that the error on a positive power measurement will remain dominated by statistical fluctuations, not systematic uncertainty.

We have not herein considered the further confirmation of the location of any positive geo-neutrino signal by neutrino direction measurement. The direction to a neutrino source was measured by the Chooz experiment team (Apollonio et al., 2000) (operating nearby a power reactor) by using the difference between the location of the initial (positron annihilation) and second (neutron capture) effective vertices, and is due to the slight neutron momentum in the direction of the incoming neutrino. The prospects for confirmation that the putative geo-reactor signal is coming from generally the direction of the center of Earth depends upon detector design (electronics and photomultiplier time resolution and scintillating oil decay lifetime). Our first estimates are not very encouraging unless the geo-reactor is at the higher end of the possible power levels. Another possibility for increasing directionality involves loading the scintillating liquid to reduce the neutron capture time and make the radiation length shorter. This possibility requires study.

We conclude that the geo-reactor measurement is easier than the mantle geo-neutrino measurement. If the geo-reactor exists at the level suggested, (and important for driving plumes) we should be able to convincingly detect it and make useful measurements of the power.

4. Recommended Detector Specifications

Based on the foregoing results, we recommend the following for a deep ocean anti-neutrino observatory:

- A detector fiducial volume of about 10 kT (some 20 times KamLAND).
- The planned live time should be at least 1 year, yielding a 10 kT-y exposure at each location.
- A depth of 4 km is sufficient to comfortably accomplish the twin geophysics goals. Depths greater than about 4 km do not make significant improvements in background levels.
- The scintillating oil must be as free of ^{210}Po as possible, with a goal of 100 times less than the initial KamLAND contamination. With testing at KamLAND already demonstrating reduction by a factor of 10^5 using a newly developed distillation process this goal should be easily attainable.
- A deep ocean (or mid-ocean island) location, far from continents, is required to resolve the mantle flux of U/Th decay neutrinos from background including neutrinos from oceanic and continental crusts.

- A location near the Hawaiian Island land mass does not add significant crust background to the measurement of U/Th neutrinos from the mantle.
- With the stated goal of 10 kT-y exposure and 4 km depth and expected background we can achieve a 25% measurement of the U/Th neutrinos from the mantle, and hence this level of global concentration.
- Again, with stated assumptions on exposure and background, we can measure a geo-reactor power to 5–22% precision for source powers in the predicted range of 10–1 TW_t. With a null result, we can set upper limits to the power of <0.5 TW_t at $>95\%$ confidence level.

5. Conclusions

In summary, we find that a 1-year deployment of a 10 kT, deep ocean anti-neutrino observatory can achieve the geophysics goals of this proposed experiment: a measurement of mantle geo-neutrinos and a definitive search for the hypothetical geo-reactor. We show that the 20% systematic uncertainty of the background from U/Th neutrinos in Earth's crusts prevents detectors at continental sites from measuring mantle geo-neutrinos and ultimately limits the measurement at an oceanic site to 8%. Subsequent deployments of Hanohano at other oceanic locations present the opportunity to measure lateral heterogeneity of U/Th concentrations in the mantle at the 25% level.

Acknowledgements

We are grateful to Robert Svoboda and Nikolai Tolich for many useful suggestions and comments. This work was partially funded by U.S. Department of Energy Grant DE-FG02-04ER41291 and by the National Defense Center of Excellence for Research in Ocean Sciences (CEROS). CEROS is a part of the Natural Energy Laboratory of Hawaii Authority (NELHA), an agency of the Department of Business, Economic Development and Tourism, State of Hawaii. CEROS is funded by the Defense Advanced Research Projects Agency (DARPA) through grants and agreements with NELHA. This work does not necessarily reflect the position or policy of the Government, and no official endorsement should be inferred.

References

Apollonio, M. et al.: 2000, *Phys. Rev. D* **61**, 012001.
Araki, T. et al.: 2005a, *Nature (London)* **436**, 499.
Araki, T. et al.: 2005b, *Phys. Rev. Lett.* **94**, 081801.
Avilez, C., Marx, G. and Fuentes, B.: 1981, *Earth as a source of antineutrinos. Phys. Rev. D* **23**, 1116.
Chen, M. C.: 2006, (this volume) doi: 10.1007/s11038-006-9116-4.
Detwiler et al.: 2002, *Phys. Rev. Lett.* **89**,191802.
Domogatski, G. et al.: 2004, hep-ph/0401221.
Eder, G.: 1966, *Nucl. Phys.* **78**, 657.
Enomoto, S., Ohtani, E., Inoue, K. and Suzuki, A.: 2005, hep-ph 0508049.
Enomoto, S.: 2005, *Neutrino Geophysics and Observation of Geo-Neutrinos at KamLAND*, Ph.D. Thesis, Tohoku University.
Fiorentini, G., Lissia, M., Mantovani, F. and Vannucci, R.: 2004, hep-ph/0401085.
Giammarchi, M. G. and Miramonti, L.: 2006, (this volume) doi: 10.1007/s11038-006-9106-6.
Herndon, J. M.: 1996, *Proc. Natl. Acad. Sci.* **93**, 646.
Hollenbach, D. F. and Herndon, J. M.: 2001, *Proc. Natl. Acad. Sci.* **98**, 11085.
Krauss, L. M., Glashow, S. L. and Schramm, D. N.: 1984, *Nature (London)* **310**, 191.
Mantovani, F. et al.: 2004, *Phys. Rev. D* **69**, 013001.
Maricic, J.: 2005, *Setting Limits on the Power of a Geo-Reactor with KamLAND*, Ph.D. Thesis, University of Hawaii.
McDonough, W. F. and Sun, S.-s.: 1995, *Chem. Geol.* **120**, 223.
Mei, D.: 2006, *Phys. Rev. D* **73**, 053004.
Pakvasa, S.: 2005, *Nucl. Phys. B* (proceedings supplement) **145**, 378.
Raghavan, R. S. et al.: 1998, *Phys. Rev. Lett.* **80**, 635.
Raghavan, R. S.: 2002, hep-ex/0208038.
Rothschild, C. G., Chen, M. C. and Calaprice, F. P.: 1998, *Geophys. Res. Lett.* **25**, 103.
Suzuki, A.: 2006, (this volume) doi: 10.1007/s11038-006-9131-5.
Tolich, N.: 2005, *Experimental Study of Terrestrial Electron Anti-neutrinos with KamLAND*, Ph.D. Thesis, Stanford University.

Earth, Moon, and Planets (2006) 99:253–264
DOI 10.1007/s11038-006-9111-9

Probing the Earth's Interior with the LENA Detector

KATHRIN A. HOCHMUTH
Max-Planck-Institut für Physik (Werner-Heisenberg-Institut), Föhringer Ring 6, 80805, München, Germany
(E-mail: hochmuth@ph.tum.de)

FRANZ V. FEILITZSCH, TERESA MARRODÁN UNDAGOITIA, LOTHAR OBERAUER, WALTER POTZEL and MICHAEL WURM
Physik Department E15, Technische Universität München, James-Franck-Strasse, 85748, Garching, Germany

BRIAN D. FIELDS
Center for Theoretical Astrophysics, Department of Astronomy, University of Illinois, Urbana, IL 61801, USA

(Received 22 May 2006; Accepted 11 July 2006)

Abstract. A future large-volume liquid scintillator detector such as the proposed 50 kton LENA (Low Energy Neutrino Astronomy) detector would provide a high-statistics measurement of terrestrial antineutrinos originating from β-decays of the uranium and thorium chains. Additionally, the neutron is scattered in the forward direction in the detection reaction $\bar{\nu}_e + p \rightarrow n + e^+$. Henceforth, we investigate to what extent LENA can distinguish between certain geophysical models on the basis of the angular dependence of the geoneutrino flux. Our analysis is based on a Monte-Carlo simulation with different levels of light yield, considering an unloaded PXE scintillator. We find that LENA is able to detect deviations from isotropy of the geoneutrino flux with high significance. However, if only the directional information is used, the time required to distinguish between different geophysical models is of the order of severals decades. Nonetheless, a high-statistics measurement of the total geoneutrino flux and its spectrum still provides an extremely useful glance at the Earth's interior.

Keywords: Low energy neutrino astronomy, geoneutrinos

1. Introduction

A future large volume liquid scintillator such as the proposed LENA detector (Oberauer et al., 2005) can obtain a high precision measurement of the geoneutrino flux, could deliver new information about the interior of the Earth, in particular its radiochemical composition, and thus give new insights on Earth and planetary formation.

Besides the geoneutrino measurement LENA will be designed for high-statistics solar neutrino spectroscopy, for spectroscopy of the cosmic diffuse supernova neutrino background, as a detector for the next galactic supernova, and to search for proton decay (Undagoitia et al., 2005). Present design studies for LENA assume 50 kt of liquid PXE scintillator that would provide a geoneutrino rate of roughly 1,000 events per year, if located on the continental crust, from the dominant

$$\bar{\nu}_e + p \rightarrow n + e^+ \tag{1}$$

inverse beta-decay reaction.

While liquid scintillator detectors do not provide direct angular information, indirectly one can retrieve directional information because the final-state neutron is displaced in the forward direction. The offset between the e^+ and the neutron-capture locations can be reconstructed, although with large uncertainties. Therefore, it is natural to study the requirements for a future large-volume liquid scintillator detector to discriminate between geophysical models of the Earth that differ both by their total neutrino fluxes and their neutrino angular distributions.

Motivated by the current design studies for LENA we will consider a 50 kt detector using a PXE-based scintillator. However, it is difficult to locate the neutron-capture event on protons because a single 2.2 MeV gamma is released that travels on average 22.4 cm before its first Compton interaction. Therefore, the event reconstruction is relatively poor.

For the geoneutrino flux we will consider a continental and an oceanic location. In each case we will use a reference model and exotic cases with an additional strong neutrino source in the Earth's core.

We begin in Section 2 with a discussion of the principles of geoneutrino detection in large-volume scintillator detectors as well as possible scintillator properties. In Section 3 we introduce our geophysical models. In Section 4 we turn to the main part of our work, a Monte-Carlo study of the power of the LENA detector for discriminating between different geophysical models. We conclude in Section 5.

2. Geoneutrino Detection

2.1. Directional information from neutron displacement

In a scintillator detector, geoneutrinos are measured by the inverse beta-decay reaction Equation (1) with an energy threshold of 1.8 MeV. The cross section is

$$\sigma = 9.52 \times 10^{-44}\text{cm}^2 \quad \frac{E_+}{\text{MeV}} \frac{p_+}{\text{MeV}} \tag{2}$$

where E_+ is the total energy of the positron and p_+ its momentum. In this equation the recoil energy (Vogel and Beacom, 1999) has been neglected, which introduces an error of ~1%. The visible energy $E_{vis} = E_+ + m_e$ always exceeds 1 MeV because the positron annihilates with an electron of the target. By measuring the visible energy one can determine the neutrino energy as $E_\nu \approx E_{vis} + 0.8$ MeV because the kinetic energy of the neutron is typically around 10 keV and thus negligible. After thermalization the neutron is captured by a nucleus, thus tagging the inverse beta decay reaction.

Kinematics implies that the neutron is scattered roughly in the forward direction with respect to the incoming neutrino (Vogel and Beacom, 1999), this being the key ingredient for obtaining directional information. The average displacement between the neutron and positron events is then theoretically found to be about 1.7 cm (Vogel and Beacom, 1999).

The reactor experiment CHOOZ, using a Gd-loaded scintillator, has measured an average neutron displacement from the e^+ event of 1.9 ± 0.4 cm (Apollonio et al., 1999). However, once the neutron has been thermalized by collisions with protons, it diffuses some distance before being captured so that the actual displacement varies by a large amount for individual events. In a PXE-based scintillator the average time interval until capture on a proton is 180 μs, leading to an uncertainty σ of the displacement of about 4 cm for the x-, y- and z-direction (Vogel and Beacom, 1999). With Gd loading σ is reduced to approximately 2.4 cm (Vogel and Beacom, 1999) because the neutron diffusion time is much shorter, on average about 30 μs (Apollonio et al., 1999).

2.2. PXE-BASED SCINTILLATOR

One option for the proposed LENA detector is to use a scintillator based on PXE (phenyl-o-xylylethane, $C_{16}H_{18}$). PXE has a high light yield, it is non hazardous, has a relatively high flashpoint of 145 °C, and a density of 0.985 g/cm^3 (Back et al., 2004, unpublished). A possible admixture of dodecane ($C_{12}H_{26}$) increases the number of free protons and improves the optical properties. A blend of 20% PXE and 80% dodecane shows a decrease in light yield of about 20% relative to pure PXE, an attenuation length of about 11 m and an increase in the number of free protons by 25% (Wurm, 2005).

In this paper we consider a detector with a total volume of about 70×10^3 m^3. This could be realized with a cylindrical detector of 100 m length and 30 m diameter. An outer water Cherenkov detector with a width of 2 m acts as a muon veto. In order to shield against external gamma and neutron radiation a fiducial volume of about 42×10^3 m^3 with a total number of 2.5×10^{33} free protons as target can be realized using a scintillator mixture as mentioned above with 20% PXE and 80% dodecane. In Monte-Carlo calculations the light yield of events in LENA has been estimated (Undagoitia et al., 2005). For events in the central detector region the yield N_{pe}, measured in photo-electrons (pe) per MeV energy deposition, can be expressed as $N_{pe} \approx 400$ pe/MeV $\times c$, where c is the optical coverage which depends on the number and aperture of the photomultiplier tubes (PMTs). A maximal coverage $c_{max} \approx 0.75$ can not be exceeded so that we assume the maximal light yield to be around 300 pe/MeV. For instance, the use of 12,000 PMTs with a diameter in aperture of 50 cm would result in an optical coverage of about 30% and a light yield $N_{pe} \simeq 120$ pe/MeV. This can be obtained either by using PMTs like in the Super-Kamiokande experiment or by smaller PMTs equipped with light concentrators as were developed for the counting test facility (CTF) at the Gran Sasso underground laboratory (Oberauer et al., 2004). For events off the axis of the cylinder the light yield would be enhanced. Hence, low-energy spectroscopy even in the sub-MeV region should be possible in LENA.

For a detection of the positron–neutron displacement the ability of the detector to locate the absorption position of both particles is crucial. The experimental reconstruction of both events is possible by analyzing the arrival times and the number of photons in each individual PMT. The position uncertainty depends on the total yield of registered photo-electrons. In the CTF, the measured position uncertainty was around 10 cm in each direction for events with 300 photo-electrons and it was shown that the uncertainty scales with the inverse square-root of that number (Alimonti et al., 1998), as the emission time dispersion of the scintillator is considerably shorter than the photon transient time through the detector. For the following discussion we assume, that the scattering length of the scintillator is considerably larger than the radius of the detector cyclinder. Therefore, we will use a Gaussian distribution for the uncertainty of the positron event reconstruction with equal width in each direction of

$$\sigma_{e^+} = 10\,\text{cm}\left(\frac{300\,\text{pe/MeV}}{N_{pe}}\frac{1\ \text{MeV}}{E_{vis}}\right)^{1/2} \tag{3}$$

where N_{pe} is the light yield and E_{vis} the visible energy released by the positron.

In PXE-based scintillators the neutron is captured by a proton with nearly 100% efficiency within an average time interval of about 180 μs, subsequently emitting a 2.2 MeV gamma. This photon has a mean free path of 22.4 cm before its first Compton scattering so that the event reconstruction is much more uncertain than for the positron event. We have simulated this case by taking into account multiple Compton scatterings of the 2.2 MeV gamma. The position of each gamma emission, representing the position of the neutron capture, is reconstructed by composing the energy-weighted sum of each Compton scattering event, taking into account the instrumental resolution. The distribution of the reconstructed position in each direction follows roughly a Lorentzian form. In Figure 1 we show the radial distribution of the reconstructed positions of these events for light yields of N_{pe} = 50, 300 and 700 pe/MeV. Increasing the light yield does not significantly narrow the distribution because its width is dominated by the large Compton mean free path of the 2.2 MeV photon. With reduced light yield the position of the maximum as well as the mean value of the distribution shifts towards larger values. This is caused by the increased uncertainty of the instrumental resolution.

2.3. BACKGROUNDS

KamLAND has reported 152 events in the energy region relevant for geo-neutrinos within a measuring time of 749 days and 3.5×10^{31} target protons. From these events 127 ± 13 are due to background (Araki et al., 2005). The most relevant background for the KamLAND site is reactor antineutrinos (80.4 ± 7.2 events). For the LENA detector positioned in the underground laboratory CUPP (Centre for Underground Physics in Pyhäsalmi) in Finland (longitude: 26° 2.709′ E, latitude: 63° 39.579′ N, 1450 m of rock (4060 m.w.e.)) this background would be reduced by a factor ≃12, as the site is far

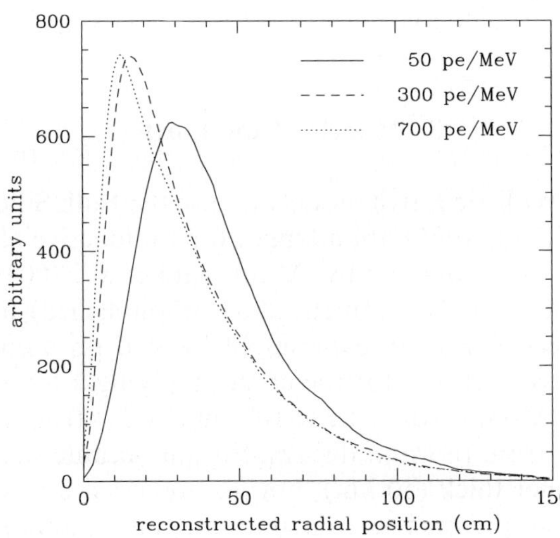

Figure 1. Monte-Carlo simulation of the radial distribution of a 2.2 MeV γ-quantum in an unloaded PXE scintillator. The curves are for light yields of N_{pe} = 50, 300 and 700 pe/MeV as indicated.

away from reactors. Hence we expect for LENA at CUPP a reactor background rate of about 687 events per year in the relevant energy window from 1.8 MeV to 3.5 MeV. Assuming a reactor run time of 100% this rate would increase by 15% to 790 events. This background can be subtracted statistically using the information on the entire reactor neutrino spectrum up to $\simeq$8 MeV.

Another important background for KamLAND is induced by radio impurities. A large concentration of the long-lived isotope ^{210}Pb is present in the KamLAND scintillator. In the decay chain of ^{210}Pb the α-emitting isotope ^{210}Po is present. Thus the reaction $^{13}C(\alpha, n)^{16}O$ can occur, mimicking the signature of geoneutrinos due to neutron scattering on protons and the subsequent neutron capture. The number of these background events in KamLAND is estimated to be 42 $\pm$ 11 (Araki et al., 2005). However, with an enhanced radiopurity of the scintillator, the background can be significantly reduced. Taking the radio purity levels of the CTF detector, where a ^{210}Po activity of 35 $\pm$ 12/m^{3}d in PXE has been observed (Back et al., 2004, unpublished), this background would be reduced by a factor of about 150 compared to KamLAND and would account for less than 10 events per year in the LENA detector.

An additional background that imitates the geoneutrino signal is due to ^{9}Li, which is produced by cosmic muons in spallation reactions with ^{12}C and decays in a β-neutron cascade. Only a small part of the ^{9}Li decays falls into the energy window which is relevant for geoneutrinos. KamLAND estimates this background to be 0.30 $\pm$ 0.05 (Araki et al., 2005). At CUPP the muon reaction rate would be reduced by a factor $\simeq$10 due to better shielding and this background rate should be at the negligible level of $\simeq$1 event per year in LENA.

3. Models of the Earth

In order to obtain realistic Earth models we use the Bulk Silicate Earth model (McDonough and Sun, 1995) abundances for radioactive elements, particularly the reference values derived by (Mantovani et al., 2003), and follow the discussion in (Fields and Hochmuch, 2004, unpublished) to generate angle dependent flux spectra. For an experiment located on a continent we have assumed a thickness of 50 km for the crust, implying a total neutrino flux in our Reference Model of 4.2×10^6 cm^{-2} s^{-1} from uranium and 4.1×10^6 cm^{-2} s^{-1} from thorium decays. For an oceanic site we have chosen the crust to be rather thick (50 km), but not included any sediments. If one wanted to determine the mantle contribution, the oceanic crust would be a background to the measurement so that the assumption of a thick oceanic crust is conservative. The neutrino fluxes in this case are 1.25×10^6 cm^{-2} s^{-1} from uranium and 0.88×10^6 cm $^{-2}$ s^{-1} from thorium decays.

Besides our reference model we get inspiration from the discussions on additional radioactivity in the core (Herndon, 1993, 2003; Labrosse et al. 2001) and consider two highly speculative models:

(A) Fully radiogenic model with additional uranium and thorium in the core, accounting for 20 TW additional heat production. (Integrated neutrino flux increase of about 32% relative to the reference model in a continental location, and 116% in an oceanic location.)
(B) Same as (A) except with 10 TW in the core. (Flux increase of 16% and 58%, respectively.)

To obtain the event rate, neutrino flavor oscillations have to be accounted for by including a global $\bar{\nu}_e$ survival-probability factor of 0.57 as measured by KamLAND (Araki et al., 2004). Matter effects for oscillations are not important because of the small geoneutrino energies. Moreover, for geoneutrino energies of 1.8–3.2 MeV and $\Delta m^2 = 7.9 \times 10^{-5}$ eV^2 the vacuum oscillation length is 57–101 km. Including distance and energy dependent survival probabilities is a negligible correction to a global reduction factor (Fiorentini et al., 2003). The annual event rates corresponding to our models, including the reduction factor, are shown in Table I for a 50-kton detector with a fiducial volume corresponding to 2.5×10^{33} protons.

Up to now we have assumed that the exotic heat source in the Earth's core is caused by uranium and thorium decays, i.e., the neutrino spectrum from this additional source was taken to be identical with the geoneutrino spectrum from the crust and mantle. However, the possibility of a natural reactor in the Earth's core ("georeactor") has been discussed in the literature (Herndon, 1993, 2003). In this case the neutrino flux could be similar to that from an ordinary power reactor with energies reaching up to about 8 MeV. With this assumption the total georeactor neutrino flux can be estimated to be $\Phi_\nu \simeq 1.9 \times 10^{23}$ s^{-1} for a thermal power of 1 TW. Taking into account neutrino oscillations, the distance to the center of the Earth, and the detection cross section we calculate an event rate of about 210 y^{-1} TW^{-1} in LENA. At Pyhäsalmi one would observe about 2,200 events per year due to

TABLE I
Annual event rates for 2.5×10^{33} target protons

Model	Continental crust	Oceanic crust
Reference model	1.02×10^3	0.29×10^3
(A) 20 TW core	1.35×10^3	0.62×10^3
(B) 10 TW core	1.19×10^3	0.45×10^3

Flavor oscillations have been included with a global reduction factor of 0.57.

neutrinos from nuclear power plants. Assuming a systematic uncertainty for the neutrino flux from the power plants of 6.5% as suggested by Araki et al. (2004), we conclude that LENA will be able to identify a georeactor of ≥2 TW after one year of measurement with a 3σ significance.

4. Monte-Carlo Study

To study the power of directional discrimination of a large liquid-scintillator detector we have performed a Monte-Carlo simulation of a large number of geoneutrino events and the corresponding directional reconstruction. We have assumed that the detector response is independent of the event location, i.e., only the spatial separation between the event $\bar{\nu}_e + p \to n + e^+$ and the location of neutron capture is relevant. However, as pointed out in Section 2, we consider a position resolution of point-like events located at the central axis of the detector. We have assumed that, on average, the neutron capture point is displaced by 1.9 cm in the forward direction relative to the e^+ event in agreement with the CHOOZ measurement (Apollonio et al., 2003). Moreover, we have assumed that neutron diffusion before capture causes a Gaussian distribution around this mean value with a width $\sigma_x = \sigma_y = \sigma_z = 4.0$ cm for an unloaded PXE-based scintillator as described in Section 2.

In addition to this distribution, the main uncertainty originates from the reconstruction of both events. For the positron event we have assumed that the reconstructed location follows a Gaussian distribution with a width given by Equation (3). The actual spread of relevant visible energies is small so that we have always used $E_{\rm vis} = 1.4$ MeV as a typical value. For an unloaded scintillator, the reconstruction of the neutron event introduces an even larger uncertainty; we have used a distribution as in Figure 1 appropriate for the given light yield.

We conclude that, given the relatively poor angular reconstruction capability of scintillator detectors, the only angular-distribution information that can be extracted is the slope of the geoneutrino distributions. Put another way, one can extract the total event rate and the dipole contribution of the angular distribution, whereas a determination of higher multipoles is unrealistic. Therefore, we write the reconstructed zenith-angle distribution in the form

$$\frac{\mathrm{d}\dot{N}}{\mathrm{d}\cos\theta} = \dot{N}\left(\frac{1}{2} + p\cos\theta\right) \tag{4}$$

where the event rate $\dot{N}$ and the coefficient p are the two numbers that characterize a given configuration of geophysical model and detector type.

The event rates for 2.5×10^{33} target protons and different geophysical models have already been reported in Table I. What remains to be determined by means of a Monte-Carlo simulation are the corresponding coefficients p and their uncertainty. In Table II we show the results for p for different cases, always assuming a light yield of 300 pe/MeV. The uncertainty σ_p of the measured p value scales with the inverse square root of the number of events N so that $s_p = \sigma_p\sqrt{N}$ is a quantity independent of N. The value of s_p can be derived analytically for $p = 0$, yielding

$$s_p = \frac{\sqrt{3}}{2} = 0.866, \tag{5}$$

which is valid for all $p \ll 1$. We have checked with our Monte Carlo that Equation (5) indeed applies to all p values of interest to us.

The number of events it takes to distinguish at the 1σ level between an isotropic event distribution ($p = 0$) and the actual coefficient is given by $N_{1\sigma} = (s_p/p)^2 = (3/4)\,p^{-2}$. For our reference model at a continental site we find $N_{1\sigma} \approx 500$ events, for an oceanic site about 200 events. In order to distinguish a geophysical model i from model j at the 1σ level, the required number of events is

$$N_{1\sigma} = \frac{2s_p^2}{(p_i - p_j)^2} = \frac{3}{2}\frac{1}{(p_i - p_j)^2}. \tag{6}$$

A detection at the $n\sigma$ level requires n^2 times more events.

In the same way as for Table II we have calculated the slope p for different light yields of the scintillator and have determined the number of events it takes to distinguish each of the exotic models from the reference case. In

TABLE II

Coefficient p for the reconstructed zenith-angle distribution for different Earth models and different detector types, always assuming a light yield of 300 pe/MeV

Model	Coefficient p for scintillator detectors
Continenal crust	
Reference model	0.0283
(A) 20 TW core	0.0377
(B) 10 TW core	0.0333
Oceanic crust	
Reference model	0.0468
(A) 20 TW core	0.0597
(B) 10 TW core	0.0560

Figure 2 we display $N_{1\sigma}$ for these cases and for both the continental- and oceanic-crust situation as a function of the light yield N_{pe}.

Of course, the time required to achieve this discriminating power depends on the detector size. For our fiducial volume with 2.5×10^{33} target protons as in LENA one needs to scale with the event rates shown in Table I. In a continental-crust location, all models produce an event rate of roughly 1,000 events per year, in full agreement with the KamLAND measurement (Araki et al., 2005). Therefore, even with optimistic assumptions a 50-kton detector would need several decades for distinguishing in a meaningful way between different geophysical models on the basis of the angular event distribution. Moreover, detector backgrounds should be included in a realistic assessment.

5. Conclusions

A future large-volume scintillator detector such as the proposed 50 kt LENA would provide a high-statistics measurement of the geoneutrino flux. The event rate would depend strongly on the detector location, notably on whether an oceanic site such as Hawaii is chosen where a reference event rate of about 300 per year (50 kt scintillator) is expected or a continental site such as the Pyhäsalmi mine in Finland where the reference rate would be about 1,000 per year. Therefore, the total geoneutrino flux could be measured with high significance and would allow one to distinguish between different Earth models.

The forward displacement of the neutron in the inverse beta decay detection reaction provides directional information on the geoneutrino flux. We have studied if this effect can be used to distinguish between different geophysical models, notably if one could diagnose a strong exotic energy source in the Earth's core under the assumption that its neutrino spectrum is identical with that emitted by the crust and mantle. While a deviation from an isotropic flux can be ascertained with high significance, we find that a 50 kt detector is too small to distinguish between different geophysical models on the basis of the directional information alone, except perhaps for extreme cases and optimistic assumptions about the detector performance.

In our study we have only used the neutrino flux from the Earth, ignoring the contribution from power reactors because it depends strongly on location. For example, in Pyhäsalmi the neutrino flux from power reactors adds roughly 60% to the counting rate in the energy window relevant for geoneutrinos. This contribution is not negligible, but it does not change our overall conclusions.

We have also estimated the sensitivity of a LENA- type detector for determining a hypothetical georeactor in the Earth's core. As a possible location the CUPP underground laboratory in Pyhäsalmi (Finland) was

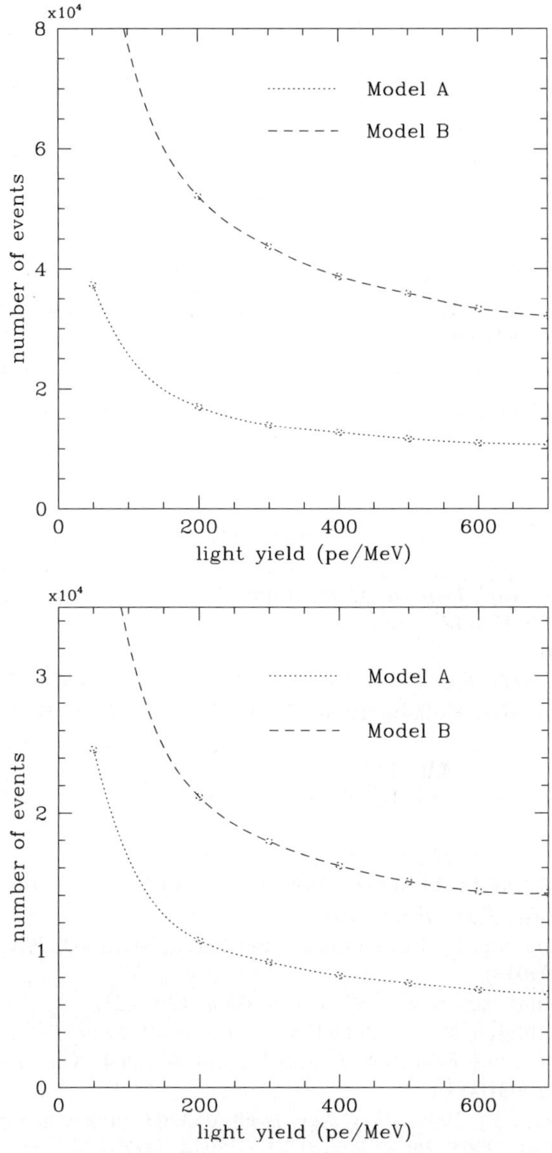

Figure 2. Number of events needed to distinguish between models A, B and the continental-crust reference model at 1σ significance. The points correspond to the values calculated with the Monte Carlo. *Upper panel:* Continental crust. *Lower panel:* Oceanic crust.

chosen and the background due to nuclear power plants was calculated. At CUPP a 2 TW georeactor could be identified at a statistical level of 3σ after only one year of measurement.

In summary, large-volume scintillator detectors of the next generation will be extremely useful to study the interior of the Earth in the "light of neutrinos." However, the prime information will be the total geoneutrino flux and its spectrum. It would be extremely challenging to use the directional information alone to distinguish between different geophysical models.

Acknowledgments

We thank E. Lisi for crucial discussions of an earlier version of this paper. Partial support by the Maier-Leibnitz-Laboratorium (Garching), the Virtual Institute for Dark Matter and Neutrinos (VIDMAN, HGF), the Deutsche Forschungsgemeinschaft under Grant No. SFB-375 and the European Union under the ILIAS project, contract No. RII3-CT-2004-506222, is acknowledged.

References

Alimonti, et al.: 1998, *Nucl. Instrum. Meth. A* **406**, 411.

Apollonio, M. et al. [CHOOZ Collaboration]: 1999, *Phys. Rev. D* **61**, 012001 (2000) [arXiv:hep-ex/9906011].

Apollonio, M. et al.: 2003, *Eur. Phys. J. C* **27**, 331 [arXiv:hep-ex/0301017].

Araki, T. et al. [KamLAND Collaboration]: 2004, *Phys. Rev. Lett.* **94**, 081801 [arXiv:hep-ex/0406035].

Araki, T. et al.: 2005, *Nature* **436**, 499.

Fiorentini, G., Lasserre, T., Lissia, M., Ricci, B. and Schonert, S.: 2003, *Phys. Lett. B* **558**, 15 [arXiv:hep-ph/0301042].

Herndon, J. M.: 1993, *J. Geomagn. Geoelectr.* **45**, 423.

Herndon, J. M.: 2003, *PNAS*, March 18, **100**(6), 3047–3050.

Labrosse, S. et al.: 2001, *Earth Planet. Sci. Lett.* **190**, 111.

Mantovani, F., Carmignani, L., Fiorentini, G. and Lissia, M.: 2003, *Phys. Rev. D* **69**, 013001 [arXiv:hep-ph/0309013].

McDonough, W. F. and Sun, S.-s.: 1995, *Chem. Geol.* **120**, 223.

Oberauer, L., von Feilitzsch, F. and Potzel, W.: 2005, *Nucl. Phys. Proc. Suppl.* **138**, 108.

Oberauer, L., Grieb, C., von Feilitzsch, F. and Manno, I.: 2004, *Nucl. Instrum. Meth. A* **530**, 453 [arXiv:physics/0310076].

Vogel, P. and Beacom, J.F.: 1999, *Phys. Rev. D* **60**, 053003 [arXiv:hep-ph/9903554].

Undagoitia, T. M. et al.: 2005, *Phys. Rev. D* **72**, 075014 [arXiv:hep-ph/0511230].

Wurm, M.: 2005, 'Untersuchungen zu den optischen Eigenschaften eines Flüssigszintillators und zum Nachweis von Supernovae Relic Neutrinos mit LENA', Diploma thesis, TU München, Germany.

Earth, Moon, and Planets (2006) 99:265–273
DOI 10.1007/s11038-006-9122-6

Neutron Background and Possibility for Shallow Experiments

TADAO MITSUI
Research Center for Neutrino Science, Tohoku University, Sendai 980-8578, Japan
(E-mail: mitsui@awa.tohoku.ac.jp)

(Received 5 June 2006; Accepted 4 September 2006)

Abstract. As a possible design of a future geoneutrino detector, a KamLAND-type, monolithic, liquid scintillator detector with a thicker veto and a method for particle identification to reject neutron and ^{9}Li background from cosmic-ray muon spallation is considered. Assuming such a detector, the possibility for geoneutrino observation at a depth of around 300 meters of water equivalent is investigated.

Keywords: Neutrino, geoneutrino, muon, spallation

1. More and More Geoneutrino Detectors

KamLAND has succeeded in the first experimental investigation of geologically produced electron antineutrinos, or geoneutrinos (Enomoto, 2006; Araki et al., 2005). Although subsequent KamLAND data will provide improved statistics with less background after purification and recirculation of the liquid scintilltor, observations at other sites are also indispensable for the full-dress "neutrino geophysics" in the near future. To this purpose, experimental sites of future observations should be chosen, in principle, based on geophysical interest rather than advantages in background environment. Hawaii is one of the mostly geologically interesting sites. By excavating deep enough or considering deep-sea experiments, like Hanohano (Dye et al., 2006), one can satisfy both conditions, i.e., geophysically interesting sites with sufficiently reduced cosmic-ray muon flux.

Here we consider a future strategy of deploying as many detectors as possible, constructed with minimal costs (especially civil engineering costs), so that a systematic measurement of Earth's power distribution is performed just like a systematic measurement of Earth's density distribution done by a large number of seismographs (Suzuki, 2005). To this purpose we consider in this report "shallow experiments" utilizing KamLAND-type detectors at a depth of around 300 meters of water equivalent (m.w.e.). For comparison KamLAND operates at a depth of 2700 m.w.e. A depth of 300 m.w.e. may still seem tough to excavate just for geoneutrino experiments. However, if existing mines are utilized, as in most current neutrino experiments, there should be many more choices at 300 m.w.e. with associated lower costs than at deeper depths. This makes it more realistic to choose geophysically favored locations.

Rejection of cosmogenic background is a key technique for implementing "cheap, shallow" experiments. For electron antineutrino ($\overline{\nu}_e$) detection via inverse beta decay on proton $\overline{\nu}_e + p \rightarrow n + e^+$, a delayed coincidence can be required between prompt positron signal and delayed gamma-ray signal from neutron capture on another proton. Although this traditional and very advantageous mode strongly rejects accidental background, correlated background remains a serious problem. The main sources of background expected in shallow experiments are (i) neutron–neutron events, in which two neutrons from the same parent muon fake the prompt–delayed pair; (ii) radioactivity–neutron events, in which radioactivity within the scintillator or an outside gamma ray fakes the prompt event and a cosmogenic neutron fakes the delayed event; (iii) fast neutron events, in which elastic scattering of a fast neutron on a proton fakes the prompt signal and the capture of that neutron on another proton fakes the delayed signal; (iv) ^{9}Li events, in which the spallation product ^{9}Li (half-life 178 ms) decays as $^9\text{Li} \rightarrow 2\alpha + n + \beta^-$ (branching ratio 50%) where the electron and neutron fake the prompt–delayed pair. In this case, the neutron and ^{9}Li are produced from spallation of carbon or other nuclei in the scintillator or surrounding materials by cosmic-ray muons. To better understand and characterize the background we perform simulations for an assumed experimental depth as described in the next section.

2. Muon and Neutron Simulation

To estimate the cosmogenic background described above, we have performed a hybrid simulation for cosmic-ray muons and fast neutrons (Araki, 2005). In this simulation, muon propagation from the ground level to the experimental site (e.g. KamLAND site or assumed shallow site) is simulated using MUSIC/MUSUN code (Antonioli et al., 1997; Kudryavtsev et al., 1999, 2000). After obtaining the energy and angular distribution of muons at the experimental site, FLUKA code (Fasso et al., 2005, Fasso et al., 2003) is employed to simulate neutron production from nuclear spallation by muons. In this step, muons obtained in the MUSIC/MUSUN simulation are injected into rocks and water around the detector, and then spallation of all nuclei in rocks and water are calculated by FLUKA. Among the various spallation products, here we record only neutrons with their energies and emitting angles. The neutron source spectrum at a given experimental site is obtained in this way. Finally, Geant4 simulation (Agostinelli et al. 2003; Allison et al., 2006) shoots those neutrons around the detector and propagates them into the detector.

Firstly we performed the case for KamLAND (depth and detector geometry). In KamLAND, fast neutron events are extracted from the real data by requiring the same delayed coincidence as neutrino events but with an outer detector (OD) hit also required. OD is a water Cherenkov detector (Enomoto, 2006; Araki et al., 2005) and in the fast neutron events the OD hits are from parent muons. By these selection criteria we have chosen fast neutron events whose parent muons passed through the OD. The observed fast neutrons produce 76 events in the prompt energy region between 2.6 and 20 MeV. The energy spectrum is almost flat. Using live time and fiducial volume, this is converted to the fast neutron rate $(1.7 \pm 0.2)\times 10^{-2}$ events d^{-1} kt^{-1} MeV^{-1}. The uncertainty includes statistical error only. In the simulated data with the same selection criteria (parent muon passes through the OD and prompt–delayed signal pair in the inner detector) we obtained the fast neutron rate $(2.0 \pm 0.2)\times 10^{-2}$ events d^{-1} kt^{-1} MeV^{-1}. Although this agreement is very good, attenuation length of neutrons obtained in the data (2.6–20 MeV) is 36.7 cm while that in the simulation is 27.8 cm. Considering that attenuation length for higher energy (2.6–700 MeV) is in good agreement (data 57.6 cm, simulation 58.0 cm), disagreement in lower energy may be because of energy error in the simulation. A likely cause is the calculation of "visible energy" in the simulation, which is very sensitive to the quenching effect of liquid scintillator. In the next section, we employ the conservative value of the attenuation length (37 cm) to estimate fast neutron background.

For verification and possible tuning of our Monte Carlo at shallow sites, we use the data of CHOOZ experiment (Lasserre, 2006; Apollonio et al., 2003). The depth of CHOOZ is around 300 m.w.e., the same as our assumed shallow site. CHOOZ reported their fast neutron rate (Apollonio et al., 2003) as 1.01 ± 0.04 (stat.) ± 0.1 (syst.) events d^{-1} for an energy region between 1.5 and 8 MeV. This corresponds to 31 events d^{-1} kt^{-1} MeV^{-1} assuming flat energy spectrum between 1.5 and 8 MeV, and fiducial mass 5 tons. In the simulation, we got 12.4 events d^{-1} kt^{-1} MeV^{-1}, which is unfortunately more than two times lower than the data. In the simulation, there should be sources of systematic errors such as quenching effect of scintillator, which is at present assumed to be the same as KamLAND scintillator. Currently this "agreement" within a factor of about two is sufficient for order of magnitude estimation, considering that we are exploring a wide range of fast neutron rates (three orders of magnitude difference between KamLAND and CHOOZ).

3. Geoneutrino Experiments at 300 m.w.e

In this section, we explore the requirements for geoneutrino experiments at 300 m.w.e. Figure 1 shows the geoneutrino and current background spectra at KamLAND. As seen, reactor neutrinos and the current $^{13}C(\alpha, n)^{16}O$

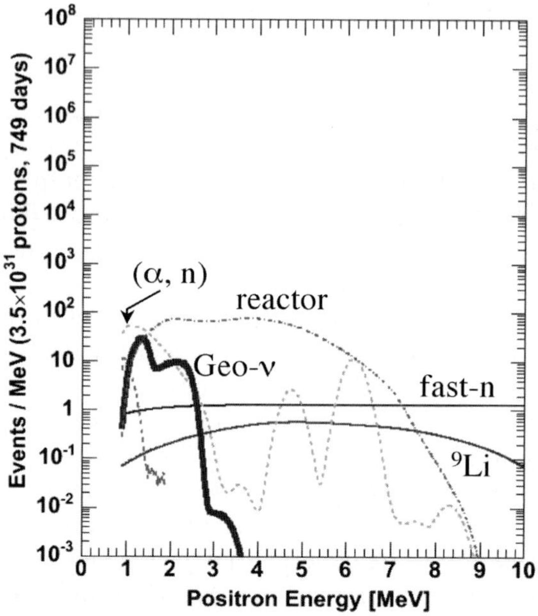

Figure 1. Prompt energy spectra for geoneutrinos and background. All curves are the result of "rate + shape" analysis of the KamLAND data (Enomoto, 2006 Araki et al., 2005).

reaction (Enomoto, 2006; Araki et al., 2005; Enomoto, 2005) are the main sources of background, which are comparable to the geoneutrino signal. Cosmogenic background due to fast neutron and ^{9}Li is not serious. Figure 2 shows the expected spectra with the same detector at 300 m.w.e. This demonstrates that KamLAND placed at 300 m.w.e. does not work because of neutron–neutron, fast neutron, and ^{9}Li background. Figure 3 shows the spectra expected if the detector is modified for operation at shallow depths as described below. Note that this final figure is just a simple estimation as explained in the following paragraphs. This should be a starting point for thinking about the possibility for deploying numerous shallow geoneutrino experiments.

In Figure 2, background sources are estimated as follows. Fast neutron rate is estimated based on the neutron source spectrum at 300 m.w.e. obtained in the simulation as described in the previous section. For the production rate of ^{9}Li, analyses and simulation are being performed (Svoboda, 2006) for the background estimation for Double Chooz (Lasserre, 2006; Double Chooz Collaboration). Here, for simplicity, the production rate of ^{9}Li is assumed to be proportional to muon flux and scaled from KamLAND data. We have to assume that ^{9}Li rejection using track and time correlation with muons is almost useless, because the muon rate is very

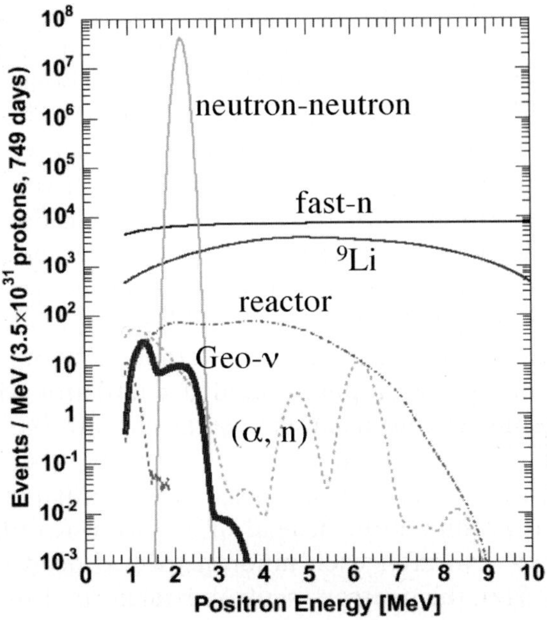

Figure 2. Expected geoneutrino and background spectra if KamLAND is located at 300 m.w.e. Assumed exposure is the same as Figure 1.

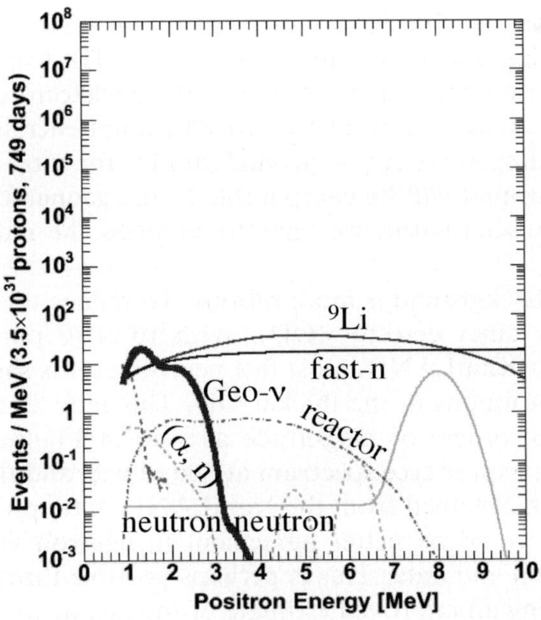

Figure 3. Expected geoneutrino and background spectra (300 m.w.e.) for a KamLAND-type detector but with thicker outer detector and less scintillation light (see text).

high (about 300 Hz) for 3 kt, including "buffer oil" (Enomoto, 2006; Araki et al., 2005) at this depth. Assuming a maximum allowable dead-time fraction of 0.15 sets the muon spallation veto at about 500 μs after each muon, which hardly works against long-lived ^{9}Li ($\tau_{1/2}$ = 178 ms, Q = 13.6 MeV). As seen in Figure 2, another large background is neutron–neutron events because of high muon rates and shorter veto after each muon.

We have to develop methods to reject background. For neutron–neutron background, a realistic method is to make neutron lifetime shorter by loading the scintillator with nuclei whose neutron absorption cross section is large. For example, by loading 0.1% (in mass) of Gd (natural Gd), 0.15% of ^{6}Li, or 0.1% of ^{10}B, one can obtain a neutron lifetime of about 30 μs. There are challenges to be met in developing a loaded scintillator with short neutron capture time including chemical stability, sufficient transparency, radiopurity, and acceptable quenching. Figure 3 shows an example of 0.1% Gd-loaded sintillator, assuming that those properties of the scintillator are satisfied. The neutron lifetime is around 30 μs instead of 200 μs in KamLAND. With this lifetime and 500 μs veto after each muon (expected dead time is 15% for a muon rate of 300 Hz), the neutron–neutron background should be negligible as shown in Figure 3. Assuming Gd-loaded sintillator, the surviving neutron–neutron background should have a peak around 8 MeV (the total energy of gammas from Gd that captured a neutron) with a possible low-energy tail (10% flat tail is assumed in Figure 3). Actually, there should be a subdominant peak at 2.2 MeV due to neutrons captured by protons (Lasserre, 2005). With muon rate 300 Hz, veto of 500 μs, and 30 μs neutron lifetime, the rate of surviving neutron events will be 2×10^{-5} Hz. Coincidences of these neutron events, or other neutrons from outside the detector, are possible with preceding radioactivity at around 1–2 MeV (prompt energies of geoneutrino events). If the radioactivity rate is around 10 kHz for total mass of 1 kt, this coincident background will be comparable to the geoneutrino signal. Then, in developing the scintillator, we have to suppress the radioactivity to less than 0.01 Bq/kg.

Fast neutron background is more serious. To reduce this background, we assume a thicker outer detector (OD). With 10 kt of pure water for OD, instead of 3 kt for KamLAND, most fast neutron events should be identified by tagging parent muons in the thicker OD. This reduces the surviving fast neutrons by three orders of magnitude as seen in Figure 3, based on the simulated fast neutron source spectrum at 300 m.w.e. and the 37-cm neutron attenuation length obtained from the KamLAND data (previous section). A clear refinement would be better agreement in neutron attenuation length between simulation and data. This is perhaps possible through tuning of the simulation allowing an improved estimate of the fast neutron background at 300 m.w.e. in the near future. Another factor is the inefficiency of the OD. At present, we assume 100% efficiency for the 10 kt OD. The efficiency actually needed is about 99.9% because we need fast neutron reduction by three orders of magnitude as seen in Figure 3. This level will be very tough unless we employ a monolithic OD like that of KamLAND. In KamLAND the OD efficiency is found to be 99.8%, which could be improved by lower OD noise rates, because the inefficiency of KamLAND OD is due in part to high rates of PMT dark noise affecting the selection criteria.

Finally, the most difficult background will be ^{9}Li. As described above, a simple veto after each muon does not work against this spallation background at this depth. A 500 μs veto after each muon is too short to reject ^{9}Li (half-life 178 ms) events. However, in KamLAND, about 80% of ^{9}Li events are found after "showering muons" that are defined as muons that deposit more energy than expected from $\mathrm{d}E/\mathrm{d}x$ of relativistic single charged particles. The fraction of showering muons is only about 1/30 of total muons. Assuming the same fraction of showering muons and ^{9}Li events after those muons at 300 m.w.e., a possible effective veto is, for example, 300 ms after each showering muon with only a region of a cylinder of radius 2 m around that muon track. This will result in ~30% of dead time and ~60% rejection of ^{9}Li events based on simple scaling from the KamLAND data. This is still a tough situation. A potential remedy is particle identification (PID) of the prompt events, i.e., possibility to distinguish the prompt signal of neutrino events (e^+) from ^{9}Li events (e^- and proton recoil from neutron scattering). Although a segmented detector is one choice for PID, cost and difficulty of construction at the 1 kt scale could be prohibitive. Also the inner detector should be in simple structure to facilitate the large, monolithic OD that is needed for fast neutron rejection as described above.

Here we consider PID by imaging Cherenkov rings in scintillator. The visible energy of a positron, which includes two 0.511-MeV annihilation gamma rays, should have a smaller fractional Cherenkov contribution than the visible energy of an electron. To observe the Cherenkov contribution, a detector with less scintillation light is considered. The light yield of KamLAND scintillator is about 500 photoelectrons (p.e.) per MeV (photo coverage of 34% including 20" PMT's (Enomoto, 2006). Cherenkov contribution is calculated at about 30 p.e./MeV for relativistic electrons above 3 MeV, although it has not been directly measured because Cherenkov rings are embedded in the isotropic scintillation light. This 30 p.e./MeV is an asymptotic value for relativistic particles. For low-energy positrons with a visible energy 1.4 MeV, which corresponds to a neutrino energy 2.2 MeV being on the lower-energy peak of geoneutrino spectrum (Enomoto, 2006; Araki et al., 2005), the calculated Cherenkov light yield is only 3 p.e./MeV. For this positron, the kinetic energy is 0.4 MeV with the other 1.0 MeV coming from the two 0.511-MeV annihilation gamma rays, which generate few electrons above Cherenkov threshold. For an electron (from ^{9}Li) with the same visible energy 1.4 MeV, the calculated Cherenkov light is 20 p.e./MeV, being as much as 2/3 of the asymptotic value. Since we have to distinguish these 3 and 20 p.e./MeV of Cherenkov contributions, the contribution of scintillation light should be comparable (30 p.e./MeV) with the new scintillator. Then simple statistical consideration shows that the expected rejection power is nearly a factor 100, assuming the solid angle for Cherenkov light emission is 20%. The expected surviving background is shown in Figure 3.

For ^{9}Li decay scheme, a neutron and two alpha particles should also be contributing to the visible energy (Enomoto 2005). If these contributions are very large, the above method for particle identification will not work. This decay mode should then be studied more. The development of scintillator with less light yield is another study. KamLAND scintillator consists of 80% dodecane ($C_{12}H_{26}$), 20% pseudocumene (1,2,4-$(CH_3)_3$-C_6H_3), and 1.5 g/l PPO (2,5-Diphenyloxazole, 2,5-$(C_6H_5)_2$-C_3HNO) (Enomoto, 2006; Araki et al., 2005; Enomoto, 2005). By decreasing the concentration of PPO, less light yield will be achieved. However, 30 p.e./MeV is substantially less than the light yield of KamLAND scintillator (about 1/20). There may be a better solution in a different type of scintillator.

With less light yield, energy resolution is of course worse as demonstrated by the geoneutrino signal spectrum shown in Figure 3. Then it becomes more difficult to distinguish background and signal by using energy spectra. In such a situation background should be further reduced. In Figure 3, a reactor background a factor of 1/100 of the KamLAND level is presented. This should be possible through site selection (1/10 may be sufficient). Also for $^{13}C(\alpha, n)^{16}O$ background, 1/100 of current KamLAND level is assumed in Figure 3, because we believe we have learned a lesson from KamLAND and will never allow ^{222}Rn to sneak into the detector.

4. Research and Development

Geoneutrino observation at 300 m.w.e. is not an easy task. Although it is one order of magnitude shallower, i.e. from 2700 to 300 m.w.e., it corresponds to three orders of magnitude in muon intensity. For research and development of a detector for such shallow experiments, utilization of reactor neutrinos is a possible step. Using the techniques to reject cosmogenic background as described above, reactor neutrino observation could be possible at sites as shallow as a few m.w.e. This could be useful for diagnosis of reactors and measurement of reactor neutrino spectrum before oscillation. From the experience with detectors for reactor physics, one develops, at the same time, techniques for shallow geoneutrino experiments. This would be a realistic and useful step for the future "mass production" of geoneutrino detectors.

Acknowledgements

I am grateful to professors John Learned and Stephen Dye for arranging agenda to insert my talk and kindly waiting for my manuscript, also for carefully reading it. A lot of important things have been pointed out by an unknown referee, including necessary care for radioactivity–neutron background. I appreciate all the organizers for a great opportunity to attend this conference where physicists and geologists gathered for a new science that has just begun.

References

Agostinelli, S. et al.: 2003, *Nucl. Instr. Method A* **506**, 250.
Allison, J. et al.: 2006, *IEEE Trans. Nucl. Sci.* **53**, 270, http://geant4.web.cern.ch/geant4/.
Antonioli, P. et al.: 1997, *Astropart. Phys.* **7**, 357.
Apollonio, M. et al.: 2003, *Eur. Phys. J.* **C 27**, 331.
Araki, T.: 2005, Master thesis (in Japanese), Tohoku University.
Araki, T. et al. (KamLAND collaboration): 2005, *Nature* **436**, 499.
Dye S. et al.: 2006, (this volume) doi: 10.1007/s11038-006-9129-z.
Double Chooz collaboration, http://doublechooz.in2p3.fr/.
Enomoto, S.: 2006, (this volume) doi: 10.1007/s11038-006-9120-8.
Enomoto, S.: 2005, Ph.D. thesis, Tohoku University.
Fasso, A., et al.: 2005, CERN-2005-10, INFN/TC_05/11, SLAC-R-773.
Fasso, A., et al.: 2003, Computing High Energy Nucl. Phys. (CHEP2003), arXiv:hep-ph/0306267, http://pcfluka.mi.infn.it/.
Kudryavtsev, V. A. et al.: 1999, *Phys. Lett. B* **471**, 251.
Kudryavtsev, V. A. et al.: 2000, *Phys. Lett. B* **494**, 175.
Lasserre, T.: 2006, (this volume) doi: 10.1007/s11038-006-9128-0.
Lasserre, T.: 2005, pointed out in this conference.
Suzuki, A.: 2005, in press release at Sendai "More geoneutrino detectors like seismographs in the future".
Svoboda, R.: 2006, for Double Chooz, talk Office High Energy Phys. http://www.science.doe.gov/hep/, see also FERMILAB-Pub-04/180 (2004) (arXiv:hep-ex/0402041).

Earth, Moon, and Planets (2006) 99:275–284
DOI 10.1007/s11038-006-9128-0

Scintillating Oils and Compatible Materials for Next Generation of Electron Anti-neutrino Detectors, After Double Chooz

THIERRY LASSERRE
CEA/Saclay, DSM/DAPNIA/SPP & APC, Paris, France
(E-mail: thierry.lasserre@cea.fr)

(Received 9 March 2006; Accepted 9 October 2006)

Abstracts. The Double Chooz reactor neutrino experiment will be built in the forthcoming years. Eventhough not dedicated to geo-neutrino detection, it is based on similar experimental methods. By pushing current technology to the limits an unprecedented precision will be reached due to careful reduction and control of systematic errors below the percent level. The experience and technical innovation achieved by this project could be valuable for future geo-neutrino experiments. After discussing the Double Chooz detector design we focus on progress achieved on scintillating oils and compatible materials.

Keywords: neutrino, reactor, scintillator

1. From Geoneutrinos to Reactor Neutrinos

Nuclear reactors produce only $\bar{\nu}_e$ through beta decays of the fragments of the fissionable materials. For instance, the fuel for light water reactors (BWR and PWR) mainly consists of ^{235}U and ^{239}Pu, which undergo thermal neutron fission. The dominant natural uranium isotope, ^{238}U, is fissile only for fast neutrons (threshold of 0.8 MeV) but it also generates fissile ^{239}Pu by thermal neutron capture. The ^{241}Pu isotope is produced in a manner similar to ^{239}Pu. Fuel composition evolves with time (burn-up).

Reactor antineutrinos are detected through the inverse neutron decay $\bar{\nu}_e + p \rightarrow e^+ + n$ (threshold of 1.8 MeV), with a cross section $\sigma(E_{e^+}) \simeq (2\pi^2\hbar^3)/(m_e^5 f\tau_n) p_{e^+} E_{e^+}$, where p_{e^+} and E_{e^+} are the momentum and the energy of the positron, τ_n is the neutron lifetime and f is the neutron decay phase space factor. For typical PWR averaged fuel composition (^{235}U (55.6%), ^{239}Pu (32.6%), ^{238}U (7.1%), ^{241}Pu (4.7%)) and a thermal power P_{th} (GW$_{th}$), the number of fissions per second N_f is given by $N_f = 3.06 \cdot 10^{19} s^{-1} P_{th}[GW]$. The energy-weighted cross section amounts to $<\sigma>_{fis} = 5.825 \cdot 10^{-43}$ cm^2 per fission. The event rate at a distance L from the source, assuming no oscillations, is given by $R_L = N_f <\sigma>_{fis} n_p \cdot 1/(4\pi L^2)$, where n_p is the number of (free) protons in the target.

Experimentally one detects the very clear signature of the coincidence signal of the prompt positron followed in space and time by the delayed neutron capture. The same technique is used to detect geo-neutrinos. This allows a strong rejection of accidental background. The visible energy seen in the detector is given by $E_{vis} = E_{e^+} + 511$ keV. Reactor neutrino experiments measure the survival probability $P_{\bar{\nu}_e \to \bar{\nu}_e}$ of the $\bar{\nu}_e$ emitted by nuclear power stations. This effect has to be considered to properly interpret geo-neutrino data. The $\bar{\nu}_e$ survival probability can be written

$$
\begin{aligned}
P_{\bar{\nu}_e \to \bar{\nu}_e} = {} & 1 - 4\sin^2\theta_{13}\cos^2\theta_{13}\sin^2\frac{\Delta m_{31}^2 L}{4E} \\
& - \cos^4\theta_{13}\sin^2(2\theta_{12})\sin^2\frac{\Delta m_{21}^2 L}{4E} \\
& + 2\sin^2\theta_{13}\cos^2\theta_{13}\sin^2\theta_{12} \\
& \left(\cos\frac{(\Delta m_{31}^2 - \Delta m_{21}^2)L}{2E} - \cos\frac{\Delta m_{31}^2 L}{2E}\right),
\end{aligned}
\tag{1}
$$

where the first two terms in Equation (1) contain respectively the atmospheric ($\Delta m_{31}^2 = \Delta m_{atm}^2$) and solar driven ($\Delta m_{21}^2 = \Delta m_{sol}^2$) oscillations. The third term is an interference between solar and atmospheric contributions, which has a detectable influence only in a small region of the space of mass and mixing parameters. For the considered case the first term dominates, which leads to a pure θ_{13} measurement.

2. Introduction to Double Chooz

The goal of the Double Chooz experiment (Ardellier et al., 2004; Berridge et al., 2004) is to search for a non-vanishing value of the θ_{13} neutrino mixing angle. This is the last step to accomplish prior to moving towards a new era of precision measurements in the lepton sector. The most stringent constraint on this mixing angle comes from the Chooz reactor neutrino experiment with $\sin^2(2\theta_{13}) < 0.2$ (90% C.L.). Double Chooz will explore the range of $\sin^2(2\theta_{13})$ from 0.2 to 0.03–0.02 (90% C.L.), within 3 years of data taking. The improvement of the CHOOZ result requires an increase in the statistics, a reduction of the systematic error below one percent, and a careful control of background. Therefore, Double Chooz will use two identical detectors, one at 300 m and another at 1.05 km distance from the Chooz nuclear cores. In addition, we will use the near detector to investigate the potential of neutrinos for monitoring civil nuclear power plants. The Double Chooz collaboration is composed of institutions from France, Germany, Italy, Spain, Russia and USA. The plan is to start operation in 2008 with the far detector and to have both detectors operating by the end of 2009. With such a scenario Double Chooz will reach a $\sin^2(2\theta_{13})$ sensitivity of 0.07 after one year of operation with one detector (far), and 0.03–0.02 after 3 years of operation with both detectors. These estimates are based on the assumptions that the relative normalization error between the near and far detectors could be kept at 0.6%, and that the background subtraction error at both sites amounts to about 1%.

To achieve such a sensitivity, all state-of-the-art technologies of the field have to be used. In the article we review the Double Chooz detector design and technologies, focusing on the scintillating oils and compatible materials. Such developments could benefit future geo-neutrino detectors.

3. Systematic Uncertainties and Backgrounds

In the first CHOOZ experiment, the total systematic error amounted to 2.7%. A summary of the CHOOZ systematic errors is given in Table I (Apollonio et al., 2003). The right column resents the new experimental goals. Entries 1, 4, and 5 correspond to systematic uncertainties related to the reactor flux and the cross section of neutrinos on the target protons. These errors become negligible if one uses two antineutrino detectors located at different baselines. The analysis cuts are potentially important sources of systematic error. In the first CHOOZ experiment, this amounted in total to 1.5% (Apollonio et al., 2003). The goal of the new experiment is to reduce this error by a factor of three. The CHOOZ experiment used seven analysis cuts to select the $\bar{\nu}_e$. In Double Chooz we plan to reduce the number of selection cuts to three (one of them will be very loose, and may not even be used). This can be achieved because of the reduction of the number of accidental background events, only possible with the new detector design thanks to the addition of the non-scintillating buffer region. Such a buffer is very important for reducing the singles rate, especially in a large scintillator detector. To select $\bar{\nu}_e$ events we have to identify the prompt positron followed by the delayed neutron (delayed

TABLE I
Overview of the systematic errors of the CHOOZ and Double Chooz experiment

	CHOOZ (%)	Double Chooz (%)
Reactor cross section	1.9	–
Number of protons	0.8	0.2
Detector efficiency	1.5	0.5
Reactor power	0.7	–
Energy per fission	0.6%	–

in time and separated in space). The trigger will require two local energy depositions of more than 500 keV in less than 200 μs.

4. The Detector

The Double Chooz detectors will consist of a target cylinder of 1150 mm radius and 2800 mm height, providing a volume of 10.3 m^3. The near and far detectors will be identical inside the PMTs supporting structure. This will allow a relative normalization systematic error of 0.6%. Starting from the center of the target the detector elements are as follows (see Figure 1).

4.1. Target and gamma-catcher vessels

Target and gamma-catcher vessels will be built with acrylic plastic material, transparent to photons with wavelengths above 400 nm. Both vessel are designed to contain the target and gamma-catcher aromatic liquids with a long term tightness (no leak for 10 years) and stability. The strongest constraint is the chemical compatibility between the vessel and the scintillating liquids of the target and gamma-catcher, for at least

Figure 1. The Double Chooz far detector, at the Chooz underground site. The detector is located in the tank used for the CHOOZ experiment (7 m high and 7 m in diameter) that is still available. 10.3 m^3 of a dodecane + PXE-based liquid scintillator doped with gadolinium is contained in a transparent acrylic cylinder surrounded by the gamma-catcher region (22.6 m^3) and the buffer (114.2 m^3). The design goal is to achieve a light yield of about 200 pe/MeV which requires an optical coverage of about 13%, provided by the surrounding PMTs. The PMTs are mounted on a cylindrical structure which separates optically the outer part of the detector, which is used as a muon veto (90 m^3).

5 years. We tolerate neither a modification of the liquid properties (scintillation, absorbency) nor a degradation of the acrylic material (breaking or crazing over more than a few percent of the acrylic surface area). The gamma-catcher vessel also has to be chemically compatible with the mineral oil of the buffer region, which is known to be a weaker constraint. Target and gamma-catcher vessels will be made of cast acrylic. The assembly will be done by gluing pieces according to the manufacturer expertise. The target vessel is a cylinder 2800 mm in height, 2300 mm in diameter, and 8 mm thick. It has a mass of 350 kg and contains a volume of 10.3 m^3 (without the chimney). A 55-cm buffer of non-loaded liquid scintillator with the same optical properties as the $\bar{\nu}_e$ target (light yield) surrounds the target vessel. This scintillating buffer around the target is necessary to measure gammas from neutron capture on Gd, to measure positron annihilation, and to reject background from fast neutrons (Ardellier et al., 2004). The double acrylic vessel is displayed in Figure 2. Simulation has shown that the truck transportation phase is hazardous for a double acrylic vessel completely assembled by glue (vibration eigen frequencies of 8 Hz). To avoid any resonance problem, the eigen frequency of the structure must be greater than 10 Hz (decoupling). This problem can be solved by transporting the target and gamma-catcher vessels separately, integrating them in the pit, and gluing the gamma-catcher top lid and the chimney in the neutrino laboratory. During the 24 h of the polymerization of the glue, the temperature has to be kept

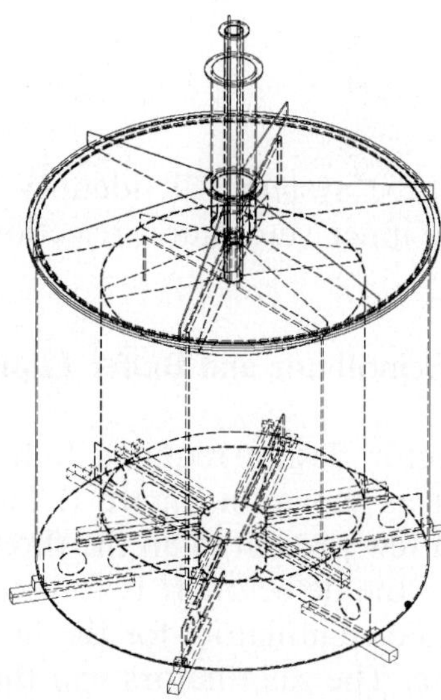

Figure 2. Viewgraph of the double acrylic vessel (8-mm thick) that encloses the neutrino target filled with 10.3 m^3 of a Gd-doped PXE/dodecane scintillator, and the gamma-catcher region that contains 22.6 m^3 of a Gd-free PXE/dodecane scintillator.

above 25 °C and the relative humidity at 40%, whereas nominal conditions in the tunnel are 14 °C and 100% humidity.

4.2. Non-scintillating buffer

A 105-cm region of non-scintillating liquid outside the gamma-catcher acts as a buffer zone to reduce the level of accidental background (mainly the contribution from photomultiplier tubes radioactivity). This region is crucial to keep the singles rate below 10 Hz in the sensitive region (target + gamma-catcher).

4.3. Buffer vessel and PMT support structure

This vessel is made of 3-mm thick stainless steel sheets and stiffeners. It contains 534 (8-in.) photomultiplier tubes mounted uniformly on the surface.

4.4. Inner veto system

The inner veto system is a 50-cm thick region region filled with liquid scintillator for the far detector, and a slightly larger thickness (~60 cm) for the near detector.

4.5. Outer veto system

A tracker proportional tube system will identify and locate "near-miss" muons. This improves the inner veto muon rejection by a factor 20.

5. Scintillator and Buffer Liquids

The Double Chooz detector design requires four different liquids in the separate detector volumes as shown in Figure 1. The densities of the liquids should be similar within a few percent in all the three inner volumes to avoid strong buoyancy forces in the detector. It is foreseen to do the purification and the mixing of the liquid scintillators for the target region as well as for the gamma-catcher off-site. The scintillators will then be transported to the Chooz site in transport tanks. On-site the tanks will be hooked up to dedicated filling systems. All three volumes have to be filled simultaneously, with a level control of a few centimeters. Material compatibility tests have been performed especially for the Gd-LS. All materials in contact with the Gd-LS in the detector or the liquid handling systems have to be checked thoroughly. To test the material compatibility with acrylic a mock up of the Double Chooz detectors was built (1/5th scale). The inner volume of this mock up contains about 110 l of Gd-LS (see Figure 3). The selection of the organic liquids are guided by physical and technical requirements, as well as by safety considerations. We review below the four different volumes.

The inner most volume of 10.3 m^3, the $\bar{\nu}_e$-target, contains a proton rich liquid scintillator mixture loaded with gadolinium (Gd-LS) at a concentration of approximately 1 g/l. The solvent is a PXE (phenyl-xylylethane)/dodecane mixture at a volume ratio of 20:80. The admixture of the dodecane reduces the light yield, but it improves the chemical compatibility with the acrylic and increases the number of free protons in the target. Besides technical requirements the solvent mixture was selected for safety considerations. In particular, both components have high flash points (PXE: fp 145 °C, dodecane: fp 74 °C). The scintillation yield of the unloaded PXE-based scintillator was measured as a function of dodecane concentration (see Figure 4). A scintillation yield of 78% with respect to pure PXE is observed at a volume fraction of 80% dodecane. As primary and secondary fluor we use PPO (6 g/l) and bis-MSB (ca. 20 mg/l). The achieved light yield is sufficient for the requirements of the Double Chooz experiment. Metal loading of liquid scintillators has been comprehensively studied at MPIK (Hartmann and Naumann, 1985, 1992; Buck et al., 2003; Hartmann, 2003) and LNGS/INR (Danilov et al., 2003; Cattadori, submitted) for several years. Research

Figure 3. Double Vessel acrylic mockup, scale 1/5th of the Double Chooz detector.

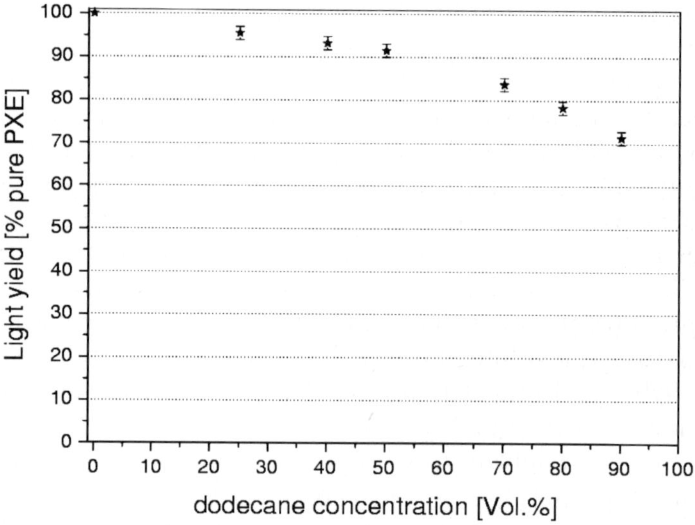

Figure 4. Scintillation light yield of PXE/dodecane mixture with varying dodecane concentration. The PPO concentration is kept constant at 6 g/l.

with gadolinium-loaded scintillators in both institutes indicates that suitable scintillators can be produced. Two scintillator formulations are investigated, one based on carboxylic acids and the other on Gd-β-diketonates. Both systems show good performance and are viable candidate liquid scintillators for the $\bar{\nu}_e$-target. Improvements concerning the light yields appear to be possible by optimization of the fluor choice and concentration. Attenuation lengths of more than 10 m were measured for Gd-doped solutions around 430 nm, the region of the scintillator emission.

The adjacent region, the gamma-catcher, has a volume of 22.6 m^3 and is filled with a Gd-free scintillator to detect gammas resulting from the neutrino interaction that escape from the target volume. Similar requirements as for the target liquid concerning material compatibility, density and optical properties have to be applied. The gamma-catcher scintillator has to match the density of the target scintillator which is about 0.80 within 1%. Since the Gd-loading does not affect the density of the liquid significantly, the same PXE/dodecane ratio in the gamma-catcher as in the target would fulfill the density requirements. Furthermore, the light yield of the gamma-catcher scintillator should be similar to the light yield of the Gd-scintillator. There is some quenching in the Gd-scintillator compared to the unloaded version and the light yield of the gamma-catcher scintillator would therefore be 10–20% higher than in the target. One option to adjust the light yield is to lower the aromatic fraction in the gamma-catcher. The attenuation length in the gamma-catcher in the wavelength region of interest is comparable to or higher than the target and the stability of the optical properties is not viewed as a problem, since this scintillator has no metal loading.

The photomultipliers are immersed in a non-scintillating buffer in order to shield the active volume from gamma rays. The volume of the buffer liquid is 114.2 m^3. This volume should shield the active volume from gamma rays emitted by the photomultipliers, external gammas, and inner material radioactivity. Since there is a thin acrylic vessel between the gamma-catcher and the buffer this liquid has to match the density of the scintillators. Additional requirements are material compatibility with acrylic and the photomultipliers and high transparency in the region of the scintillator emission. It is foreseen to use pure mineral oil in this region. Optically pure mineral oil with lower density has to be found or the mineral oil has to be mixed with dodecane (density 0.749).

Last, an instrumented volume of approximately 90 m^3 encloses the whole setup serving as a shield against external radiation and as a muon veto system. The veto volume contains low-scintillating organic liquid viewed by about 100 PMTs.

Table II summarizes the liquid inventory of a single detector system.

6. Material Compatibility

The three inner volumes, target, gamma-catcher and buffer are in contact with acrylic, therefore the material compatibility of the respective liquids with the acrylic used in the experiment is crucial. The compatibility of several PXE/dodecane mixtures, mineral oils (mixture of alkanes), and of both Gd-scintillator versions with acrylic have been tested. Two kinds of tests have been performed: acrylic without stress in contact with liquids for 30 days at 20 and 50 °C, and acrylic under stress (flexion, up to 30 MPa) for 24 h in order to determine the constraint at which crazing appears. Results indicate that flexion up to a few MPa is tolerable for the experiment. These measurements confirmed the compatibility with mineral oil, as well as the improving compatibility with increasing dodecane concentrations. This was

TABLE II
Overview of liquid inventory for a single detector

Labeling	Volume (m^3)	Type
$\bar{\nu}_e$-Target	10.3	Gd loaded LS (0.1%)
Gamma-catcher	22.6	Unloaded LS
Buffer	114.2	Non-scintillating organic liquid
Veto	110	Low-scintillating organic liquid

expected because pure PXE is known to induce hard crazing in acrylic materials. The PXE/dodecane ratio of 20:80 by volume is a good compromise providing sufficient material compatibility and scintillator light yield. The PXE concentration in the target scintillator could be lowered to about 15% to improve the safety margin, since the loss of light would be only a few percent (light output decreases, but transparency of oil is better than that of PXE). No significant compatibility difference was observed between the different Gd-loaded scintillators. There are also other materials in contact with the scintillators during the measurement (calibration system), handling and purification. Steel surfaces should be avoided for the target scintillator.

Acrylic-scintillator compatibility would have been demonstrated for a 'single piece' acrylic vessel. However the selection of the glue as well as the gluing technique to be used, on and off site, are critical, especially at the weak points of the vessel mechanical structure (chimney, lids).

7. Conclusion

Double Chooz will lay the foundation for future experiments through the development of innovative technologies (scintillators, material compatibility, detector inter-calibration, etc.). These developments will contribute to the development of applied neutrino physics, as the geo-neutrino detection.

References

Apollonio, M. et al. (CHOOZ Collaboration): 2003, *Eur. Phys. J.* **C27**, 331.

Ardellier, F. et al. (Double Chooz Collaboration): 2004, hep-ex/0405032.

Berridge, S. et al. (Double Chooz Collaboration): 2004, hep-ex/0410081.

Buck. C., Hartmann, F. X. , Motta, D., Schönert, S. and Schwan, U.: 2003, Metal beta-diketone scintillators, Presentation at the workshop on Future Low Energy Neutrino Experiments. TU Munich, Munich 9–11 October.

Danilov, N. A., Krylov, Yu. S. and Cattadori, C.: 2003, *Radiochemistry* **45**(2), 128.

Hartmann, F. X.: 2003, Low level scintillators and gadolinium, Presentation at the Workshop on Future Low Energy Neutrino Experiments. TU Munich Munich 9–11 October.

Hartmann, F. X. and Naumann, R. A.: 1992, *Nucl. Inst. Methods Phys. Res.* **A313**, 237.

Hartmann, F. X. and Naumann, R. A.: 1985, *Phys. Rev. C. (Rapid Comm.)* **31**, 1594–1596.

Earth, Moon, and Planets (2006) 99:285–307
DOI 10.1007/s11038-006-9101-y

Neutrino Tomography – Learning About The Earth's Interior Using The Propagation Of Neutrinos

WALTER WINTER
Institute for Advanced Study, School of Natural Sciences, Einstein Drive, Princeton, NJ, 08540, USA
(E-mail: winter@ias.edu)

(Received 2 February 2006; Accepted 28 June, 2006)

Abstract. Because the propagation of neutrinos is affected by the presence of Earth matter, it opens new possibilities to probe the Earth's interior. Different approaches range from techniques based upon the interaction of high energy (above TeV) neutrinos with Earth matter, to methods using the MSW effect on the oscillations of low energy (MeV to GeV) neutrinos. In principle, neutrinos from many different sources (sun, atmosphere, supernovae, beams etc.) can be used. In this talk, we summarize and compare different approaches with an emphasis on more recent developments. In addition, we point out other geophysical aspects relevant for neutrino oscillations.

Keywords: Matter effects, neutrino absorption, neutrino attenuation, neutrino oscillations, neutrino tomography

1. Introduction

Neutrinos are elementary particles coming in three active (i.e. weakly interacting) flavors. Since the cross sections for neutrino interactions are very small, neutrinos practically penetrate everything. However, one can compensate for these tiny cross sections by just using enough material in the detector. Depending on neutrino energy and source, the detector has to be protected from backgrounds such that the neutrino events cannot be easily mixed up with different particle interactions. Neutrinos are produced in detectable numbers and with detectable energies by nuclear reactions in the sun, by cosmic ray interactions in the Earth's atmosphere, in nuclear fission reactors, in supernova explosions, in the Earth's crust and mantle, and possibly by astrophysical sources. In addition, accelerator-based neutrino sources specifically designed to produce a high-intensity neutrino beam have been successfully operated (such as K2K (Aliu et al., 2005) or MINOS) or are planned. Thus, there are neutrinos from various different sources with different energies.

One of the most recent exciting discoveries in neutrino physics is neutrino oscillations, i.e., neutrinos change flavor while traveling from source to detector. This quantum mechanical phenomenon implies that neutrinos mix, i.e., the eigenstates of the weak interaction are not the same as the mass eigenstates, and at least two out of the three have non-vanishing masses. This is probably the most direct evidence today for physics beyond the standard model of elementary particle physics. Recent neutrino oscillation experiments, especially SNO (Ahmad et al., 2002), KamLAND (Eguchi et al., 2003), Super-Kamiokande (Fukuda et al., 1998), and CHOOZ (Apollonio et al., 1999) have helped to quantify this picture. Unlike in quark mixing, two out of the three mixing angles are large, and one is even close to maximal. In addition, the oscillation frequencies have been fairly precisely measured. For one of the mixing angles θ_{13}, however, only an upper bound exists, and several parameters (the arrangement of masses, i.e., mass hierarchy, and one complex phase δ_{CP} which allows neutrinos and antineutrinos to oscillate differently) are still unknown. Future experiments will probe these parameters starting with the Double Chooz (Ardellier et al., 2004), T2K (Itow et al., 2001), and NOνA (Ayres et al., 2004) experiments (for the prospects for the next decade, see, e.g., Huber et al., 2004).

For neutrino tomography the relevant aspect is the sensitivity to Earth matter. Since it is well known that the cross sections with matter rise at least until 10 TeV (Quigg et al., 1986), the probability of matter interactions can be increased by higher neutrino energies. Neutrino absorption tomography uses this effect to infer on the matter structure. For neutrino oscillations, we know that the so-called MSW matter effect (Wolfenstein, 1978; Mikheev and Smirnov, 1985, 1986) is the most plausible explanation for the solar neutrino deficit (Fogli and Lisi, 2004). This, however, implies that neutrino oscillations in the Earth have to experience this effect, too. Neutrino oscillation tomography uses the MSW effect to study the matter structure.

2. Tomography Using the Propagation of Neutrinos

Tomography using the propagation of neutrinos (Nedyalkov, 1981a, b) assumes a neutrino source with a well-known flux and flavor composition, a well-understood neutrino detector, and a specific neutrino propagation model between source and detector. The key ingredient to any such tomography is a considerable dependence of the propagation model on the matter structure between source and detector. Compared to the detection of geoneutrinos, the object of interest is not the neutrino source, but the material along the baseline (path between source and detector). If the matter structure along the baseline is (partly) unknown, the information from counting neutrino events at different energies by the detector can be used to infer on the matter profile. Two accepted propagation models could be used for neutrino tomography:

Neutrino absorption. Because the cross section for neutrino interactions increases proportional to the energy, neutrino interactions lead to attenuation effects. Useful neutrino energies for a significant attenuation are $E_\nu \gtrsim 1$ TeV.

Neutrino oscillations. The MSW effect (Wolfenstein, 1978; Mikheev and Smirnov, 1985, 1986) in neutrino oscillations (coherent forward scattering in matter) leads to a relative phase shift of the electron flavor compared to the muon and tau flavors. This phase shift depends on the electron density. Useful neutrino energies require substantial contributions from the MSW effect as well as large enough oscillation amplitudes. Depending the relevant Δm^2 (see Eq. 2 below), neutrino energies between 100 and 35 GeV are optimal for studying the Earth's interior.

Beyond these two models, at least small admixtures of non-standard effects have not yet been excluded. Some of these non-standard effects are sensitive to the matter density, too. Examples are mass-varying neutrinos with acceleron couplings to matter fields (Kaplan et al., 2004), non-standard neutrino interactions (see Huber et al., 2002b and references therein), and matter-induced (fast) neutrino decay (Giunti et al., 1992). Because there is not yet any evidence for such effects, we do not include them in this discussion.

Given the above neutrino energies, there are a number of potential sources which could be used for neutrino propagation tomography. For neutrino oscillations, solar neutrinos, supernova neutrinos, atmospheric neutrinos, and neutrino beams (such as superbeams or neutrino factories) are potential sources. For neutrino absorption, high-energy atmospheric neutrinos, a possible high-energy neutrino beam, or cosmic sources are possible sources.

As far as potential geophysics applications are concerned, neutrinos may be interesting for several reasons:

1. Neutrinos propagate on straight lines. The uncertainty in their path (direction) is only as big as the surface area of the detector.
2. Neutrinos are sensitive to complementary quantities to geophysics: Neutrino absorption is directly sensitive to the matter density via the nucleon density. Neutrino oscillations are sensitive to the electron density which can be converted to the matter density by the number of electrons per nucleon (for stable "heavy" materials about two). On the other hand, seismic wave geophysics needs to reconstruct the matter density by the equation of state from the propagation velocity profile.
3. Neutrinos are, in principle, sensitive to the density averaged over the baseline, whereas other geophysics techniques are, in principle, less sensitive towards the innermost parts of the Earth. For example, seismic shear waves cannot propagate within the outer liquid core, which means that a substantial fraction of the energy deposited in seismic waves is

reflected at the mantle-core boundary. Other direct density measurements by the Earth's mass or rotational inertia are less sensitive towards the innermost parts, because they measure volume-averaged quantities.

Given these observations, there may be interesting geophysics applications exactly where complementary information is needed. Possible applications range from the detection of density contrasts in the Earth's upper or lower mantle, to the measurement of the average densities of the outer and inner core by independent methods.

3. Neutrino Absorption Tomography

Here we discuss tomography based on attenuation effects in a neutrino flux of high enough energies, which we call, for simplicity, "neutrino absorption tomography." After we have introduced the principles, we will discuss possible applications with respect to tomography of the whole Earth as well as specific sites.

3.1. Principles

"Neutrino absorption tomography" uses the attenuation of a high-energy neutrino flux as a propagation model. In this case, weak interactions damp the initial flux by the integrated effect of absorption, deflection, and regeneration. For example, muons produced by a muon neutrino interaction are absorbed very quickly in Earth matter, whereas tauons produced by tau neutrinos tend to decay before absorption (and some of the decay products are again neutrinos). Only the integrated effect leads to attenuation of the flux. The magnitude of the attenuation effect can be estimated from the cross section

$$\frac{\sigma}{E} \sim 10^{-35} \frac{\text{cm}^2}{\text{TeV}} \tag{1}$$

to be of the order of several percent over the Earth's diameter for $E_\nu = 1$ TeV. The interaction cross section rises linearly up to about 10 TeV (Quigg et al., 1986), whereas the behavior above these energies is somewhat more speculative. The energies are usually as high as standard neutrino oscillations do not develop within the Earth. Since the neutrinos interact with nucleons, the attenuation is directly proportional to the nucleon density. Therefore, neutrino absorption is a very directly handle on the matter density with an extremely tiny remaining uncertainty from composition and the difference between neutron and proton mass.

As far as potential neutrino sources are concerned, Eq. (1) requires very high neutrino energies. The existence of corresponding neutrino sources is plausible and will be tested by upcoming experiments commonly referred to as "neutrino telescopes." These neutrino telescopes could also serve as prototypes for the detectors useful for neutrino absorption tomography.

The only detected source so far is atmospheric neutrinos produced by the interaction of cosmic rays in the atmosphere. Unfortunately, the atmospheric neutrino flux drops rapidly with energy, which means that statistics are limited at the relevant energies $E_\nu > 1$ TeV (see, e.g., Gonzalez-Garcia et al., 2005 for specific values). Other potential candidates include many different possible astrophysical objects, as well as particle physics mechanisms such as decays of dark matter particles. Since we know that the Earth is hit by cosmic rays of very high energies, it might be inferred that astrophysical mechanisms exist which accelerate particles (for example, protons) to these high energies. It is plausible that such mechanisms also produce neutrinos. Potential mechanisms could either produce discrete fluxes from individual objects, or their integrated effect could lead to a diffuse flux over the whole sky. Eventually, one could think about a neutrino beam producing high-energy neutrinos. If, for instance, one used the protons from LHC (7 TeV) to hit a target, the decaying secondaries (pions, kaons) would produce a neutrino flux peaking at about 1 TeV.

3.2. Whole earth tomography

For possible applications of neutrino absorption tomography, there exist two different directions in the literature: Either one could "X-ray" the whole Earth ("Whole Earth tomography"), or one could think about the investigation of specific sites in the Earth's mantle. We summarize in Figure 1 different approaches to "Whole Earth tomography." In case (a) (isotropic flux) a neutrino flux from many directions is detected by a detector with good directional resolution. For instance, a possible neutrino source would be a cosmic diffuse flux (Jain et al., 1999) (related work: Reynoso and Sampayo, 2004) or the high-energy tail of atmospheric neutrinos (see, e.g., Gonzalez-Garcia et al., 2005). This application could be very interesting because it might be available at no additional experimental effort. However, if one wants to study the innermost parts of the Earth, it is (except from sufficient directional resolution and flux isotropy) a major challenge that the fraction of the sky which is seen through the Earth's inner core are very small (~1%), which means that the statistics for this specific goal are very low. Very good precisions may, on the other hand, be obtained for the mantle (see Figure 4 of Jain et al., 1999). In case (b) (high-energy neutrino beam)

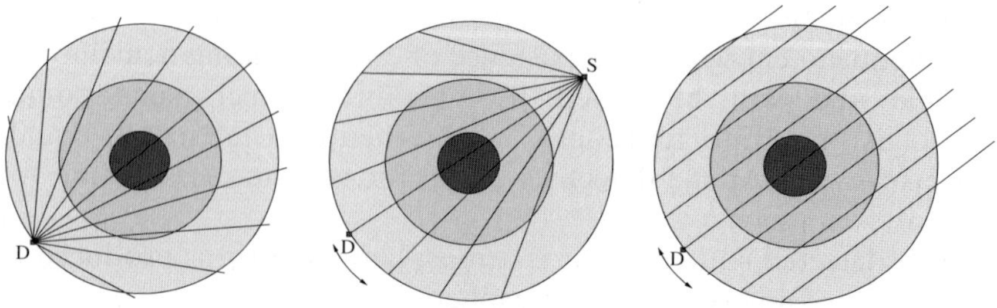

a) Isotropic flux **b)** High-energy neutrino beam **c)** Cosmic point source

Figure 1. Three different approaches to "Whole Earth Tomography" using neutrino absorption. The lines refer to different baselines.

(De Rujula et al., 1983; Wilson, 1984; Askar'yan, 1984; Borisov et al., 1986; Borisov and Dolgoshein, 1993) the detector is moved to obtain many baselines, whereas the source is kept fixed. In this case, high precision could be obtained (De Rujula et al., 1983). However, a major challenge might actually be the operation of a high-energy neutrino beam with a moving decay tunnel. Note that such a beam could not only be used for whole Earth tomography, but also for local searches (see below). In case (c) (cosmic point source) (Wilson, 1984; Kuo et al., 1995), the flux from a single object is used for the tomography of the Earth. In this case, the flux has to be constant in time to be detected either by a moving detector, or by one detector using many baselines by the rotation of the Earth. Note that the second mechanism cannot be used for the currently largest planned neutrino telescope "Ice-Cube" (Ahrens et al., 2004) because it is residing at the south pole.

3.3. Specific site tomography

Compared to "whole Earth tomography," a different direction is the investigation of individual sites, such as in the Earth's mantle. For example, De Rujula et al. (1983) extensively reviews techniques based on a high-energy

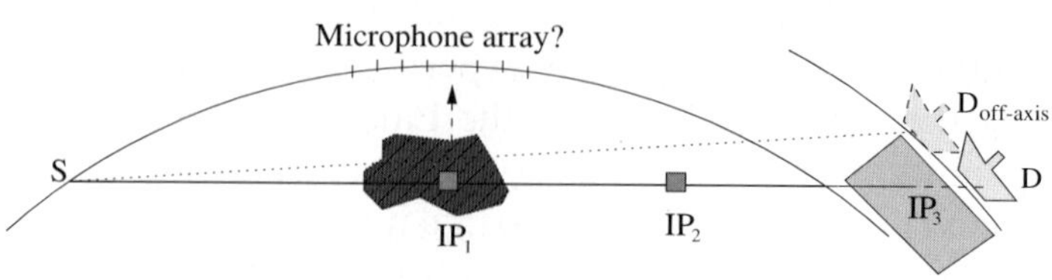

Figure 2. Different possibilities for neutrino tomography using a high-energy neutrino beam. The labels "IP" refer to possible interaction points. See text for more details.

neutrino beam. We summarize some of these in Figure 2. The neutrinos, produced by the source "S," may interact at several possible interaction points IP. If, for example, the site of interest is the dark-shaded cavity, an interaction at IP_1 could create a particle shower leading to sound production, which may be detected by a microphone array at the surface. In addition, the final neutrino flux detected at "D" would be damped depending on the material density in the cavity. An interaction at IP_2 just below the surface (200 m) would produce muons which could still be detected at the surface (such as possibly by a muon detector on a truck). A variation of this flux detected by a moving muon detector could point towards heavy materials. Eventually, a neutrino interaction at IP_3 within the sea water below a moving muon detector would indicate that the initial neutrino has arrived. Since the neutrino energy decreases rapidly by moving the detector out of the beam axis by kinematics, attenuation effects also decrease and the initial flux could be measured by the "off-axis" technology. Comparing this flux to the on-axis flux reveals the attenuation along the path and therefore some information on the matter structure.

In summary, there are many potential applications of neutrino absorption tomography. The coming years, especially the operation of IceCube, will reveal the possible existence of high-energy cosmic neutrino fluxes. Operating a high-energy neutrino beam may be a major technical challenge, which definitely needs further investigation.

4. Neutrino Oscillation Tomography

In this section, we discuss neutrino tomography using oscillations. First, we introduce the principles of neutrino oscillation tomography: Neutrino oscillations in vacuum and matter, numerical approaches to neutrino oscillation tomography, as well as conceptual (mathematical) problems. Then, we show applications related to solar and supernova neutrinos, and we discuss tomography with neutrino beams.

4.1. Principles

Neutrino oscillation tomography uses neutrino oscillations in matter as propagation model. Possible neutrino sources include "natural" ones (e.g., sun, supernova, atmosphere), as well as "man-made" ones (e.g., superbeam, β-beam, neutrino factory). The detection technology depends on the neutrino energy and ranges from water Cherenkov detectors (lower energies), to liquid scintillators (medium energies), to iron calorimeters (high energies), just to mention some examples.

4.1.1. *Neutrino oscillation phenomenon*

Neutrino oscillations are a quantum mechanical phenomenon with two prerequisites: First, the weak interaction eigenstates have to be different from the propagation/mass eigenstates (flavor mixing). Second, the neutrino masses have to be different from each other, which implies that at least two of the active neutrinos have to have non-zero mass (Bilenky and Pontecorvo, 1978). In the limit of two flavors, the flavor transition probability $\nu_\alpha \rightarrow \nu_\beta$ in vacuum can be written as

$$P_{\alpha\beta} = \sin^2(2\theta)\sin^2\left(\frac{\Delta m^2 L}{4E}\right), \tag{2}$$

where θ is the mixing angle of a 2×2 rotation matrix U, $\Delta m^2 \equiv m_a^2 - m_b^2$ is the mass-squared difference describing the oscillation frequency, L is the baseline (distance source-detector), and E is the neutrino energy. Note that the quotient L/E determines the oscillation phase. Similarly, the flavor conservation probability $\nu_\alpha \rightarrow \nu_\alpha$ is given by $P_{\alpha\alpha} = 1 - P_{\alpha\beta}$ from conservation of unitarity. Practically, $P_{\alpha\beta}$ is measured as function of E (convoluted with the neutrino flux and cross sections) for a fixed baseline since the detector cannot be moved. Since we do know that we deal with three active flavors, the complete picture is somewhat more complicated. Three-flavor neutrino oscillations can be described by six parameters (three mixing angles, one complex phase, and two mass squared differences), which decouple into two-flavor oscillations, described by two parameters each, in certain limits (see Fogli et al., 2005 for a recent review). In summary, we have two almost decoupled two-flavor oscillations described by two very different frequencies and large mixing angles, often referred two as "solar" (Δm_{21}^2, θ_{12}) and "atmospheric" (Δm_{31}^2, θ_{23}) oscillations. Those could be coupled by θ_{13}, for which so far only an upper bound $\sin^2(2\theta_{13}) < 0.1$ (Apollonio et al., 1999) exists. In addition, we do not yet know anything about the complex phase δ_{CP}, which could lead to sub-leading effects, and the sign of Δm_{31}^2 ("mass hierarchy"). These parameters will be probed by neutrino oscillation experiments in the coming years. In this section, we concentrate on the two-flavor case for pedagogical reasons.

4.1.2. *Matter effects in neutrino oscillations*

Key ingredient to neutrino tomography are matter effects in neutrino oscillations (Wolfenstein, 1978; Mikheev and Smirnov, 1985, 1986). Since Earth matter contains plenty of electrons, but no muons or tauons, charged-current interactions of the electron neutrino flavor through coherent forward scattering lead to a relative phase shift compared to the muon and tau neutrino flavors. In the Hamiltonian in two flavors, the matter term enters as the second term in

$$\mathcal{H}(n_e) = U \begin{pmatrix} 0 & 0 \\ 0 & \frac{\Delta m_{21}^2}{2E} \end{pmatrix} U^\dagger + \begin{pmatrix} A(n_e) & 0 \\ 0 & 0 \end{pmatrix} \tag{3}$$

in flavor space, where $A(n_e) = \pm\sqrt{2}G_F n_e$ is the matter potential as function of the electron density n_e and the coupling constant G_F, and the different signs refer to neutrinos (plus) and antineutrinos (minus). Assuming that the number of electrons per nucleon is approximately 0.5 for stable "heavy" (considerably heavier than hydrogen) materials, the electron density can be converted into the matter density as $n_e = 0.5\ \rho/m_N$ with m_N the nucleon mass. In this case, there is some material dependence of this factor 0.5 ("electron fraction"), which, however, might also be used to obtain additional information on the composition. In two flavors and for constant matter density, Eq. (2) can bc easily re-written by a parameter mapping between vacuum and matter parameters:

$$P_{\alpha\beta} = \sin^2(2\tilde{\theta})\sin^2\left(\frac{\Delta\tilde{m}^2 L}{4E}\right), \tag{4}$$

where

$$\Delta\tilde{m}^2 = \xi \cdot \Delta m^2, \quad \sin(2\tilde{\theta}) = \frac{\sin(2\tilde{\theta})}{\xi}, \tag{5}$$

with

$$\xi \equiv \sqrt{\sin^2(2\theta) + (\cos(2\theta) - \hat{A})^2}, \tag{6}$$

$$\hat{A} \equiv \frac{2\sqrt{2}G_F n_e E}{\Delta m^2}. \tag{7}$$

One can easily read-off these formulas that for $\hat{A} \to \cos(2\theta)$ the parameter ξ in Eq. (6) becomes minimal, which means that the oscillation frequency in matter becomes minimal and the effective mixing maximal (cf., Eq. (5)). This case is often referred to as "matter resonance," where the condition $\hat{A} \to \cos(2\theta)$ evaluates to

$$E_{\text{res}} \sim 13,200\cos(2\theta)\frac{\Delta m^2[\text{eV}^2]}{\rho[\text{g/cm}^3]}. \tag{8}$$

This condition together with the requirement of a large oscillation phase $\sin^2(\Delta m^2 L/(4E)) = \mathcal{O}(1)$ leads to the "ideal" energies for neutrino oscillation tomography depending on the considered Δm^2:

$$\Delta m_{21}^2 : E \sim 100\,\text{MeV to} 1\,\text{GeV},$$
$$\Delta m_{31}^2 : E \sim \text{ few GeV to} 35\,\text{GeV}.$$

If the neutrino energy is far out of this range, either the matter effects or the overall event rate from oscillations will be strongly suppressed. However, there are also possible applications. Since, for instance, for solar neutrinos $E \ll E_{\text{res}}$, one can use the absence of the resonance for analytical simplifications, as we will discuss later.

4.1.3. *Numerical evaluation and conceptual problems*

In order to numerically study neutrino oscillation tomography, a commonly used method is the "evolution operator method" (cf., e.g., Ohlsson and Snellman, 2000). This method assumes that the matter density profile be discretized into layers with constant density (cf., Figure 3 for an example). The initial state $|\nu_\alpha\rangle$ is then propagated through the different matter density layers with depths x_j with the evolution operators

$$\mathcal{V}(x_j, \rho_j) = \mathrm{e}^{-i\mathcal{H}(\rho_j)x_j} \tag{9}$$

while the Hamiltonian within each layer $\mathcal{H}$ (cf., Eq. (3)) is assumed to have no explicit time-independence (it is given in constant density ρ_j). The transition probability is then obtained as

$$P_{\alpha\beta} = \left|\langle\nu_\beta|\mathcal{V}(x_n, \rho_n)\ldots\mathcal{V}(x_1, \rho_1)|\nu_\alpha\rangle\right|^2. \tag{10}$$

In practice, Eq. (10) is evaluated by diagonalizing the Hamiltonian for each density step, i.e., by calculating the mass eigenstates in each matter layer. Note that in general

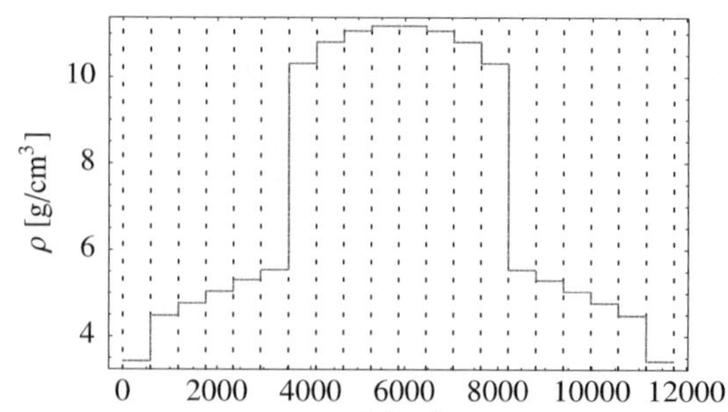

Figure 3. Example for a REM- ("Reference Earth Model") based matter density profile for a baseline of 12,000 km as used for the numerical evaluation of the neutrino propagation (20 steps). The matter density within each layer is assumed to be constant.

$$[\mathcal{V}(x_i, \rho_i), \mathcal{V}(x_j, \rho_j)] \neq 0 \quad \text{for } \rho_i \neq \rho_j, \tag{11}$$

which means that the evolution operators of different layers do not necessarily commute. This already implies that the information from a single baseline must be somehow sensitive towards the arrangement of the matter density layers. This is very different from X-ray or neutrino absorption tomography which do not have positional information from one baseline.

An important conceptual problem in neutrino oscillation tomography is the matter profile inversion problem (Ermilova et al., 1986, 1988). Assume that a matter density profile such as in Figure 3 is given. For a specific experimental setup, it is then fairly easy to compute the corresponding transition probabilities or event rates as function of energy. However, the reverse problem is theoretically generally unsolved: Assume that the transition probability is known up to infinite energies, then it would be very useful to be able to compute the matter profile from that. So far, there have been several attempts to solve this problem using simplifications, such as

- Simple models using only very few discrete steps (see, e.g., Nicolaidis, 1988; Nicolaidis et al., 1991; Ohlsson and Winter, 2002).
- Linearization in a low density medium (solar, supernova neutrinos) (Akhmedov et al., 2005).
- Discretization of a more complex profile using non-deterministic algorithms to fit a large number of parameters (Ohlsson and Winter, 2001).

Below, we will discuss some of these approaches in greater detail.

4.2. Neutrino oscillation tomography with solar and supernova neutrinos

Solar and supernova neutrinos are theoretically very interesting for neutrino tomography because matter effects are introduced off the resonance in Earth matter, i.e., the neutrino energy $E \ll E_{\mathrm{res}}$ (cf., Eq. (8) for $\Delta m^2 = \Delta m^2_{21}$), or equivalently $\hat{A} \ll 1$. This means that one does not expect strong matter effects in Earth matter as opposed to within the sun. However, this limit is theoretically very useful to study tomography because it allows for perturbation theory and other simplified approaches. It is often referred to as neutrino oscillations in a "low density medium" (Ioannisian and Smirnov, 2004; Ioannisian et al., 2005) because the density in the Earth is much lower than in the sun.

4.2.1. *Detecting a cavity*

We show in Figure 4 a possible setup for neutrino tomography using solar neutrinos following Ioannisian and Smirnov (2002). In this setup, the detector is fixed while the Earth is rotating, which means that the cavity with density ρ is "exposed" (in line of sight sun-detector) a time $0 < t_{\mathrm{exp}} < 24$ h per day. The change in the oscillation probability during this time is, depending on geometry and density contrast, $\lesssim 0.1\%$. This leads to a required detector mass $M \gtrsim 130\,\mathrm{Mt}/(t_{\mathrm{exp}}[\mathrm{h}])$, which has a lower limit of 5 Mt at the poles. Thus, from the statistics point of view, this approach is very challenging, and backgrounds might be an important issue. In addition, for such large detectors, the detector surface area might be of the order of the cavity size. There are, however, interesting theoretical results from such a discussion. Let us define the oscillation phases in the individual steps x_1, d, and x_2 as (Ioannisian and Smirnov, 2002)

$$\Phi_i \equiv \frac{\Delta m^2_{21} x_i}{2E} \sqrt{\sin^2 2\theta_{12} + (\cos 2\theta - \hat{A}_i)^2} \tag{12}$$

with the corresponding matter potentials $\hat{A}_i$ (cf., Figure 4). One can show that if mass eigenstates arrive at the surface of the Earth (solar and supernova neutrinos), the change in probability ΔP (cavity exposed-not exposed) only depends on Φ_2, but not on Φ_1. In addition, there is a damping of contributions from remote distances x_2, which means that solar neutrinos are less sensitive to the deep interior of the Earth than to structures close to the detector.

4.2.2. *Matter density inversion problem*

A further application of the low density limit is to theoretically solve the matter profile inversion problem. Following Akhmedov et al. (2005), the Earth matter effect on solar or supernova neutrinos is fully encoded in the quantity ("day–night regeneration effect")

$$P^{\mathrm{night}}_{2e} - P^{\mathrm{day}}_{2e} = \frac{1}{2} \cos^2\theta_{13} \sin^2 2\theta_{12} f(\delta) \tag{13}$$

with

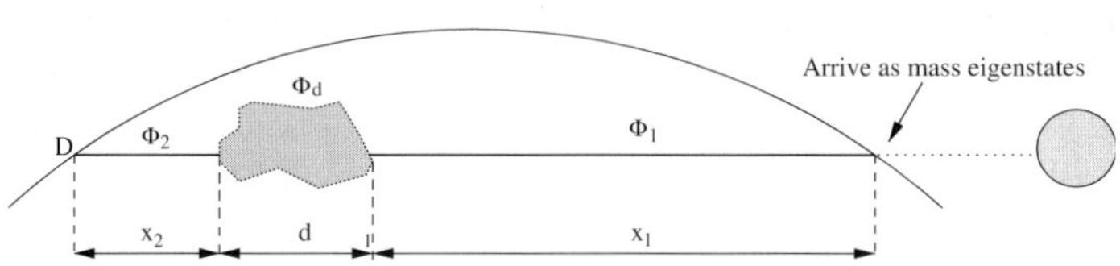

Figure 4. Neutrino oscillation tomography using solar neutrinos for the investigation of a cavity in the Earth's mantle.

$$f(\delta) = \int_0^L dx A(x) \sin\left[2 \int_x^L \omega(x') dx'\right], \tag{14}$$

$$\omega(x) = \sqrt{(\delta \cos 2\theta_{12} - A(x)/2)^2 + \delta^2 \sin^2 2\theta_{12}}, \tag{15}$$

$$A(x) = \sqrt{2} G_F n_e(x), \quad \delta \equiv \frac{\Delta m_{21}^2}{4E}. \tag{16}$$

This implies that the measured quantity is $f(\delta)$, i.e., a function of energy, which needs to be inverted into the matter profile $A(x)$. Especially, the double integral in Eq. (14) is quite complicated to invert. However, using the low density limit $A \ll 2\delta$ (or equivalently $\hat{A} \ll 1$) as well as $AL \ll 1$ ($L \ll$ 1700 km), one can linearize Eq. (14) in order to obtain

$$f(\delta) = \int_0^L dx A(x) \sin[2\delta(L - x)]. \tag{17}$$

This is just the Fourier transform of the matter density profile, i.e.,

$$A(x) = \frac{4}{\pi} \int_0^\infty f(\delta) \sin(2\delta(L - x)) d\delta, \tag{18}$$

and the matter density profile inversion problem is solved. One problem is very obvious from Eq. (18): One needs to know $f(\delta)$ in the whole interval $0 \leq E < \infty$ which is practically impossible. The authors of Akhmedov et al., (2005) suggest an iteration method to solve this problem. Additional challenges are statistics and a finite energy resolution, which is "washing out" the edges in the profile. One interesting advantage of using solar or supernova neutrinos is the sensitivity to asymmetric profiles, i.e., for mass-flavor oscillations there is no degeneracy between one profile and the time-inverted version, which otherwise (for flavor–flavor oscillations) can only be resolved by suppressed three-flavor effects.

4.2.3. *Supernova neutrinos to spy on the Earth's core*

Unlike solar neutrinos, which are limited to energies below 20 MeV, supernova neutrinos from a possible galactic supernova explosion have a

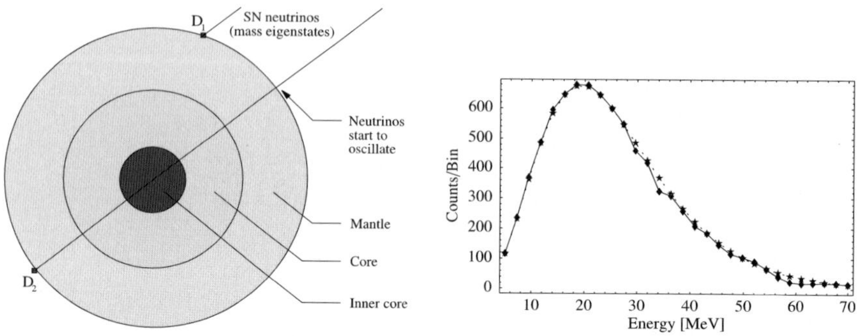

Figure 5. Illustration for the tomography of the Earth's core using supernova neutrinos (left) and difference between the event rate spectra (electron antineutrinos) of D_1 and D_2 for a Super-Kamiokande-like detectors (right). The difference between these spectra corresponds to $\Delta\chi^2 \simeq 35$ for the Earth matter effects, i.e., it is highly significant. Figures from Lindner et al. (2003).

high-energy tail which is closer to the Earth matter resonance energy. This effect is illustrated in Figure 5 (right) which compares the energy spectrum between two Super-Kamiokande-like detectors with and without Earth matter effects. It is obvious from this figure that the difference between the spectra around the peak at ~20 MeV is tiny, whereas statistically significant deviations can be found at larger energies.[1] Such a scenario could happen if supernova neutrinos were detected by two similar-sized detectors, one on the Earth's surface and with the Earth's core in the line of sight (cf., Figure 5, left). Note that the supernova neutrinos are detected within a very short time frame $\ll 24$ h, which means that one would actually obtain a "snapshot" of the Earth's interior. As it has been demonstrated in Lindner et al. (2003), for a galactic supernova in the distance $D = 10$ kpc with an energy release of $E = 3\ 10^{53}$ ergs, two megaton-size water Cherenkov detectors could measure the density of the Earth's core at the percent level with a number of challenges: First, the Earth's mantle density is assumed to be known at the 2% level. Second, the solar oscillation parameters have to be known at the 0.2% level. Third, too similar supernova fluxes for the different flavors (similar temperatures) and deviations from energy equipartition are unfavorable. And fourth, one has to have some knowledge on the flavor composition of the flux, possibly from detection of different flavors.

[1] For example, at around 34 and 60 MeV deviations between the two curves in Figure 5 (right) can be identified. The difference between these spectra corresponds to $\Delta\chi^2 \simeq 35$ for the Earth matter effects, i.e., it is highly significant.

4.3. Neutrino oscillation tomography with neutrino beams

We now discuss neutrino oscillation tomography with the "man-made" neutrino beams. Neutrino beams are planned or future neutrino sources using accelerators, where the neutrino beam is produced by pion/kaon decays (superbeams, see, e.g., Itow et al., 2001; Ayres et al., 2004), by muon decays (neutrino factory, see, e.g., Geer, 1998; Apollonio et al., 2002; Albright et al., 2004), or by the decay of unstable nuclei (β-Beam, see, e.g., Zucchelli, 2002; Bouchez et al., 2004; Burguet-Castell et al., 2004; Huber et al., 2006). Neutrino beams have, compared to "natural" neutrino sources, the advantage that either flux and flavor composition are well-known, or a near detector can be used to improve the knowledge on these quantities as well as on the interaction cross sections. There is, however, one major obstacle common to all of these experiments: Matter effects especially enter in the $\nu_e \leftrightarrow \nu_\mu$ flavor transition which is suppressed by the small mixing angle $\sin^2(2\theta_{13})$. Up to now, this mixing angle is unknown and only an upper bound exists (Apollonio et al., 1999). Experiments within the coming 10 years will reveal if $\sin^2(2\theta_{13})$ is suitably large for the applications discussed here (for a summary, see, e.g., Huber et al., 2004). Therefore, the experiment performance has always to be evaluated as function of $\sin^2(2\theta_{13})$. In this section, we split the discussion into conceptual areas linked to tomography with neutrino beams.

4.3.1. *Positional information for a single baseline*

Interesting questions are discussed in Ohlsson and Winter (2002): Assume we have a beam crossing a cavity with a specific density contrast compared to the surrounding matter. Then one wants to know

- How large has the cavity to be detected?
- Can the position of the cavity be measured and if so, how precisely?

In Ohlsson and Winter (2002) a 500 MeV superbeam is assumed with very luminous 200,000 events in total. The density in the cavity is assumed to be 1 g/cm^3 (water), the baseline $L = 1000$ km, and $\sin^2(2\theta_{13}) = 0.03$, where a smaller number of events can be compensated by a larger $\sin^2(2\theta_{13})$. It turns out that the cavity has to be longer than about 100 km to be found and its size can be measured to about $\pm$ 50 km. The most important result is that the position of the cavity can be reconstructed $\pm$ 100 km from a single baseline, which is very different from X-ray or absorption tomography. However, there is a degeneracy in the position between x and $L - x$ which can be only resolved by suppressed three-flavor effects. This example demonstrates already one of the basic principles of neutrino oscillation tomography: Positional information is available already from a single baseline.

4.3.2. *Resolution of structures and edges*

One can learn about the resolution of structures and edges from the numerical solution of the matter density inversion problem. In Ohlsson and Winter (2001) a (symmetrized) REM ("reference Earth Model") profile is reconstructed from a single baseline crossing the outer core with 14 degrees of freedom using a genetic algorithm. Naturally, there are many degenerate profiles close to the 1σ, 2σ, and 3σ contours, and one cannot show a contour in 14-dimensional parameter space. Therefore, we show in Figure 6 several "typical" representatives close to the 1σ, 2σ, and 3σ contours for a neutrino factory, where the total number of oscillated events is for $\sin^2(2\theta_{13}) = 0.1$ only about a factor of four above currently discussed luminosities (see, e.g., Huber et al., 2002a). From Figure 6, one can easily read-off that such an experiment could, in principle, reconstruct the mantle-core-mantle structure of the Earth. However, structures smaller than several hundred kilometers cannot be resolved. In addition, the mantle-core boundary cannot be resolved at a sufficiently high confidence level from a single baseline. Analytically, it has been demonstrated in Ohlsson and Winter (2001) that structures much smaller than the oscillation length in matter cannot be resolved – as one would naturally expect similar to other quantum mechanical phenomena. In conclusion, neutrino oscillations in matter are

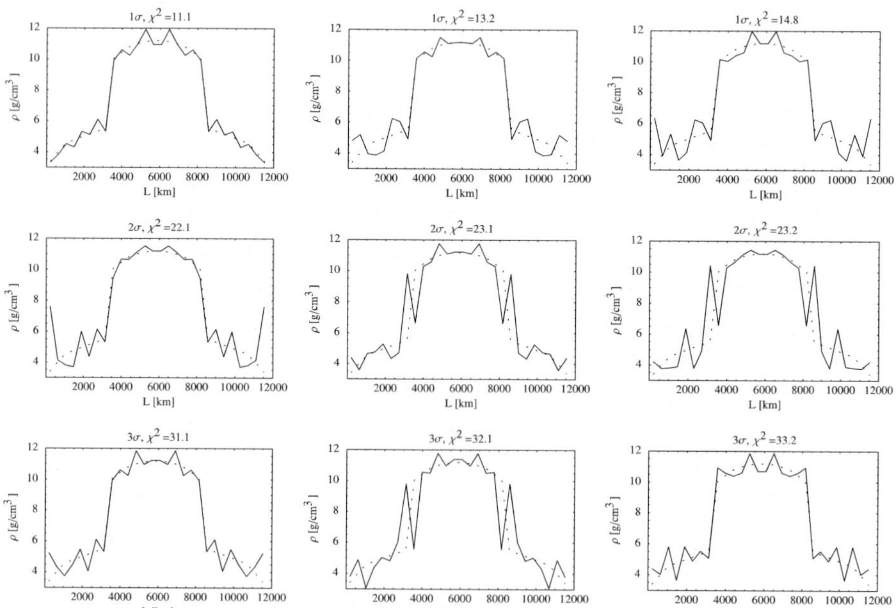

Figure 6. Examples for reconstructed (symmetric) REM profiles from an extremely luminous neutrino factory close to the 1σ (upper row), 2σ (middle row), and 3σ (lower row) contours for $L = 11{,}736$ km (14 d.o.f.). Figure from Ohlsson and Winter (2001).

very sensitive towards average densities and the arrangement of structure on the length scale of the oscillation length. However, neither can edges nor small structures be precisely resolved.

4.3.3. *Density measurement*

Since we know that neutrino oscillations measure more or less the baseline-averaged densities $\bar{\rho}_i^L = 1/L \int_0^L \tilde{\rho}(l)dl$ over long distances plus some suppressed interference effects, we can use this to discuss possible applications. For example, let us assume that we want to perform a simple one-parameter measurement of the Earth's inner core density. Because the Earth's mass is fixed, we need to correct the average mantle or outer core density for any shift of the inner core density. Note, however, that it is the volume-averaged density to be corrected, which means that large shifts in the Earth's inner core density cause only very small density corrections in the mantle. This example illustrates already one potential strength of neutrino oscillation tomography: Since neutrinos from a "vertical" baseline travel similar distances in mantle, core, and inner core, there should be no *a priori* disadvantage for the innermost parts of the Earth. In Winter (2005b) a neutrino factory setup from Huber et al. (2002a) with currently anticipated luminosities was chosen to test this hypothesis for realistic statistics. In order to measure the oscillation parameters, the experiment with $L = 2R_E$ was combined with a $L = 3000$ km. The precision of the measurement can be found in Figure 7 as function of $\sin^2(2\theta_{13})$. One case easily read-off that a percent level measurement is realistic for $\sin^2(2\theta_{13}) \gtrsim 0.01$. Most importantly, the application survives the unknown oscillation parameters and the performance is already close to the optimum (dashed curves). For smaller values of $0.001 \lesssim \sin^2(2\theta_{13}) \lesssim 0.01$, the correlations would be much worse without the $L = 3000$ km baseline. For large values of $\sin^2(2\theta_{13}) \gtrsim 0.01$, the vertical baseline alone is hardly affected by correlations with the oscillation parameters: As illustrated in Winter (2006), CP effects are suppressed for very long baselines. Since there is only a number of potential high-energy laboratories around the world which could host a neutrino factory, we show in Figure 8 some examples and the corresponding outer and inner core crossing baselines. Obviously, there are potential detector locations for some of the laboratories, which are, however, not exactly on the $L = 2R_E$-axis. Relaxing this baseline constraint somewhat, one can show that one can find detector locations for a small drop in precision (Winter, 2005b). In summary, this application illustrates that a density measurement could be performed with (a) reasonable statistics, (b) including the correlations with the oscillation parameters, and (c) reasonably small values of $\sin^2(2\theta_{13})$. In the future, it has to be clarified how large the additional effort for such a facility (the vertical storage ring) would be. Note, however, that there are plenty of other applications of a "very long" neutrino factory baseline, such as the "magic

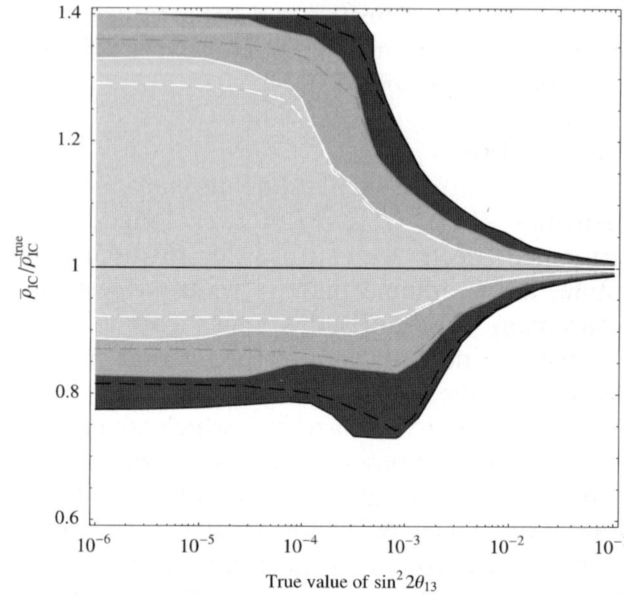

Figure 7. The measurement of $\bar{\rho}_{IC}$ (inner core density) as function of the true value of $\sin^2(2\theta_13)$ at the 1σ, 2σ, and 3σ confidence levels (from light to dark shaded regions). For the baselines, $L = 2{\cdot}R_E$ combined with a shorter baseline $L = 3000$ km to reduce correlations is used. The dashed curves correspond to fixing the oscillation parameters, i.e., to not taking into account correlations and degeneracies. Figure from Winter (2005b).

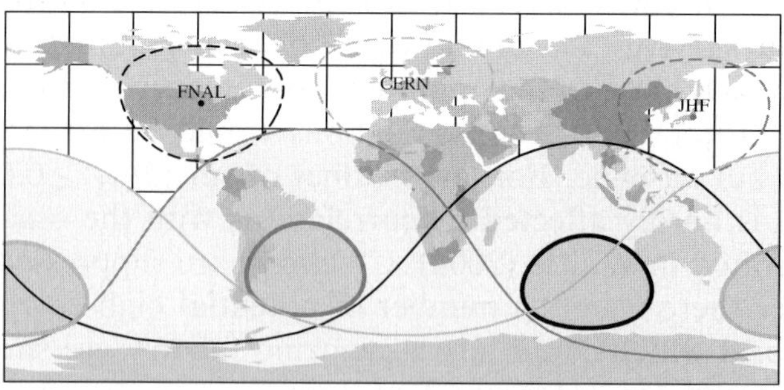

Figure 8. Positions of three of the major potential neutrino factory laboratories, (typical) $L = 3000$ km detector sites (dashed curves), as well as potential detector sites with outer core crossing baselines (below thin solid curves), and inner core crossing baselines (within thick solid curves). The grayscales of the curves represent the different laboratories. Figure from Winter (2005c).

baseline" to resolve degeneracies (Huber and Winter, 2003) ($L \sim 7500$ km), the test of the MSW effect for $\sin^2(2\theta_{13})=0$ (Winter, 2005a) ($L \gtrsim 5500$ km), the mass hierarchy measurement for $\sin^2(2\theta_{13})=0$ (de Gouvea et al., 2005;

de Gouvea and Winter, 2006) ($L \sim 6000$ km), and the test of the "parametric resonance" (Akhmedov, 1999; Petcov, 1998) ($L > 10{,}665$ km).

In addition to the described neutrino sources, note that tomography comparing the neutrino and antineutrino disappearance information from atmospheric neutrinos might, in principle, be possible as well (Geiser and Kahle, 2002).

5. Other Geophysical Aspects of Neutrino Oscillations

It is well known that matter density uncertainties spoil the extraction of the oscillation parameters from the measurements (see Jacobsson et al., 2002; Geller and Hara, 2001; Shan et al., 2002; Fogli et al., 2001; Ota and Sato, 2003; Shan et al., 2003; Kozlovskaya et al., 2003; Ohlsson and Winter, 2003 and references therein). In particular for baselines sensitive to δ_{CP}, such as $L\sim 3000$ km at a neutrino factory, the additional correlation with the matter density affects the precision measurements of $\sin^2(2\theta_{13})$ and δ_{CP}, and the CP violation sensitivity. This effect is illustrated in Figure 9 for the precision of δ_{CP} and different matter density uncertainties $\Delta\rho^*$. Especially for large $\sin^2(2\theta_{13})$, any uncertainty larger than about 1% affects the precision severely. Note that the baseline used for Figure 9 is $L = 3000$ km, which means that the neutrinos travel in an average depth of ~120 km up to a maximum depth of ~180 km. In these depths, the uncertainty among geophysics models is currently at the level of 5% (Geller and Hara, 2001). Since the matter density uncertainties may affect the competitiveness of a neutrino factory with a superbeam (operated at shorter baselines) for large values of $\sin^2(2\theta_{13})$, improved knowledge for specifically chosen baselines would be very helpful.

6. Summary and Conclusions

In summary, neutrino tomography might be a very complementary approach to geophysical methods. For example, neutrinos travel on straight lines with almost no uncertainty in their path. Furthermore, neutrino tomography is either sensitive to the nucleon density (absorption tomography) or electron density (oscillation tomography). In comparison, the paths of seismic waves are curved, and there is some uncertainty in them. In addition, the matter density has to be reconstructed from the propagation velocity profile by the equation of state. This means that neutrino tomography might be a more "direct" handle on the matter density and could be very useful to investigate specifically localized regions, such as in the lower mantle. Moreover, there is no principle reason to prevent neutrinos from penetrating the Earth's core,

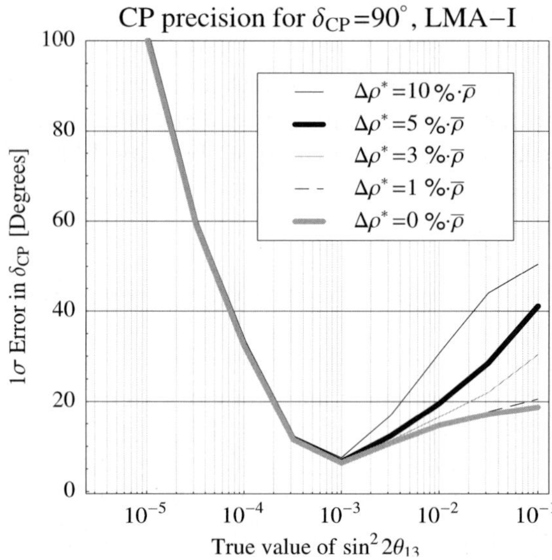

Figure 9. The precision of the measurement of δ_{CP} for a neutrino factory and the simulated value $\delta_{CP} = 90°$ as a function of the true value of $\sin^2(2\theta_{13})$ at the 1σ confidence level. The different curves correspond to different allowed matter density uncertainties $\Delta\rho^*$ as described in the plot legend, especially the thick curves correspond to no matter density uncertainty (light thick curve) and the often used uncertainty $\Delta\rho^* = 5\% \cdot \bar{\rho}$ (dark thick curve). Figure from Ohlsson and Winter (2003).

whereas seismic waves are partially reflected at the mantle-core and outer-inner core boundaries. Note that though the most precise information on deviations from the REM in the Earth's mantle comes from seismic waves, there are other geophysical methods which might be more directly sensitive towards the matter density, such as normal modes, mass, and rotational inertia of the Earth. Nevertheless, none of those could provide a measurement along a very specific path.

The main challenges for neutrino tomography might be the existence of high-energy neutrino sources for absorption tomography, and the statistics for oscillation tomography. For example, neutrino oscillation tomography could, in principle, reconstruct the matter density profile along a single baseline due to interference effects among different matter density layers. Note, however, that neutrino oscillations are to first order sensitive towards densities averaged over the scale of the oscillation length, which means that such sophisticated applications require extremely large luminosities (detector mass×source power×running time) and might be very challenging. On the other hand, very simple questions, such as a one-parameter measurement of the average density along the path or the discrimination between two very specific degenerate geophysical models might be feasible within the next decades. For example, the achievable precision for the inner core density of the Earth with a neutrino factory experiment might be quite comparable ($\pm$ 0.23 g/cm^3 for $\sin^2(2\theta_{13})$ = 0.01 and $\pm$ 0.06 g/cm^3 for $\sin^2(2\theta_{13})$ = 0.1 Winter, 2005b) to current precisions given for the density jump at the inner-core boundary from geophysics (e.g., $\pm$ 0.18 g/cm^3 in Masters and Gubbins, 2003). We therefore conclude that it will be important that the right and simple questions be asked by discussions between neutrino physicists and geophysicists.

Acknowledgements

I would like to thank Evgeny Akhmedov, Tommy Ohlsson, and Kris Sigurdson for useful comments and discussions. In addition, I would like to thank the organizers of the workshop "Neutrino Sciences 2005: Neutrino geophysics" for an excellent workshop with vivid discussions. This work was supported by the W. M. Keck Foundation and NSF Grant PHY-0503584.

References

Ahmad, Q. R. et al.: 2002, *Phys. Rev. Lett.* **89**, 011, 301, nucl-ex/0204008.

Ahrens, J. et al.: 2004, *Astropart. Phys.* **20**, 507–532 astro-ph/0305196.

Akhmedov, E. K.: 1999, *Nucl. Phys. B* **538**, 25–51 hep-ph/9805272.

Akhmedov, E. K., Tortola, M. A. and Valle, J. W. F.: 2005, *JHEP* **06**, 053, hep-ph/0502154.

Albright, C. et al.: 2004, The neutrino factory and beta beam experiments and development physics/0411123.

Aliu, E. et al.: 2005, *Phys. Rev. Lett.* **94**, 081, 802, hep-ex/0411038.

Apollonio, M. et al.: 1999, *Phys. Lett. B* **466**, 415–430 hep-ex/9907037.

Apollonio, M. et al.: 2002, Oscillation physics with a neutrino factory. hep-ph/0210192.

Ardellier, F. et al.: 2004, Letter of intent for Double-CHOOZ: A search for the mixing angle theta(13). hep-ex/0405032.

Askar'yan, G. A.: 1984, *Usp. Fiz. Nauk.* **144**, 523--530, [*Sov. Phys. Usp.* **27** (1984) 896].

Ayres, D. S. et al.: 2004, NOvA proposal to build a 30-kiloton off-axis detector to study neutrino oscillations in the Fermilab NuMI beamline. hep-ex/0503053.

Bilenky, S. M. and Pontecorvo, B.: 1978, *Phys. Rept.* **41**, 225–261.

Borisov, A. B., Dolgoshein, B. A., and Kalinovsky, A. N.: 1986, *Yad. Fiz.* **44**, 681–689.

Borisov, A. B. and Dolgoshein, B. A.: 1993, *Phys. Atom. Nucl.* **56**, 755–761.

Bouchez, J., Lindroos, M., and Mezzetto, M.: 2004, *AIP Conf. Proc.* **721**, 37–47 hep-ex/0310059.

Burguet-Castell, J., Casper, D., Gomez-Cadenas, J. J., Hernandez, P., and Sanchez, F.: 2004, *Nucl. Phys. B* **695**, 217–240 hep-ph/0312068.

De Rujula, A., Glashow, S. L., Wilson, R. R., and Charpak, G.: 1983, *Phys. Rept.* **99**, 341 .

Eguchi, K. et al.: 2003, *Phys. Rev. Lett.* **90**, 021, 802, hep-ex/0212021.

Ermilova, V. K., Tsarev, V. A. and Chechin, V. A.: 1986, *Pis'ma. Zh. Eksp. Teor. Fiz.* **43**, 353, [*JETP Lett.* **43** (1986) 453].

Ermilova, V. K., Tsarev, V. A. and Chechin, V. A.: 1988, *Bull. Lebedev. Phys. Inst.* NO. **3**, 51.

Fogli, G. L., Lettera, G. and Lisi, E.: 2001, Effects of matter density variations on dominant oscillations in long baseline neutrino experiments. hep-ph/0112241.

Fogli, G. and Lisi, E.: 2004, *New J. Phys.* **6**, 139.

Fogli, G. L., Lisi, E., Marrone, A. and Palazzo, A.: 2005, Global analysis of three-flavor neutrino masses and mixings. hep-ph/0506083.

Fukuda, Y. et al.: 1998, *Phys. Rev. Lett.* **81**, 1562–1567 hep-ex/9807003.

Geer, S.: 1998, *Phys. Rev. D* **57**, 6989–6997.

Geiser, A. and Kahle, B.: 2002, Tomography of the earth by the oscillation of atmospheric neutrinos: A study of principle. Poster presented at Neutrino 2002 Munich.

Geller, R. J. and Hara, T.: 2001, *Nucl. Instrum. Meth. A* **503**, 187–191 hep-ph/0111342.

Giunti, C., Kim, C. W., Lee, U. W., and Lam, W. P.: 1992, *Phys. Rev. D* **45**, 1557–1568.

Gonzalez-Garcia, M. C., Halzen, F. and Maltoni, M.: 2005, *Phys. Rev. D* **71**, 093,010, hep-ph/0502223.

de Gouvea, A. and Winter, W.: 2006, *Phys. Rev. D* **73**, 033003, hep-ph/0509359.

de Gouvea, A., Jenkins, J. and Kayser, B.: 2005, Neutrino mass hierarchy, vacuum oscillations, and vanishing ue3 hep-ph/0503079.

Huber, P., Lindner, M., and Winter, W.: 2002a, *Nucl. Phys. B* **645**, 3–48 hep-ph/0204352.

Huber, P., Schwetz, T. and Valle, J. W. F.: 2002b, *Phys. Rev. D* **66**, 013,006, hep-ph/0202048.

Huber, P. and Winter, W.: 2003, *Phys. Rev. D* **68**, 037,301, hep-ph/0301257.

Huber, P., Lindner, M., Rolinec, M., Schwetz, T. and Winter, W.: 2004, *Phys. Rev. D* **70**, 073,014, hep-ph/0403068.

Huber, P., Lindner, M., Rolinec, M. and Winter, W.: 2006, *Phys. Rev. D* **73**, 053002, hep-ph/0506237.

Ioannisian, A. N. and Smirnov, A. Y.: 2002, Matter effects of thin layers: Detecting oil by oscillations of solar neutrinos. hep-ph/0201012.

Ioannisian, A. N. and Smirnov, A. Y.: 2004, *Phys. Rev. Lett.* **93**, 241,801, hep-ph/0404060.

Ioannisian, A. N., Kazarian, N. A., Smirnov, A. Y. and Wyler, D.: 2005, *Phys. Rev. D* **71**, 033,006, hep-ph/0407138.

Itow, Y. et al.: 2001, *Nucl. Phys. Proc. Suppl.* **111**, 146–151 hep-ex/0106019.

Jacobsson, B., Ohlsson, T., Snellman, H., and Winter, W.: 2002, *Phys. Lett. B* **532**, 259–266 hep-ph/0112138.

Jain, P., Ralston, J. P., and Frichter, G. M.: 1999, *Astropart. Phys.* **12**, 193–198 hep-ph/9902206.

Kaplan, D. B., Nelson, A. E. and Weiner, N.: 2004, *Phys. Rev. Lett.* **93**, 091,801, hep-ph/0401099.

Kozlovskaya, E., Peltoniemi, J. and Sarkamo, J.: 2003, The density distribution in the earth along the CERN-Pyhaesalmi baseline and its effect on neutrino oscillations. hep-ph/0305042.

Kuo, C., Crawford, H. J., Jeanloz, R., Romanowicz, B., Shapiro, G., and Stevenson, M. L.: 1995, *Earth Plan. Sci. Lett.* **133**, 95 .

Lindner, M., Ohlsson, T., Tomas, R., and Winter, W.: 2003, *Astropart. Phys.* **19**, 755–770 hep-ph/0207238.

Masters, G. and Gubbins, D.: 2003, *Phys. Earth Planet Int.* **140**, 159–167.

Mikheev, S. P. and Smirnov, A. Y.: 1985, *Sov. J. Nucl. Phys.* **42**, 913–917.

Mikheev, S. P. and Smirnov, A. Y.: 1986, *Nuovo. Cim. C* **9**, 17–26.

Nedyalkov, I. P.: 1981a, *Bolgarskaia Akademiia Nauk* **34**, 177–180.

Nedyalkov, I. P.: 1981b, On the study of the earth composition by means of neutrino experiments JINR-P2-81-645.

Nicolaidis, A.: 1988, *Phys. Lett. B* **200**, 553 .

Nicolaidis, A., Jannane, M., and Tarantola, A.: 1991, *J. Geophys. Res.* **96**, 21,811–21,817.

Ohlsson, T. and Snellman, H.: 2000, *Phys. Lett. B* **474**, 153–162 hep-ph/9912295.
Ohlsson, T. and Winter, W.: 2001, *Phys. Lett. B* **512**, 357–364 hep-ph/0105293.
Ohlsson, T. and Winter, W.: 2002, *Europhys. Lett.* **60**, 34–39 hep-ph/0111247.
Ohlsson, T. and Winter, W.: 2003, *Phys. Rev. D* **68**, 073,007, hep-ph/0307178.
Ota, T. and Sato, J.: 2003, *Phys. Rev. D* **67**, 053,003, hep-ph/0211095.
Petcov, S. T.: 1998, *Phys. Lett. B* **434**, 321–332 hep-ph/9805262.
Quigg, C., Reno, M. H., and Walker, T. P.: 1986, *Phys. Rev. Lett.* **57**, 774 .
Reynoso, M. M. and Sampayo, O. A.: 2004, *Astropart. Phys.* **21**, 315–324 hep-ph/0401102.
Shan, L. Y., Young, B. L. and Zhang, X. M.: 2002, *Phys. Rev. D* **66**, 053,012, hep-ph/0110414.
Shan, L. Y. et al.: 2003, *Phys. Rev. D* **68**, 013,002, hep-ph/0303112.
Wilson, T. L.: 1984, *Nature* **309**, 38–42.
Winter, W.: 2005a, *Phys. Lett. B* **613**, 67–73 hep-ph/0411309.
Winter, W.: 2005b, *Phys. Rev. D* **72**, 037,302, hep-ph/0502097.
Winter, W.: 2006, *Nucl. Phys. B* (Proc. Suppl.) **155**, 197–198, hep-ph/0510025.
Wolfenstein, L.: 1978, *Phys. Rev. D* **17**, 2369 .
Zucchelli, P.: 2002, *Phys. Lett. B* **532**, 166–172.

Earth, Moon, and Planets (2006) 99:309–330
DOI 10.1007/s11038-006-9110-x

Far Field Monitoring of Rogue Nuclear Activity with an Array of Large Anti-neutrino Detectors

EUGENE H. GUILLIAN

Department of Physics, University of Hawaii, Manoa, 2505 Correa Road, Honolulu, HI 96822, USA

Department of Physics, Queen's University, Stirling Hall, Kingston, ON, K7L 3N6, Canada (E-mail: guillian@owl.phy.queensu.ca)

(Received 28 February 2006; Accepted 11 July 2006)

Abstract. The result of a study on the use of an array of large anti-neutrino detectors for the purpose of monitoring rogue nuclear activity is presented. Targeted regional monitoring of a nation bordering large bodies of water with no pre-existing legal nuclear activity may be possible at a cost of about several billion dollars, assuming several as-yet-untested schemes pan out in the next two decades. These are: (1) the enabling of a water-based detector to detect reactor anti-neutrinos by doping with $GdCl_3$; (2) the deployment of a KamLAND-like detector in a deep-sea environment; and (3) the scaling of a Super-Kamiokande-like detector to a size of one or more megatons. The first may well prove feasible, and should be tested by phase-III Super-Kamiokande in the next few years. The second is more of a challenge, but may well be tested by the Hanohano collaboration in the coming decade. The third is perhaps the least certain, with no schedule for construction of any such device in the foreseeable future. In addition to the regional monitoring scheme, several global, untargeted monitoring schemes were considered. All schemes were found to fail benchmark sensitivity levels by a wide margin, and to cost at least several trillion dollars.

Keywords: nuclear fission, anti-neutrino, anti-neutrino detector, anti-neutrino detector array, arms control, nuclear non-proliferation, rogue nuclear activity, nuclear reactor, fission bomb

1. Introduction

The human race first tapped into nuclear energy with the success of the Manhattan project. Ever since, the practical know-how regarding the use of this source of energy has expanded and spread, and, so far as civilization as we know it continues to exist, this, no doubt, will continue to be the case. The spread of practical know-how in this area, however, presents a threat to peace, since there will always be desperate characters among the world's political leaders, and it is a matter of time before one such leader gets access to this know-how and decides to use it indiscriminately against his enemies.

Monitoring regimes exist to guard against the uncontrolled spread of nuclear technology and the detonation of nuclear bombs. The International Atomic Energy Agency (IAEA) works under the auspices of the United Nations to make sure that nations that use nuclear energy do so only for peaceful purposes.[1] Another monitoring regime is the Comprehensive Test Ban Treaty (CTBT), which is an agreement among nations to ban all nuclear explosions.[2] As recent world events (the detonation of a fission bomb by Pakistan in 1998, and the current political crisis involving nuclear activities in North Korea and Iran) have shown, however, neither regime has proved sufficient to curb the spread of nuclear technology nor the detonation of bombs. Clearly, the flaws in the regimes are mostly political. For instance, the detonation of nuclear bombs by Pakistan in 1998 was not against the CTBT because Pakistan is not a signatory. Also, the recent events in North Korea and Iran have little to do with monitoring techniques, but, rather, with flaws in the political process that allows headstrong political leaders to use nuclear threats as political bargaining chips.

Although much of the problem with today's monitoring regime is political, some of the political problems are abetted by insufficiencies in the monitoring techniques. For instance, in 2002, after mounting tensions with the United States and her allies, North Korea expelled United Nations inspectors and threatened to restart its nuclear facilities in Yongbyon[3] Once the inspectors were ousted, it was impossible to tell whether or not the North Koreans had actually carried through with their threat to reprocess nuclear fuel. This scenario is made possible by the fact that the IAEA monitoring technique requires the cooperation of participants. Clearly, a more robust monitoring regime requires far-field monitoring techniques that do not depend on participant cooperation.

Such techniques are already in use to monitor nuclear explosions by the CTBT (seismology, hydroacoustics, infrasound, and radionuclide monitoring),[2] but they are useless for detecting nuclear reactor operation because a reactor burns nuclear fuel at a steady rate, and it does not release redionuclides into the environment. Far-field monitoring, however, is possible in principle using anti-neutrinos produced in nuclear fission. Indeed, the KamLAND experiment (Eguchi et al., 2003) detects anti-neutrinos from nuclear reactors at an average distance of about 180 km. Anti-neutrinos are electrically neutral particles produced in nuclear fission; they interact with matter only via the *weak nuclear force*. Because of this, anti-neutrinos can easily travel through hundreds of kilometers of matter with almost no probability of interaction with the intervening material. This feature of the anti-neutrinos makes their detection very difficult; however, given a big

[1] http://www.iaea.org/.

[2] http://www.fas.org/nuke/control/ctbt/.

[3] http://www.cnn.com/2005/WORLD/asiapcf/02/10/ nkorea.timeline/index.html.

enough target, a sufficiently long exposure time, and a sufficiently low background level, they can be reliably detected.

The purpose of the study presented herein is to determine the feasibility of using anti-neutrino detection for far-field monitoring of both nuclear reactor operation and fission bomb detonations. At the most basic level, the feasibility of this technique has been established by KamLAND. However, they were helped by the very large signal due to the unusually large concentration of nuclear reactors in Japan.[4] In a realistic far-field monitoring scenario, the signal is expected to be tiny – probably much less than 100 MW_{th} (a typical commercial nuclear reactor power is about 2500 MW_{th}). We have found that a regionally targeted monitoring regime – e.g., the monitoring of nuclear reactor operations in North Korea – may be possible at a projected cost of several billion dollars, as long as several as-yet-untested schemes pan out in the coming decade. We also considered the possibility of setting up a global array of large anti-neutrino detectors to detect surreptitious nuclear fission activity anywhere in the world. This was found to miss benchmark sensitivity levels by a wide margin, and to be unrealistic because of the prohibitive projected cost on the order of trillions of dollars.

2. Anti-neutrinos Produced in Nuclear Fission

Nuclear reactors and fission bombs make use of the energy released by splitting heavy nuclei (primarily uranium and plutonium). The former are designed to keep the rate of splitting constant so that energy is released at a steady rate, whereas the latter is designed to cause the energy to be released in a very short period. In both cases, the daughter nuclei from the splitting of uranium and plutonium are unstable and undergo radioactive decay; an anti-neutrino is produced from every beta decay.

The rate of anti-neutrino production in a nuclear reactor is directly proportional to its thermal power. Each nuclear fission releases about 200 MeV (million electron volts) of thermal energy,[5] which is equal to 3.2×10^{-11} J. A typical nuclear reactor has a thermal power of one gigawatt, or 10^9 joules per second. The number of fissions per second required to produce this power, therefore, is 10^9 J/s divided by 3.2×10^{-11} J, which is equal to 3.1×10^{19}. Finally, since about 6 anti-neutrinos are produced per fission, we find that

[4] As of 2002, Japan has 16 nuclear reactor plants producing a total power of 130 GW_{th} (Bemporad et al., 2002).

[5] One electron volt is the amount of energy that an electron gains when it moves from the "−" to "+" terminal of a 1-volt battery. The energy released by nuclear fission, therefore, is hundreds of million times as large as the energy released by a typical battery used in every-day life.

1.9×10^{20} anti-neutrinos per second are produced for every one gigawatt of thermal power.

The corresponding calculation for a fission bomb proceeds similarly. The strength of a fission bomb is usually quoted in terms of its "yield", which is the mass of TNT that produces the same amount of energy. A small fission bomb has a yield of about 1000 tons of TNT, and one metric ton of TNT releases 4.18×10^9 J, so this bomb releases 4.18×10^{12} J. As in the case of a nuclear reactor, each fission releases 200 MeV, and 6 anti-neutrinos are produced per fission. Thus 1.30×10^{23} anti-neutrinos are released for every kiloton of fission bomb yield. Unlike a reactor, which releases anti-neutrinos at a steady rate, a fission bomb releases the anti-neutrino impulsively over a period of several seconds; almost no anti-neutrinos are emitted after about 10 s from the blast (Bernstein et al., 2001).

3. Detecting Anti-neutrinos

The stuff that composes the material world is responsive to the electromagnetic force. It is this force that keeps a ball from going through one's hand when caught. Anti-neutrinos, unlike ordinary stuff, are unresponsive to the electromagnetic force. Consequently, an anti-neutrino can travel through an extraordinary thickness of matter with almost zero chance of hitting stuff on its way through; the illustration of a neutrino traveling through a light-year thick block of lead is famous. They, however, do interact with matter via the weak nuclear force. This force has a strength comparable to that of the electromagnetic force, but the carriers of this force are very massive unlike photons, the massless particles that transmit the electromagnetic force. As a result, the weak force has a very short range; an anti-neutrino interacts with matter only if it happens to pass by very close to a target particle. The chance of this happening is extraordinarily low, and this accounts for an anti-neutrino's ability to travel through a large amount of matter.

The weak interaction between an anti-neutrino and matter can take place in various ways. For instance, an anti-neutrino can hit an electron that orbits an atomic nucleus, transmitting some of its momentum. Since an electron carries an electric charge, its motion through matter is very noticeable; a sensitive particle detector can detect the effects produced by this motion. For instance, an electron carrying several million electron-volts travels faster than the speed of light in matter; this super-luminal motion creates a shock wave of electromagnetic radiation, which is referred to as the Cherenkov effect. Detectors like Super-Kamiokande (Fukuda et al., 2003) detect anti-neutrinos (and neutrinos) using this effect.

Another way that an anti-neutrino can be detected is via the inverse beta process, in which an anti-neutrino encounters a proton and comes out transformed into a positron, while the proton is transformed into a neutron. This is written symbolically as follows:

$$\overline{\nu_e} + p \rightarrow n + e^+ \qquad (1)$$

The symbols $\overline{\nu_e}$, p, n, and e^+ stand, respectively, for anti-neutrino (electron-type), proton, neutron, and positron. In the study we performed, we considered detectors that use this process for detecting anti-neurtrinos. The advantages of this technique are the relatively high probability of the occurrence of the inverse beta process, and the "double-bang" signature produced by the anti-neutrino. That is, the out-going particles e^+ and n both produce signals in the detector. First, the e^+ produces a burst of light, the amount of which is proportional to its energy (which is also closely related to the anti-neutrino's energy). This happens promptly after the transformation in Equation (1) takes place. The neutron, however, rattles around for tens to hundreds of microseconds (millionth of a second, which is a relatively long time in the present context) until it is eventually absorbed by a proton or a dopant like gadolinium in the target. This absorption is followed by the emission of gamma ray(s) of several million electron volts; a burst of light proportional to this energy is produced. In summary, then, an anti-neutrino interacting in this manner produces two bursts of light separated by a meter or so in distance, and tens to hundreds of microseconds in time. This double-bang signature is useful for picking out anti-neutrino events from the large background produced, for example, by radioactive contaminants in the detector. The background events produce random flashes of light, but they are not very likely to produce the double-bang signature (Bemporad et al., 2002).

The detection of anti-neutrinos using inverse beta decay is typically done using a liquid scintillator detector; KamLAND is an example of such a detector (Eguchi et al., 2003). Liquid scintillator is used primarily because the amount of light produced by an anti-neutrino interaction event is very large compared to Cherenkov radiation produced in water; greater light yield translates to higher sensitivity (i.e., particles with lower energy are visible) and better energy resolution. In the present context, however, the required detector size is of the order of 1 megaton, which, for liquids, is about a cube of sides 100 m. At this scale, the use of liquid scintillator becomes impractical because of the cost; water is the only economically realistic target material. By itself, however, water cannot be used to detect the inverse beta decay process because the second "bang" in the double-bang signature is below the energy threshold. In order to make the second bang visible, the detector must be doped with an element such as gadolinium, which aggressively absorbs the produced neutron and emits gamma rays above the detector energy threshold. Because of the very large absorption cross section for thermal neutrons, only a 0.2% concentration is needed to capture 90% of the neutrons. For a 1-megaton detector, this corresponds to 200 tons of the salt $GdCl_3$ (Beacom and Vagins, 2001).

The rate of detection of anti-neutrinos from a nuclear reactor in a water detector is given by the equation below:[6]

$$N = 3.04 \times 10^3 \text{ events} \times \left(\frac{T}{1\text{ year}}\right) \cdot \left(\frac{M}{1\text{ Megaton}}\right) \times \left(\frac{P}{100\text{ MW}_{\text{th}}}\right) \cdot \left(\frac{100\text{ km}}{D}\right)^2 \tag{2}$$

This equation shows that a 100 MW_{th} nuclear reactor (about the upper limit of the power expected from a rogue nuclear reactor) at 100 km from a 1-megaton detector exposed for 1 year produces about 3000 observable events. The number of anti-neutrino events from a fission bomb is given by:

$$N = 2.25 \text{ events} \times \left(\frac{M}{1\text{ Megaton}}\right) \cdot \left(\frac{100\text{ km}}{D}\right)^2 \cdot \left(\frac{Y}{1\text{ kiloton}}\right) \tag{3}$$

Unlike Equation (2), which gives the rate of anti-neutrino detection (events per year), Equation (3) gives the total anti-neutrino yield over 10 s during which most of the anti-neutrinos are released by a fission bomb. Based on this equation, one finds only 2.25 events for a 1 kiloton bomb detonated at 100 km from a 1-megaton detector. This may seem small, but since the events arrive in a 10 s window, the signal-to-background ratio is actually quite good. For instance, a 2500 MW_{th} reactor (typical power of a commercial reactor) 100 km away produces about 2.5 events in this time window, giving a signal-to-background ratio close to 1; at most locations, the ratio is much better than this.

Both Equation (2) and Equation (3) are somewhat optimistic because they were calculated assuming that the detector is sensitive to all values of

[6] This rate assumes 100% efficiency above the threshold anti-neutrino energy of 1.8 MeV. The anti-neutrino energy spectrum depends somewhat on the relative fraction of isotopes in a fission reactor; we took average values used in (Eguchi K. et al., 2003), which were 0.567, 0.078, 0.297, and 0.057 for ^{235}U, ^{238}U, ^{239}Pu, and ^{241}Pu, respectively. The rate also depends on the production cross section for the inverse beta process; the energy dependence was taken to be the same as that used in (Eguchi K. et al., 2003). Finally, it was assumed that all isotopes produce 204 MeV of thermal energy per fission; in fact, different isotopes produce somewhat different energies, but the value used here is a good-enough approximation for the present purpose.

anti-neutrino energy (the inverse beta process requires at least 1.8 MeV in anti-neutrino energy). In reality, the anti-neutrino energy probably needs to be at least 3.8 MeV to be visible by the detector. Only 58% of events have energy above this. Other data selection cuts may decrease the event rate somewhat, probably to a total efficiency of about 50%. For the sake of simplicity, we shall take the efficiency to be 100%. Any result we obtain here, therefore, will be over-optimistic by a factor of about $\sqrt{0.5}$. In other words, the actual sensitivity will be worse by about a factor of $\sqrt{2} \approx 1.4$.

4. Shielding from Cosmic Rays

Because anti-neutrinos interact very rarely with matter, extreme care must be taken to ensure that the signal is not overwhelmed by background noise. One way to deal with this is to increase the signal so much that it is comparable to the background. This is what is done in short-baseline reactor detectors (Bemporad et al., 2002) and near-field reactor monitoring detectors (Bernstein et al., 2001). For some applications, however, the distance between the anti-neutrino source and the detector must be large. Since the anti-neutrino flux is inversely proportional to the square of this distance, the signal rate is tiny in these situations. To make up for this, the detector must be large and it must have a very small level of background noise.

There are two main classes of background noise: radioactivity present in and around the detector, and radioactivity produced by cosmic ray muons. The level of the former can be reduced to very low levels, thanks to decades of experience from numerous neutrino detection experiments. The latter, however, can only be reduced by brute force: i.e., by having sufficient shielding material, such as rock or water. As a rough rule of thumb for typical anti-neutrino measurements, 2 km of water is barely enough shielding, 3 km is satisfactory, and four or more kilometers provides good shielding. Shielding by rock is very costly unless it is already present, such as inside of a pre-existing mine. Indeed, most large anti-neutrino detectors in existence today are located in commercial mines. For the purpose of nuclear monitoring, however, it is unlikely in general that commercial mines would exist in locations where the detectors need to be placed. Thus, for economic reasons, the only locations where far-field monitoring detectors could be placed are in large bodies of water (i.e., oceans, seas, and large lakes).

5. Anti-neutrino Background Sources

One of the most formidable background sources for nuclear monitoring with ant-inuetrino detectors is the flux of anti-neutrinos from commercial

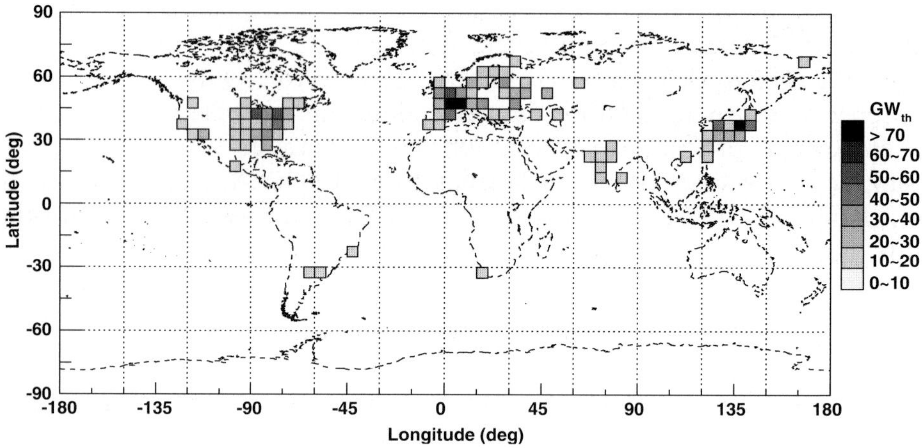

Figure 1. A map of the thermal power of commercial and research reactors in 5° × 5° cells on Earth.

and research nuclear reactors around the world. Distributed mostly around the northern hemisphere, a total thermal power of about 1 TW is produced by these reactors. The distribution of these reactors is shown in Figure 1. See Appendix A for details on the location and power of these reactors. The number of anti-neutrinos produced by these nuclear reactors detected per year by a 1-megaton detector located at various locations on Earth is shown in Figure 2a. Anti-neutrinos produced by these reactors are virtually indistinguishable from those produced by a rogue reactor; if a rogue reactor operates in a region where the anti-neutrino flux from commercial and research reactors is high, it would be very difficult to detect.

Another possible source of background is the "georeactor", which is a hypothetical natural nuclear reactor in the core of the Earth (Herndon, 1979; Herndon, 1980; Herndon, 1996; Hollenbach and Herndon, 2001; Herndon, 2003). If it exists, this reactor is expected to have a radius of several kilometers and have a thermal power of about 1–10 TW_{th}. Since commercial and research nuclear reactors world-wide produce a total power of about 1 TW_{th}, the existence of a georeactor would have a large impact on the background rate for detecting a rogue reactor. This is illustrated in Figure 2b, which is the same as Figure 2a, but with a contribution from a 3 TW_{th} georeactor. The effect of a georeactor is not particularly serious in much of the northern hemisphere because the anti-neutrino flux is already high, but it causes a serious increase in background in much of the southern hemisphere. For this reason, a measurement of this background is an important prerequisite for the anti-neutrino detector array being considered here. A preliminary measurement

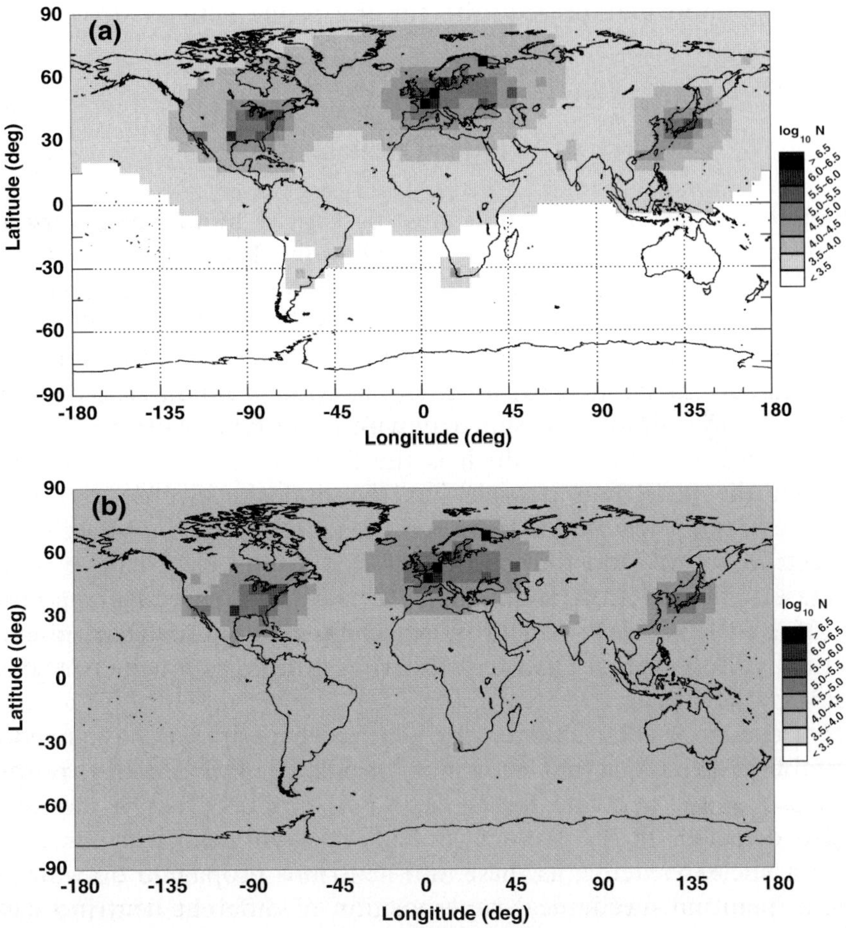

Figure 2. The number of events per year (log scale) detected by a 1-megaton anti-neutrino detector due to commercial and research reactors around the world. (**a**) Assuming no georeactor. (**b**) Assuming a 3 TW_{th} georeactor.

has already been carried out by KamLAND (Maricic, 2005), but the result is imprecise because of the large background from commercial nuclear reactors. However, a detector with size comparable to KamLAND and located far away from commercial reactors can easily make a precise measurement of georeactor power down to about 1 TW_{th}. Hanohano[7] (Hawaii Anti-Neutrino Observatory) is an example of a detector capable of making this measurement. Like the detectors in this array, Hanohano will be placed deep in the ocean. Thus, it is a prototype of the megaton detectors, and the successful implementation and operation of it

[7] http://www.phys.hawaii.edu/sdye/hnsc.html.

would be an important prerequisite for the rogue activity detector array concept.

6. Neutrino Oscillations

A potentially important detail that must be kept in mind when considering neutrino detection is neutrino flavor oscillation. The current view of the nature of neutrinos is that three "flavors" of neutrinos exist; the flavors are referred to as the "electron-type", "muon-type", and "tau-type". An electron-type neutrino turns into an electron when it interacts with the target via the charged-current electroweak interaction, while a muon-type neutrino is transformed into a muon and similarly with a tau-type neutrino. The situation with "anti-neutrinos" – which is the focus of this study – is similar, except that the out-going particle has the opposite electric charge. The importance of the foregoing discussion is the fact that the final state particles – the electron, muon, and the tau (and their anti-particles) – have very different masses. The electron, muon, and the tau have, respectively, a mass of 0.511, 105, and 1777 MeV. A neutrino can only undergo the charged-current interaction with the target particle if it carries at least as much energy as the out-going particle mass.

Recent results of solar and reactor neutrino experiments have unequivocally established the fact that neutrinos "oscillate". For practical purposes, this means that the neutrino flavor when it is produced is not the same as when it is detected. In the present context, electron anti-neutrinos are produced in a nuclear reactor; as these anti-neutrinos propagate outward, they become a quantum mechanical superposition of different neutrino flavors. Since these anti-neutrinos have energy well below 10 MeV, that part of the superposition that has turned into a muon- or tau-type neutrino cannot interact with the target because the available energy is insufficient to produce a muon or tau. The result is that the anti-neutrino detection rate is smaller than is expected in the absence of neutrino oscillations. The neutrino survival probability – defined as the fraction of detection rate compared to the rate without oscillations – is a function of distance from the reactor. For a threshold energy of 1.8 MeV, this starts out at 100% for distances of 0 to several 10s of kilometers. The probability then oscillates around an asymptotic value of 0.57 as the distance ranges from about 100–300 km. Beyond this, the amplitude of the oscillation approaches zero, and the probability is practically indistinguishable from 0.57.

In this study, we consider two cases: regional monitoring (section 7) and global monitoring (section 8). In the former, the variation of the survival probability with distance affects the result of performance studies, so this has been taken into account in all figures and results. In the latter, the effect of oscillation was implemented by simply scaling the no-oscillation rate by 0.57. This simplification is valid because we are only interested in how the detector array performs as a group spanning many thousands of kilometers. In other words, in the global scheme, we are not interested in how well several near-by detectors perform (which is covered in the regional scheme), but in how well many hundreds of widely separated detectors perform. We have established that, for the global scheme, the asymptotic approximation of the survival probability is accurate to within a fraction of a percent.

7. Regional Monitoring

As an example of the capability of an array of megaton-scale anti-neutrino detectors for the purpose of detecting rogue nuclear activity, we consider a scenario in which rogue activity is taking place in North Korea. To make the illustration concrete, it was assumed that the rogue reactor is located deep inside of North Korean territory at longitude 127.0° E and latitude 40.5° N (Figure 3a). North Korea presents a realistic test case not only because of recent events, but also because of the fact that it does not operate any nuclear reactors legally. If this were not the case, the monitoring regime presented here would be easily defeated because the rogue reactor could be placed close to a legally operated reactor, which would obscure this activity.

The choice of location of anti-neutrino detectors should be based on the sensitivity to rogue reactor detection. Figure 3a shows the number of events detected by a 1-megaton detector exposed for 1 year, assuming the rogue reactor power is 100 MW_{th}. Figure 3b shows the number of background events, mostly from commercial nuclear reactors in South Korea and Japan. The sensitivity of a detector depends on the signal S and the background B according to the formula $S/\sqrt{S+B}$. Twenty-three candidate locations were chosen based on the sensitivity contour (Figure 3c). We assumed, of course, that the detector must be located in the ocean for cosmic ray shielding. We did not consider the feasibility of the candidate locations from the point of view of political boundaries, depth, or ease of sabotage.

The general outlinc of the monitoring regime proceeds as follows. First, detectors are placed in several locations around North Korea. In our simulations, we examined array configurations with two to four detectors, the location of which was chosen from the 23 shown in Figure 3c. Of course, we do not know the rogue reactor location *a priori*, but North Korea is not such a big territory, so the exact choice of locations should not matter so long as the detectors are reasonably close to land. Second, the detectors take data for 1 year. During this exposure period, it receives background events from commercial reactors, but the expected level can be calculated accurately using data provided by the reactor operators. Finally,

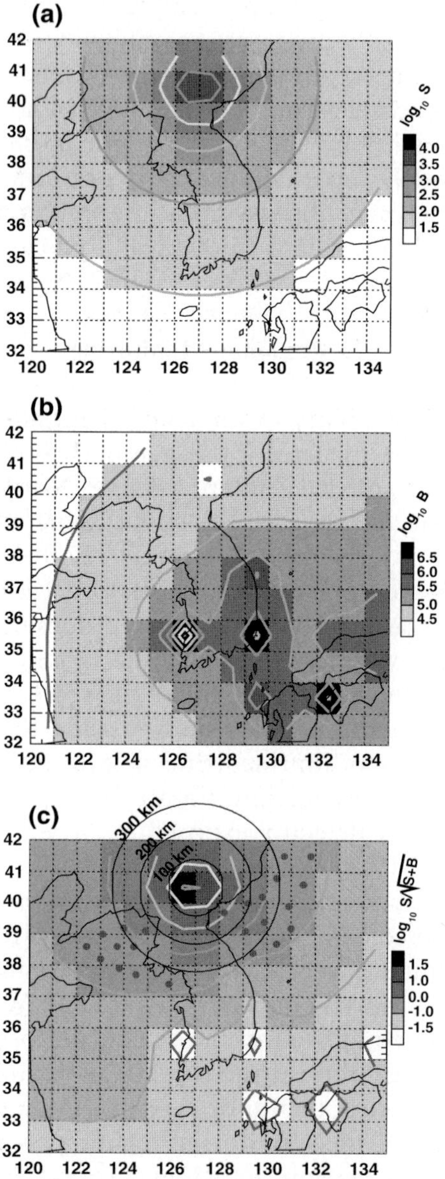

Figure 3. The signal and background from a 100 MW_{th} rogue reactor deep in North Korean territory (127.0° E longitude, 40.5° N latitude). (**a**) The signal S, defined as the number of anti-neutrino events detected by a 1-megaton detector exposed for 1 year. (**b**) The background B, defined as above, but the source of anti-neutrinos are all commercial and research reactors around the world; the vast majority of detected background comes from reactors in South Korea and Japan. (**c**) The signal significance $S/\sqrt{S+B}$. The dots are candidate locations of the 1-megaton detectors. Each location was chosen based on the value of the significance, whose contours are not circular because of distortions from the background.

one compares the observed number of events in each detector to the expected number of background events. If a significant excess is observed in any of the detectors, an alarm is raised and one would then use the data from all the detectors to try to triangulate the location of the rogue activity. At the same time, political action would commence against the rogue regime. The statistical technique used in the comparison of the data against the background expectation is described in detail in Appendix B.

The quantity P_{99} (defined in detail in Appendix B) stands for the threshold rogue reactor thermal power that triggers an alarm at the "99%" confidence level. Figure 4 shows an array configuration with four detectors for which $P_{99} = 128$ $\mathrm{MW_{th}}$. To quantify the ability of the array to pin-point the reactor, a map of $\Delta\chi^2$ was made (Figure 4). This map was made by comparing the observed number of events (sum of signal and background) in each detector with the number of expected events for a hypothetical rogue reactor at different locations and power levels. At each location, the power was varied until the χ^2 between the observed and expected set of events was

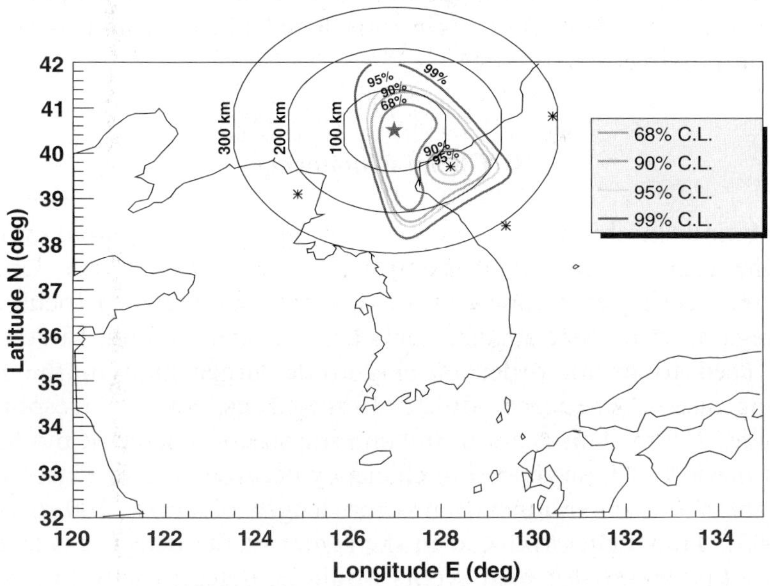

Figure 4. This map demonstrates the ability of four 1-megaton detectors (indicated with asterisks) exposed for 1 year to detect and pin-point a 128 $\mathrm{MW_{th}}$ rogue reactor (star). The power of 128 $\mathrm{MW_{th}}$ corresponds to the 99% detection threshold for this configuration. It was made by varying the hypothesized position and power of the unknown reactor and comparing the number of expected events with the mean number of events that would be observed for the true reactor position; the comparison was quantified using the χ^2 technique. At each longitude and latitude, χ^2 was minimized with respect to the rogue reactor power. The contours correspond to the 68, 90, 95, and 99% confidence level contour for two free parameters (longitude and latitude).

minimized. As one would expect, the smallest *minimized* χ^2 occurs at the true reactor location; $\Delta\chi^2$ is defined as the difference between this smallest minimized χ^2 and that at any given location in the map (by definition, $\Delta\chi^2 = 0$ at the true location). The contours shown in the $\Delta\chi^2$ map indicate the range of likely reconstructed positions at the specified confidence level. In other words, an X-% contour indicates that there is an X-% chance that the reconstructed position would lie within the contour.

There are several notable features in Figure 4. First is the fact that each detector strongly rules out a circular region of radius of several 10s of kilometers. Second is the fact that the alarm level P_{99} is determined almost completely by the closest detector. The addition of the other detectors do not lower the alarm level (i.e., they do not improve the sensitivity); their role is to help pin-point the location of the rogue reactor. To see this, note that if only the closest detector were present, the minimum χ^2 would be an annular region around it; the other detectors strongly rule out circular regions surrounding their locations, thus narrowing down the possible locations.

We finally note that nuclear reactors need to be cooled; reactors located inland are often cooled with a river or a lake. Thus the intersection of rivers and lakes with the confidence region discussed above would allow one to focus in on possible reactor sites.

8. Global Monitoring

More ambitious in objective than regional monitoring is a global monitoring regime, the goal of which is to monitor all locations on Earth. Unlike the regional monitoring case, one cannot optimize resources to focus in on a suspect region, so the size requirements are very demanding. First, detector modules need to be an order of magnitude larger than in the regional monitoring case – i.e., each module is 10 megatons, which corresponds to a cube of sides 216 m. This is about the limiting size of a detector module from the point of view of light detection efficiency because even in extremely pure water, light has a maximum attenuation length of about 100 m (Fukuda et al., 2003). Thus light produced in the center of the detector is attenuated by about a factor of 0.40; such events would be detected with low efficiency with any detector that is significantly larger. Also, the number of modules in an array needs to be on the order of 1000. As before, we took the exposure time to be 1 year. We considered three different array configurations in our study, shown in Figure 5. To measure the performance of the arrays, a map of P_{99} was made (see Figure 6 and Appendix B). In other words, a rogue reactor was assumed to exist in various locations on Earth. For each location, the rogue reactor power was varied until the reactor was detectable at the 99% confidence level. The maps show that in most costal regions

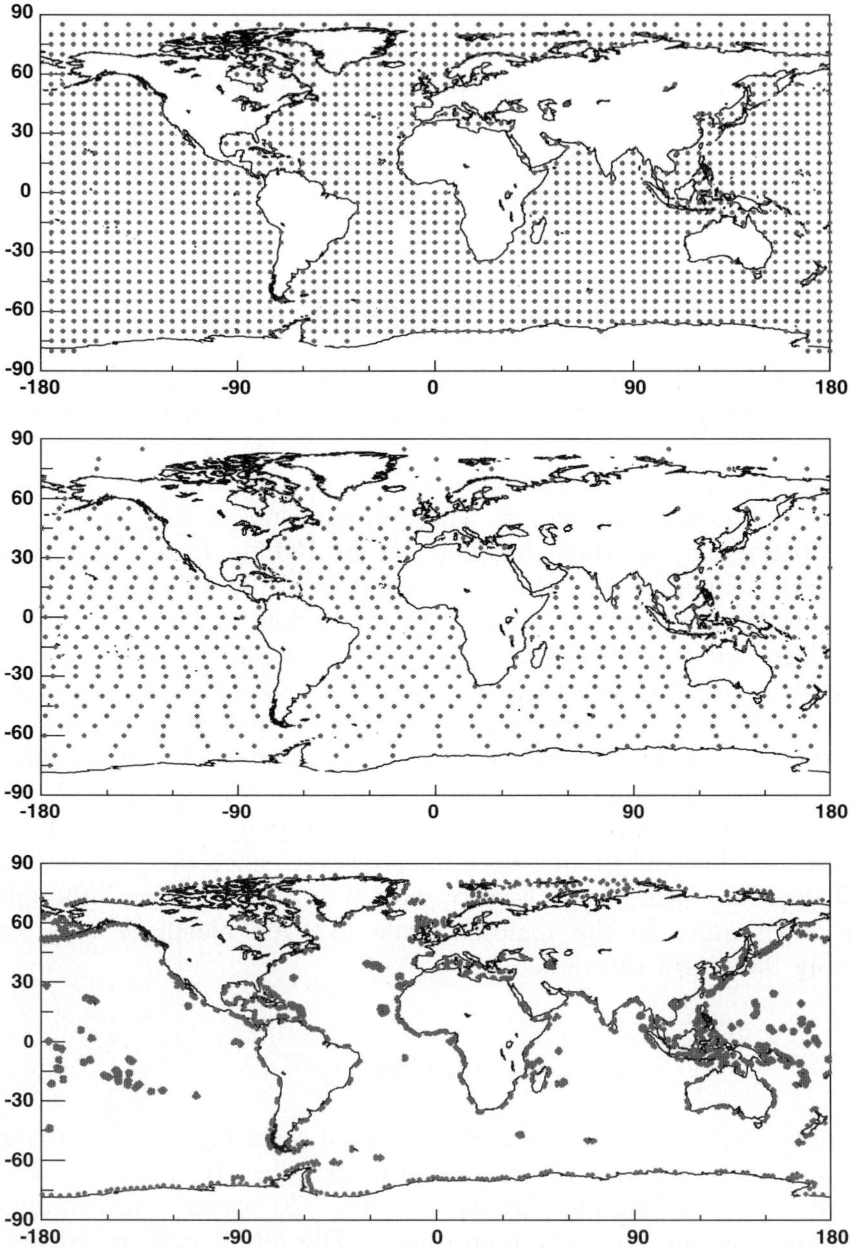

Figure 5. The three array configurations considered in the world-wide monitoring regime. Top: detector modules distributed on a $5° \times 5°$ grid in longitude and latitude. Middle: modules distributed so that they are approximately equidistant from one another. Bottom: modules distributed to hug coastlines; they are approximately 100 km from land, and 100 km from each other. The number of modules in each array are 1596, 623, and 1482. The final results were normalized so that the total detector mass is equal to 1596 modules' worth.

(i.e., within several hundred kilometers of the shore), the array is sensitive to rogue reactor power of several hundred MW_{th}. The sensitivity worsens to about 1000 MW_{th} for regions with a high level of legal nuclear activity. The sensitivity worsens yet to almost 2000 MW_{th} deep within continents. Since rogue reactors, realistically, should have a power of less than about 100 MW_{th}, it is seen that the world-wide monitoring regime considered here does not measure up well to the task at hand.

9. Monitoring of Fission Bomb Detonation

The energy production mechanism of a fission bomb is basically the same as that of a nuclear reactor. The main difference is that the latter operate in a steady-state mode, while the former releases its energy in a short burst. Most of the anti-neutrinos from a fission bomb are released in 10 s from the moment of detonation. The anti-neutrino yield from a 1 kiloton bomb observed by a 1-megaton detector at 100 km from ground-zero is 2.25 events (Equation 3). This may seem like a small number, but one must consider the fact that the amount of background is reduced greatly by the fact that the observation time is ten seconds. A study of the sensitivity of a global array was carried out as in the case for rogue reactors (section 8). In this case, a map of Y_{99} was made instead of P_{99}, where Y_{99} is the yield of a fission bomb that can be detected at the 99% confidence level, as defined in Appendix B. The result is shown in Figure 7. Unfortunately, at most places on Earth, the sensitivity is several kilotons, compared to the goal of one kiloton. However, since the goal is not too far off, one may achieve the goal by targeting certain regions (though not going down quite to the regional scale like for North Korea), or by loosening the alarm threshold.

10. Cost

Here, only a very rough estimate of the cost of the arrays will be attempted. The main costs involve: (1) photomultiplier tubes (PMTs), (2) detector material, civil engineering, transport, etc., (3) water purification, (4) gadolinium dopant, and (5) man-power. The PMT cost is fairly well-understood. If we assume the same coverage as the Super-Kamiokande detector (about 40% of the detector wall area), assuming that the same PMTs as Super-Kamiokande will be used, and assuming that the cost per PMT will be about \$1000 (assuming that bulk-discount or economy of scale reduces the price per PMT), about 120,000 PMT will be required per detector module.

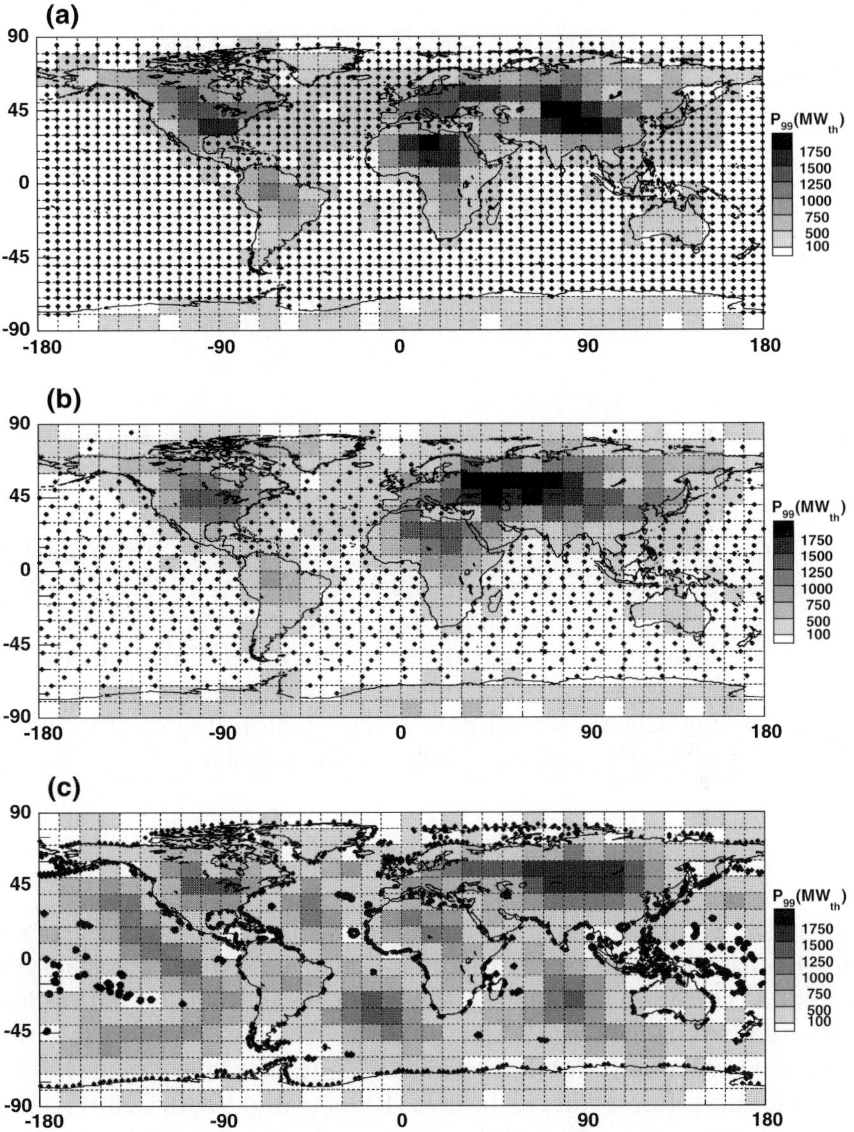

Figure 6. Map of P_{99} for the three array configurations shown in Figure 5. The power is in units of MW_{th}. The number of detector modules in each array is different between arrays, but the total mass has been scaled to 1596 times 10 megatons≈16 gigatons. The target power of $P_{99} < 100$ MW_{th} is indicated by the white areas.

This translates to 120 million dollars per detector. The material, civil engineering, transport, etc. cost is not well-known at this point, but probably several hundred million dollars per module is the right order of magnitude. The cost of water purification is also not known right now, but probably about 100 million dollars is the right order of magnitude. Several thousand

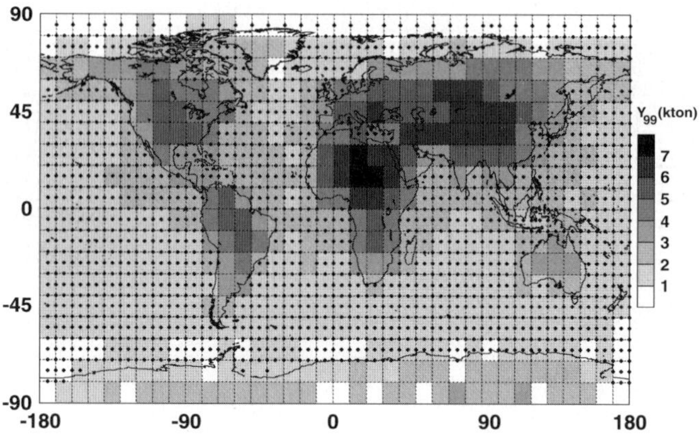

Figure 7. Map of Y_{99} for the array configuration shown in the top of Figure 5. The target sensitivity is less than 1 kiloton, which is indicated by the white boxes.

tons of $GdCl_3$ per module is required; at $3 per kilogram, this translates to several million dollars, which is negligible compared to the total cost. Manpower and everything else is quite vague, but something on the order of $100 million is probably the right order of magnitude. In total, then, each detector module will probably cost about $1 billion. This implies that a regional monitoring scheme would cost several billion dollars. In contrast, a world monitoring regime will cost several trillion dollars.

11. Conclusion

Our study shows that targeted regional monitoring of rogue nuclear reactor activity in a nation without pre-existing legally operated nuclear reactors may be done at a cost of several billion dollars, provided that the nation has significant coastlines facing large bodies of water. We note, however, that the cost accounting is very rough, and that several key features of the scheme have yet to be proved feasible. For instance, the idea (Beacom and Vagins, 2001) of doping a water-based detector with gadolinium to make it sensitive to reactor antineutrinos is promising, but yet unproven. The verdict should be out in the next several years as Super-Kamiokande starts its third experimental phase this year (2006). Another unproven scheme is the deploying of a KamLAND-like detector in a deep-sea environment. Hanohano appears to be the single experiment that will test this idea in the coming decade. Finally, the idea of deploying a megaton-scale detector is unproven. Ideas to construct detectors on this scale exist (e.g., Hyper-K and UNO), but there are no schedules for building any. Moreover, these detectors are to be deployed on land, which simplifies matters considerably compared to deploying them deep in the ocean.

The more ambitious idea of an untargeted world-wide array was studied, but it did not reach the target sensitivity of about 100 MW_{th} for reactors nor the 1 kiloton yield for fission bombs. This was at a prohibitive cost of more than trillion dollars. Thus it is concluded that an untargeted world-wide monitoring scheme is unrealistic. A viable monitoring scheme must focus in on some region in order to optimize resources.

Acknowledgments

I would like to thank Prof. John Learned for his leadership in initiating at the University of Hawaii a program exploring the use of anti-neutrinos for monitoring nuclear activities. I feel honored to have been invited to join the research effort, and am indebted to him for his kind support throughout the time I was there. I would like to thank Prof. Stephen Dye for his dedication to Hanohano and related research activities. I would also like to thank Jelena Maricic for working with me on various aspects of research in anti-neutrino detection. I thank all members of the University of Hawaii Hanohano team who worked hard to keep up the momentum for an exceptionally worthy goal. Thanks also to the members of the Neutrino Physics Group and all others in the Physics Department for maintaining a high-quality research environment, and to the Department Staff whose daily support was invaluable.

Appendix A: A List of Nuclear Reactor Location and Power

This appendix gives a list of the location (longitude and latitude) and the nominal thermal power of registered nuclear reactors throughout the world. The list was obtained in 2003 from the International Nuclear Safety Center.[8] The list may not be up to date, and the positions are only approximate. Our results, however, do not depend sensitively on the exact world-wide distribution of nuclear reactors, nor on the relatively small change in the total nuclear power world wide since 2003, so this should be sufficient for the purpose of this study. A total of 433 reactors are in the list, and the total thermal power is 1.06 TW_{th}. This list will be provided in ASCII format upon request to ehguillian@gmail.com.

[8] The International Nuclear Safety Center, operated by the Argonne National Laboratory, maintains a list of registered nuclear reactors worldwide. Prof. John Learned of the University of Hawaii, Manoa obtained a text version of the list through private channels.

Appendix B: The Statistical Technique Used to Compare the Observed Number of Events against the Expected Number

Let us say that there are N_{det} detectors in the array. In the absence of any rogue reactor, detector number i detects b_i events per year; these are from commercial and research reactors, and (possibly) from the georeactor in Earth's core. We denote the set of observed number of events $b_1, b_2, \ldots b_{N\text{det}}$ by the set notation used in mathematics, $\{b_i\}$.

The actually observed number of events in each detector is represented as $\{n_i\}$. In the absence of rogue activity, the numbers in $\{n_i\}$ should agree with those in $\{b_i\}$. If, however, rogue activity is taking place, the predicted numbers $\{b_i\}$ is incorrect, and it should be replaced with $\{b_i + s_i\}$, where s_i represents the number of events in detector number i due to the rogue activity.

The method used to detect rogue activity starts with the assumption that no rogue activity is taking place, so that the predicted number of events at each detector is given by $\{b_i\}$. The set of observed number of events is compared against the observed number $\{n_i\}$ using a *likelihood function*; as the name suggests, this function provides information about how likely a set of numbers $\{n_i\}$ is to have resulted from the predicted set $\{b_i\}$, given statistical and systematic uncertainties. For this report, we considered only statistical uncertainty, in which case the logarithm of the likelihood function (the log-likelihood function) is defined as follows:

$$\mathcal{L} = -b_i + n_i \ln b_i - \ln \Gamma(n_i + 1) \tag{B1}$$

The last term $\ln \Gamma\ (n_i + 1)$ is the logarithm of the Gamma function.

The value of $\mathcal{L}$ for a given measurement (lasting 1 year) is not known *a priori* because of statistical fluctuations, although the mean expected value $\langle\mathcal{L}\rangle$ is (Figure 8a). The mean expected value depends on the power P of the rogue reactor; we denote this dependence as $\langle\mathcal{L}\rangle(P)$. As the power increases, the assumption that no rogue reactor exists becomes increasingly inconsistent with observations; this inconsistency causes $\langle\mathcal{L}\rangle$ to be biased to lower values. When P is small, the slightly biased distribution of $\langle\mathcal{L}\rangle(P)$ largely overlaps the distribution of $\langle\mathcal{L}\rangle(0)$, which implies that the detector array is not sensitive enough to detect such a low value of P (Figure 8 b). However, as the power is raised, a point is reached where the two distributions are different enough that the rogue reactor can be judged to exist with great confidence (Figure 8c).

For the purpose of identifying a rogue reactor, a threshold level of the log-likelihood value for triggering an alarm is necessary. If this is set too close to $\langle\mathcal{L}\rangle(0)$, the observed value of $\langle\mathcal{L}\rangle$ would easily trigger a false-positive alarm just from statistical fluctuations. For the purpose of the present study, we decided to tolerate a 1% false-positive probability (Figure 8c). Once this threshold is set, we can talk about the sensitivity of an array. We quantified

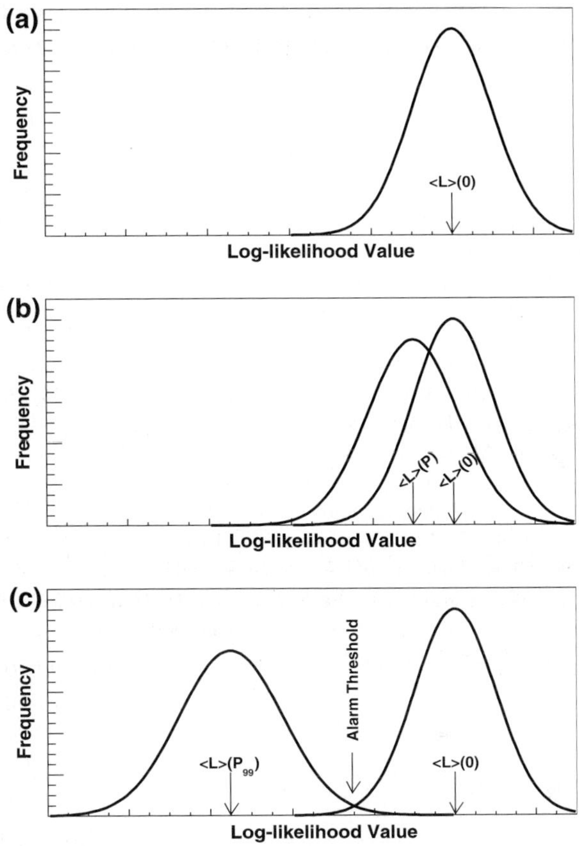

Figure 8. Illustrating how the log-likelihood distribution changes with rogue reactor power. (**a**) When no rogue reactor exists, the data agree well with the assumption, so that the mean log-likelihood value $\langle\mathcal{L}\rangle(0)$ is high. Any given measurement is distributed around the mean due to statistical fluctuations. (**b**) As the rogue reactor power increases to P, the mean value $\langle\mathcal{L}\rangle(P)$ decreases. However, P is small so the distribution at this power largely overlaps with the distribution at zero power, which means that the detector array is not sensitive enough to confidently detect the rogue reactor. (**c**) When the rogue reactor power is sufficiently large, the overlap between the distributions become very small, and the existence of the reactor can be confirmed with great confidence. The power P_{99} is defined as the power above which there is 99% chance that the likelihood value will be above the alarm threshold, which is defined as the log-likelihood value below which there is only 1% chance for a "false positive". We note that the width of the distribution increases slowly with power.

this with P_{99}, which is the rogue reactor power that has a 99% chance of clearing the alarm threshold. Since rogue reactors are not likely to be much larger than 100 MW_{th}, a promising detector array should have P_{99} at this level. Of course, the tolerance for false-positives and -negatives chosen here are arbitrary; looser tolerance would result in sensitivity to lower power, but at the cost of greater chance of mis-identification and missing an actually existing reactor.

A list of registered nuclear reactors worldwide is available as supplementary material in the online version of this article at doi:10.1007/s11038-006-9110-x and is accessible for authorized users.

References

Eguchi K. et al.: 2003, *Phys. Rev. Lett.* **90**, 021802, URL http://www.citebase.org/cgi-bin/citations?id = oai:arXiv.org:hep-ex/0212021.

Bernstein, A., West, T. and Gupta, V.: 2001, *An assessment of antineutrino detection as a tool for monitoring nuclear explosions, Science & Global Security*, Vol. 9, Taylor and Francis, pp. 235-255.

Fukuda, S. et al.: 2003, *Nucl. Instrum. Meth. A* **501**, 418462.

Bemporad, C., Gratta, G. and Vogel, P.: 2002, *Rev. Mod. Phys.* **74**, 297, http://www.citebase.org/cgi-bin/citations?id = oai:arXiv.org:he p-ph/0107277.

Beacom, J. F. and Vagins, M. R.: 2001, *Phys. Rev. Lett.* **93**, 171101, http://www.citebase.org/cgi-bin/citations?id = oai:arXiv.org:he p-ph/0309300.

Herndon, J. M.: 1979, *Proc. R. Soc. London A* **368**, 495.

Herndon, J. M.: 1980, *Proc. R. Soc. London A* **372**, 149.

Herndon, J. M.: 1996, *Proc. Natl. Acad. Sci. USA* **93**, 646.

Hollenbach, D. F. and Herndon, J. M.: 2001, *Proc. Natl. Acad. Sci. USA* **98**, 11085.

Herndon, J. M.: 2003, *Proc. Natl. Acad. Sci. USA* **100**, 3047.

Maricic, J.: 2005, Ph.D. thesis, University of Hawaii, Manoa.

Earth, Moon, and Planets (2006) 99:331–341
DOI 10.1007/s11038-006-9105-7

Neutrinos and Non-proliferation in Europe

MICHEL CRIBIER
CEA/Saclay, DAPNIA/SPP and APC, Paris, France
(E-mail: mcribier@cea.fr)

(Received 10 March 2006; Accepted 28 June 2006)

Abstract. The main effort in Europe to evaluate the interest for IAEA of neutrinos detectors close to nuclear power stations is made within the Double Chooz experiments. Specific simulation of diversion scenarios as well as new experimental measurements of neutrinos emitted are underway.

Keywords: neutrino, non-proliferation, nuclear reactors

The International Atomic Energy Agency (IAEA) is the United Nations agency in charge of the development of peaceful use of atomic energy. In particular IAEA is the verification authority of the Treaty on the Non-Proliferation of Nuclear Weapons (NPT). To do that job inspections of civil nuclear installations and related facilities under safeguards agreements are made in more than 140 states.

IAEA uses many different tools for these verifications, like neutron monitor, gamma spectroscopy, but also bookeeping of the isotopic composition at the fuel element level before and after their use in the nuclear power station. In particular it verifies that weapon-origin and other fissile materials that Russia and USA have released from their defense programmes are used for civil application.

The existence of an antineutrino signal sensitive to the power and to the isotopic composition of a reactor core, as first proposed by Mikaelian et al. and as demonstrated by the Bugey (Achkar et al., 1995) and Rovno experiments, (Vidyakin et al., 1994), could provide a means to address certain safeguards applications. Thus the IAEA recently asked member states to make a feasibility study to determine whether antineutrino detection methods might provide practical safeguards tools for selected applications. If this method proves to be useful, IAEA has the power to decide that any new nuclear power plants built have to include an antineutrino monitor.

Within the Double Chooz collaboration, an experiment (Lasserre, these proceeding) mainly devoted to studying the fundamental properties of neutrinos, we thought that we were in a good position to evaluate the interest of using antineutrino detection to remotely monitor nuclear power stations.

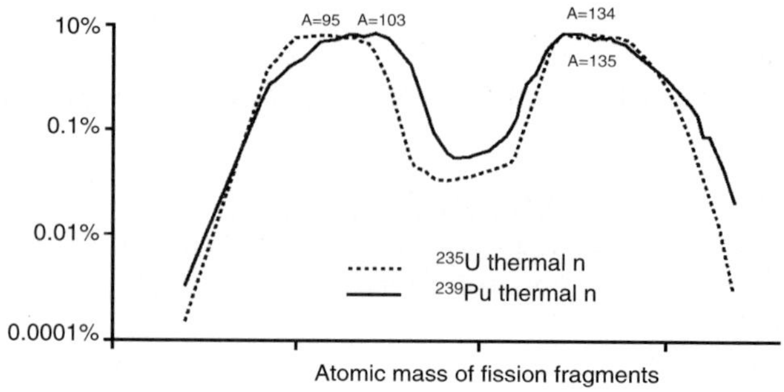

Figure 1. The statistical distribution of the fission products resulting from the fission of the most important fissile nuclei ^{235}U and ^{239}Pu shows two humps, one centered around masses 100 and the other one centered around 135. The low mass hump is at higher mass in ^{239}Pu fission than in ^{235}U, resulting in different nuclei and decays.

This effort in Europe, supplemented by the US effort (Ardellier et al., 2006), will constitute the basic answer to IAEA of the neutrino community.

The high penetration power of antineutrinos and the detection capability might provide a means to make remote, non-intrusive measurements of plutonium content in reactors (Bernstein et al., 2002). The antineutrino flux and energy spectrum depends upon the thermal power and on the fissile isotopic composition of the reactor fuel. Indeed, when a heavy nucleus (uranium, plutonium) experiences a fission, it produces two unequal mass fission fragments (and a few free neutrons); the statistical distribution of the atomic masses is depicted in Figure 1. All these nuclei immediately produced are extremely unstable – they are too rich in neutrons – and thus β decay toward stable nuclei with an average of 6 β decays. All these process involving several hundreds of unstable nuclei, with their excited states, makes it very difficult to understand details of the physics. Moreover, the most energetic antineutrinos, which are detected more easily, are produced in the very first decays involving nuclei with typical lifetimes smaller than a second.

Based on predicted and observed β spectra, the number of antineutrinos per fission from ^{239}Pu is known to be less than the number from ^{235}U, and the energy released bigger by 5%. Hence an hypothetical reactor able to use only ^{235}U would induce in a detector an antineutrino signal 60% higher than the

	^{235}U	^{239}Pu
Released energy per fission	201.7 MeV	210.0 MeV
Mean energy of ν	2.94 MeV	2.84 MeV
ν per fission > 1.8 MeV	1.92	1.45
Average inter. cross section	$\approx 3.2 \times 10^{-43}$ cm^2	$\approx 2.76 \times 10^{-43}$ cm^2

same reactor producing the same amount of energy but burning only ^{239}Pu (see Table). This offers a means to monitor changes in the relative amounts of ^{235}U and ^{239}Pu in the core. If made in conjunction with accurate independent measurements of the thermal power (with the temperature and the flow rate of cooling water), antineutrino measurements might provide an estimate of the isotopic composition of the core, in particular its plutonium inventories. The shape of the antineutrino spectrum can provide additional information about core fissile isotopic composition.

Because the antineutrino signal from the reactor decreases as the square of the distance from the reactor to the detector a precise "remote" measurement is really only practical at distances of a few tens of meters if one is constrained to "small" detectors of a few cubic meters in size.

1. Simulations

1.1. Magnitudes of some effects

In our group, the development of detailed simulations using professional reactor codes started (see below), but it seems wise to use less sophisticated methods in order to evaluate already, with some flexibility, the magnitude of some effects. To do that we started from the set of Bateman equations, as depicted graphicaly in Figure 2, which describe the evolution of fuel elements in a reactor. The gross simplification in such treatment is the use of average cross section, depending only on three groups (thermal neutron, resonance region, fast neutrons), and moreover the fact that the neutron flux is imposed and not calculated.

Given this we use for each isotope under consideration, the cross section for capture, fission, and also plug in the parameters of the decays. Then it is rather easy (and fast) to simulate the evolution of a given initial core composition; in the same way, it is possible to "make a diversion" by manupulating the fuel composition at a choosen moment. As an example, Figure 3 shows the evolution of a fresh core composed of uranium enriched at 3.5 % in ^{235}U: the build up of ^{239}Pu and ^{241}Pu is rather well reproduced.

Knowing the amount of fissions at a given time, it is straight forward to translate a given antineutrino flux using the parametrisation of Huber and Schwetz (2004), and finally using the interaction cross section for inverse β decay reaction, to produce the recorded signal in a given detector placed at a suitable location from the reactor under examination (Figure 4).

As an example of this type of computation, we show in Figure 5, the effect of the modification of fuel composition after 100 days: here the operator, clever enough, knows that he cannot merely remove plutonium from the core without changing the thermal power which will be immediatly noticed. Hence

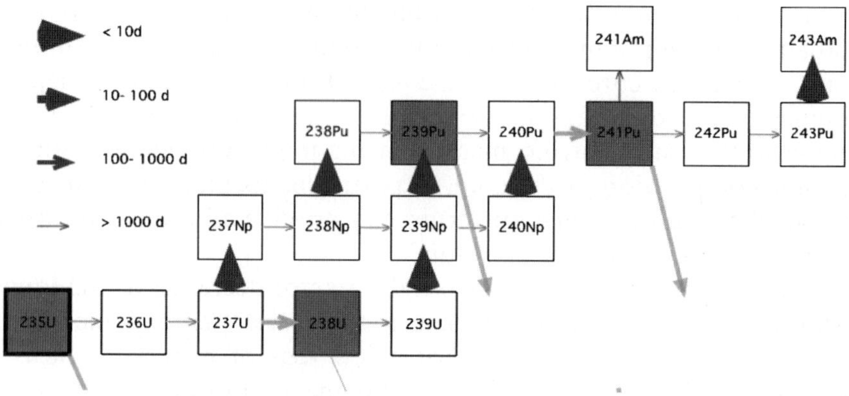

Figure 2. The Bateman equations are the set of differential equations which described all transformations of the nuclei submitted to a given neutron flux: capture of neutrons are responsible to move at Z constant (rightward arrow), β-decay are responsible to increase the atomic mass by one unit (upward arrow), and fission destroy the heavy nuclei and produce energy (sideward arrows).

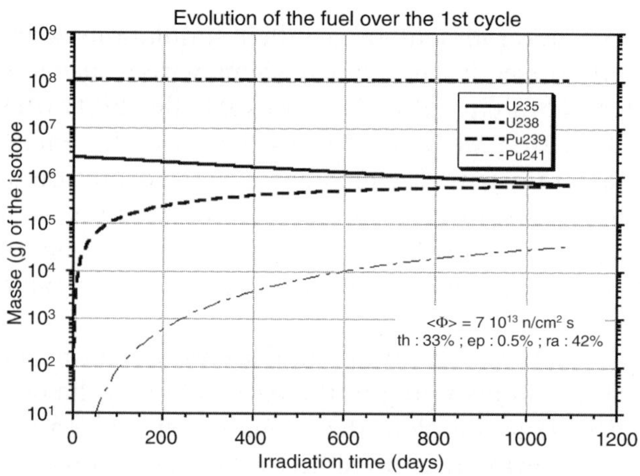

Figure 3. In a new reactor the initial fuel consists of enriched uranium rods, with an ^{235}U content typically at 3.5%, the rest is ^{238}U. As soon as the reactor is operating, reactions described by Bateman equations produce ^{239}Pu (and ^{241}Pu), which then contribute to the energy production, at the expense of ^{238}U.

he takes the precaution to add 28 kg of ^{235}U at the same time where he removes 20 kg of ^{239}Pu: although the thermal power is kept constant, the imprint on the antineutrino signal, although modest, is such that, after 10 days, there is an increase of more than 1σ in the number of interactions recorded. Such a diversion is clearly impossible in a Pressurized Water Reactor (PWR) or in a Boiling Water Reactor (BWR), but more easy in a Candu-type reactor, and even more so in a molten salt reactor.

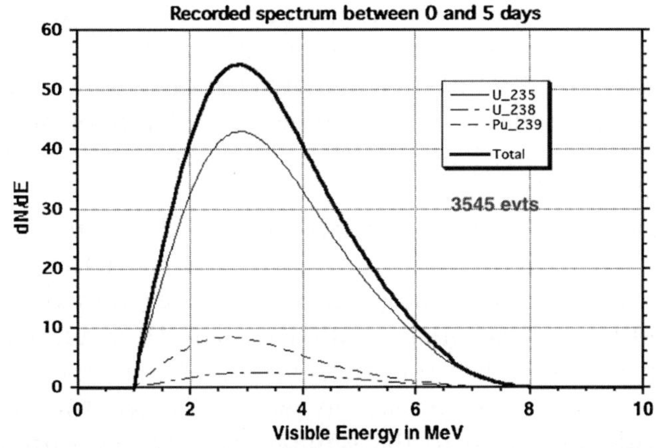

Figure 4. Positron spectrum recorded in an typical antineutrino detector (10 tons of target) placed 150 m from a nuclear reactor (1000 MW_{el}). Positrons results from the inverse β-decay reaction used in the detection of anti-neutrino. The signal is the superposition of several components whose spectra exhibit small but sizeable differences, especialy at high energy.

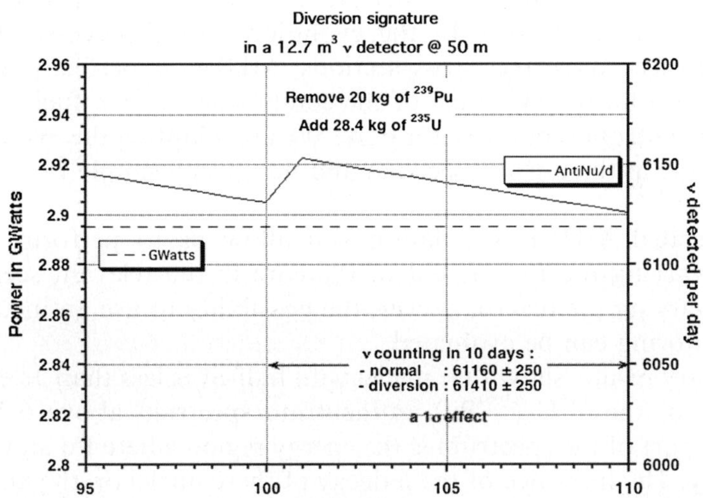

Figure 5. An hypothetical diversion scenario where an exchange of ^{239}Pu with ^{235}U is made such that the power does not change, but the antineutrino signal recorded by the monitor is slightly increased, giving some evidence of an abnormal operation.

1.2. Simulations of diversion scenarios

The IAEA recommends the study of specific safeguards scenarios. Among its concerns are the confirmation of the absence of unrecorded production of fissile material in declared reactors and the monitoring of the burn-up of a reactor core. The time required to manufacture an actual weapon estimated by the IAEA (conversion time), for plutonium in partially irradiated or spent fuel, lies between 1 and 3 months. The significant quantity of Pu is 8 kg, to be compared with the 3 tons of ^{235}U contained in a PWR of power 900 MWe enriched to 3%. The small magnitude of the researched signal requires a carefull feasability study.

The proliferation scenarios of interest involve different kinds of nuclear power plants such as light water or heavy water reactors (PWR, BWR, Candu...), it has to include isotope production reactors of a few tens of MWth, and future reactors (e.g., PBMRs, Gen IV reactors, accelerator-driven sub-critical assemblies for transmutation, molten salt reactors). To perform these studies, core simulations with dedicated Monte-Carlo codes should be provided, coupled to the simulation of the evolution of the anti-neutrino flux and spectrum over time.

We started simulation work using the widely used particle transport code MCNPX (Monte Carlo), coupled with an evolution code solving the Bateman equations for the fission products within a package called MURE (MCNP Utility for Reactor Evolution, 2005). This package offers a set of tools, interfaced with MCNP or MCNPX, that allows to define easily the geometry of a reactor core. In the evolution part, it accesses the set of evaluated nuclear data and cross sections. MURE is perfectly adapted to simulate the evolution with time of the composition of the fuel, taking into account the neutronics of a reactor core. We are adapting the evolution code to simulate the antineutrino spectrum and flux, using simple Fermi decay as starting point.

The extended MURE simulation will allow us to perform sensitivity studies by varying the Pu content of the core in the relevant scenarios for IAEA. By varying the reactor power, the possibility to use antineutrinos for power monitoring can be evaluated.

Preliminary results show that nuclei with half-lives less than 1s emit about 70% (50%) of the ^{235}U (^{239}Pu) antineutrino spectrum above 6 MeV. The high energy part of the spectrum is the energy region where Pu and U spectra differ mostly. The influence of the β decay of these nuclei on the antineutrino spectrum might be preponderant also in scenarios where rapid changes of the core composition are performed, e.g. in reactors such as Candu, refueled on line.

The appropriate starting point for this scenario is a representative PWR, like the Chooz reactors. For this reactor type, simulations of the evolution of the antineutrino flux and spectrum over time will be provided and compared to the accurate measurement provided by the near detector of Double Chooz. This should tell the precision on the fuel composition and of an independent thermal power measurement. An interesting point to study is at the time of the partial refuelling of the core, thanks to the fact that reactors like Chooz (N4-type) do not use MOX fuel.

Without any extra experimental effort, the near detector of the Double Chooz experiment will provide the most important data set of antineutrinos detected (5×10^5 ν per year) by a PWR. The precise neutrino energy spectrum recorded at a given time will be correlated to the fuel composition and to the thermal power provided by electrical company. This valuable dataset will constitute an excellent experimental basis for the above feasibility studies of potential monitoring and for bench-marking fuel management codes; it is expected that individual components due to fissile elements (^{235}U, ^{239}Pu) could be extracted with some modest precision and serve as a benchmark of this technique.

To fulfil the goal of non-proliferation additional lab tests and theoretical calculations should be performed to more precisely estimate the underlying neutrino spectra of plutonium and uranium fission products, especially at high energies. Contributions from decays to excited states of daughter nuclei are mandatory to reconstruct the shape of each spectrum. Following the conclusion of Huber and Schwetz (2004) to achieve this goal a reduction of the present errors on the antineutrino fluxes of about a factor of three is necessary. We will see that such improvement needs an important effort.

2. Experimental Effort

The precise measurement of β-decay spectra from fission products produced by the irradiation of a fissile target can be performed at the high flux reactor at Institut Laue Langevin (ILL) in Grenoble, where similar studies performed in the past (Schreckenbach et al., 1985) are the basis of the actual fluxes of antineutrinos used in this reactor neutrino experiment. The ILL reactor produces the highest neutron flux in the world: the fission rate of a fissile material target placed close to the reactor core is about 10^{12} per second. It is possible to choose different fissile elements as targets in order to maximize the yield of the nucleus of interest. Using the LOHENGRIN recoil mass spectrometer (ILL Instrument, 2004/2005, measurement of individual β-spectra from short-lived fission products are possible; in the same irradiation channel, measurements of integral β-spectra with the Mini-INCA detectors (Marie et al., 2006) could be envisaged to study the evolution with time of the antineutrino energy spectrum of a nuclear power plant.

2.1. Experiments with Lohengrin

The LOHENGRIN recoil mass spectrometer offers the possibility to measure β-decays of individual fission products. The fissile target (^{235}U, ^{239}Pu,

^{241}Pu, ...) is placed into a thermal neutron flux of 6×10^{14}n/cm^2/s, 50 cm from the fuel element. Recoil fission products are selected with a dipolar magnetic field followed by an electrostatic condenser. At the end the fragments could be implanted in a moving tape, and the measurement of subsequent β and γ-rays are recorded by a β-spectrometer (Si-detector) and Ge-clover detectors, respectively. Coincidences between these two quantities could also be made to reconstruct the decay scheme of the observed fission products or to select one fission product. Fragments with half-lives down to 2 μs can be measured, so that nuclei with large Q_β (above 4 MeV) can be measured.

The LOHENGRIN experimental objectives are to complete existing β-spectra of individual fission products (Tengblad et al., 1989) with new measurements for the main contributors to the detected ν-spectra and to clarify experimental disagreements between previous measurements. This ambitious experimental programme is motivated by the fact – noted by C. Bemporad et al. (2002) – that unknown decays contribute as much as 25% of the antineutrinos at energies > 4 MeV. Folding the antineutrino energy spectrum over the detection cross-section for inverse beta decay enhances the contribution of the high energy antineutrinos to the total detected flux by a factor of about 10 for $E_\nu > 6$ MeV. The focus of these experiments will be on neutron-rich nuclei with yields very different in ^{239}Pu and ^{235}U fission. In the list: ^{86}Ge, $^{90\text{-}92}$Se, ^{94}Br, $^{96\text{-}98}$Kr, ^{100}Rb, $^{100\text{-}102}$Sr, $^{108\text{-}112}$Mo, $^{106\text{-}113}$Tc, $^{113\text{-}115}$Ru... contribute to the high energy part of the spectrum and have never been measured.

2.2. Irradiation tests in summer 2005

A preliminary experiment was performed during two weeks in summer 2005. The isobaric chains $A = 90$ and $A = 94$ were studied where some isotopes possess a high Q_β energy, contributing significantly to the high energy part of the antineutrino spectra following ^{235}U and ^{239}Pu fissions and moreover produced with very different fission yields after ^{235}U and ^{239}Pu fission (England and Rider, 1993). The well-known nuclei, such as ^{90}Br, will serve as a test of the experimental set-up, while the beta decay of more exotic nuclei such as ^{94}Kr and ^{94}Br will constitute a test case for how far one can reach in the very neutron-rich region with this experimental device. The recorded data (Figure 6) will validate the simulation described in the previous section, in particular the evolution over time of the isobaric chains beta decay spectra.

2.3. Integral β spectra measurements

In complement to individual studies on LOHENGRIN, more integral studies can be envisaged using the so called "Mini-INCA chamber" at ILL (Marie

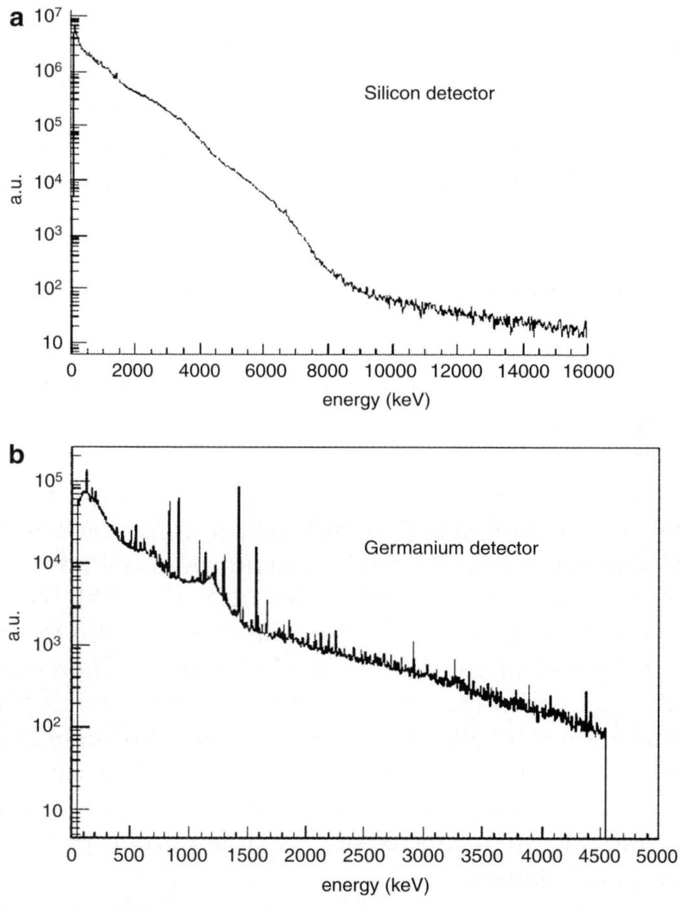

Figure 6. Beta energy spectrum (a) recorded with the silicon detector corresponding to β decay of fission products with mass $A = 94$. The fission products arising from the LOHENGRIN spectrometer were implanted on a mylar tape of adjustable velocity in front of the silicon detector. The highest velocity was selected in order to enhance shorter-lived nuclei such as ^{94}Kr and ^{94}Br. The gamma energy spectrum (b) obtained with the germanium detector corresponding to the same runs is displayed also.

and Letourneau et al., 2006) in return for adding a β-spectrometer (to be developed). The existing α- and γ-spectroscopy station is connected to the LOHENGRIN channel and offers the possibility to perform irradiations in a quasi-thermal neutron flux up to 20 times the nominal value in a PWR. Moreover, the irradiation can be repeated as many times as needed. It offers then the unique possibility to characterize the evolution of the β spectrum as a function of the irradiation time and the irradiation cooling. The expected modification of the β spectrum as a function of the irradiation time is connected to the transmutation induced by neutron capture of the fissile and fission fragment elements. It is thus related to the "natural" evolution of the spent-fuel in the reactor. The modification of the β spectrum as a function of the cooling time is connected to the decaying chain of the fission products and is then a means to select the emitted fragments by their livetime. This information is important because long-lived fission fragments accumulate in the core and after a few days mainly contribute to the low energy part of the antineutrino spectra.

Due to the mechanical transfer of the sample from the irradiation location to the measurement station an irreducible delay time of 30 mn is imposed leading to the loss of short-live fragments.

2.4. Prospect to study fission of ^{238}U

The integral beta decay spectrum arising from ^{238}U fission has never been measured. All information relies on theoretical computations (Vogel and Engel, 1989). Some experiments could be envisaged using few MeV neutron sources in Europe (Van de Graaf in Geel, SINQ in PSI, ALVARES or SAMES accelerators at Valduc, ...). Here the total absence of experimental data on the β emitted in the fission of ^{238}U changes the context of this measurement compared to the other isotopes. Indeed any integral measurements performed could be used to constrain present theoretical estimations of the antineutrino flux produced in the fission of ^{238}U. In any case it seems rather difficult to fulfil the goal of a determination of the isotopic content from antineutrino measurements as long as an important part of the energy spectrum is so poorly known.

3. Conclusions

After the preliminary studies, some conclusions can already be made. A realistic diversion ($\approx$10 kg Pu) has an imprint in the antineutrino signal which is very small. The present knowledge of antineutrino spectra emitted in fissions is not precise enough to allow a determination of the isotopic content in the core sensitive to such diversion.

On the other hand, the thermal power measurement is a less difficult job. Neutrinos sample the whole core, without attenuation, and would bring valuable information on the power with totally different systematics than present methods.

Even if its measurement is not dissuasive by itself, the operator cannot hide any stops or change of power, and in most cases, such a record made with an external and independent device, virtually impossible to fake, will act as a strong constraint.

In spite of the uncertainty mentioned previously, we see that the most energetic part offers the best possibility to disentangle fission from ^{235}U and ^{239}Pu. The comparison between the cumulative numbers of antineutrinos as a function of antineutrino energy detected at low versus high energy is an efficient observable to distinguish pure ^{235}U and ^{239}Pu.

IAEA seeks also monitoring large spent-fuel elements. For this application, the likelihood is that antineutrino detectors could only make measurements on large quantities of beta-emitters, e.g., several cores of spent fuel. In the time of the experiment the discharge of parts of the core will happen and the Double Chooz experiment will quantify the sensitivity of such monitoring.

More generally the techniques developed for the detection of antineutrinos could be applied for the monitoring of nuclear activities at the level of a country. Hence a KamLAND-type detector deeply submerged off the coast of the country, would offer the sensitivity to detect a new underground reactor located at several hundreds of kilometers. All these common efforts toward more reliable techniques, remotely operated detectors, not to mention undersea techniques will automatically benefit both safeguards and geo-neutrinos.

References

Achkar, B. et al.: 1995, *Nucl. Phys.* **B434,** 503–532.

Ardellier, F. et al. hepex/0606025.

Bemporad, C. et al.: 2002, *Rev. of Mod. Phys.* **74**, 297–328.

Bernstein, A. et al.: 2002, *J. Appl. Phys.* **91**, 4672–4676.

England, T.R. and Rider, B.F.: 1993, Los Alamos National Laboratory 3106.

Huber, P. and Schwetz, Th.: 2004, *Phys. Rev.* **D70**, 053011.

T. Lasserre, these proceeding.

ILL Instrument Review, 2004/2005.

Mikaelian, L.A. Neutrino laboratory in the atomic plant, Proc. Int. Conference Neutrino-77, v. 2, pp. 383–387.

Marie, F. et al.: 2006, *Nucl. Instr. and Meth.* **A556**, 547–555.

Monte Carlo N-Particle eXtended, LA-UR-05–2675, J.S.Hendricks et al.

MURE: MCNP Utility for Reactor Evolution—Description of the methods, first applications and results. MÃl' plan O., Nuttin A., Laulan O., David S., Michel-Sendis F. et al. In Proceedings of the ENC 2005 (CD-Rom) (2005), 1–7.

Schreckenbach, K., Colvin, G., Gelletly, W. and Feilitzch, F.v.: 1985, *Phys. Lett.* **B160**, 325–330.

Tengblad, O. et al.: 1989, *Nuclear Physics* **A503**, 136–160.

Vidyakin, G.S. et al.: 1994, *JETP* **59**, 390–393.

Vogel, P. and Engel, J.: 1989, Phys. Rev. **D39**, 3378–3383.

Earth, Moon, and Planets (2006) 99:343–358
DOI 10.1007/s11038-006-9103-9

Strategy for Applying Neutrino Geophysics to the Earth Sciences Including Planetary Habitability

NORMAN H. SLEEP
Department of Geophysics, Stanford University, Stanford, CA, 94305, USA
(E-mail: norm@pangea.stanford.edu)

(Received 7 February 2006; Accepted 28 June 2006)

Abstract. Antineutrino data constrain the concentrations of the heat producing elements U and Th as well as potentially the concentration of K. Interpretation is similar to but not homologous with gravity. Current geoneutrino physics efficiently asks simple questions taking advantage of what is already known about the Earth. A few measurements with some sites in the ocean basins will constrain the concentration of U and Th in the crust and mantle and whether the mantle is laterally heterogeneous. These results will allow Earth science arguments about the formation, chemistry, and dynamics of the Earth to be turned around and appraised. In particular, they will tell whether the Earth accreted its expected share of these elements from the solar nebula and how long radioactive heat will sustain active geological processes on the Earth. Both aspects are essential to evaluating the Earth as a common or rare habitable planet.

Key words: Bulk silicate earth, neutrino, planetary habitability, potassium, solar nebula, thorium, uranium

1. Introduction

Physicists now have the capability to detect antineutrinos from the decay of U and Th in the Earth's interior (Araki et al., 2005). They may eventually be able to detect the products of potassium decay. I will concentrate on the fruits of geoneutrino research, as interpretation of the X-ray mode is straightforward.

For the foreseeable future, neutrino detectors of all types will be bulky and expensive. To be successful, neutrino geophysicists must ask simple questions and take advantage of the fact that a lot is already known about the Earth. Essentially, they will turn around historical geochemical and cosmochemical arguments made to constrain the absolute abundances of antineutrino-producing elements. In the process, they will validate and refine or refute the underlying assumptions of these arguments. I show that these antineutrino studies will be a major advance to these sciences and geodynamics.

I keep my arguments general and cite review papers where available. My intention is to avoid obscuring this review with details that arise once antineutrino data for an actual site are in hand. Because of the scarcity of detectors, I concentrate on global issues. The total amounts of U and K constrain models of the Earth's accretion from the solar nebula. The antineutrino producing elements, U and Th are also the major heat producing elements in the Earth's interior; their abundances calibrate information on the thermal history of the Earth and on the expected duration of active tectonic processes on terrestrial planets. In turn, these endeavors provide information on the abundance and dynamics of habitable planets.

I do not discuss a "georeactor" at the center of the Earth (Herndon, 2003; Schuiling, this volume). The reality of this feature will be evident once antineutrino data are available at a site far from man-made reactors. Except for the slight geometrical effects of the Earth's ellipticity and topography, there will be no lateral variation in flux if this feature is in fact at the center of the Earth. It is conceivable, however, that georeactors exist within the deep mantle within subducted U-rich organic sediments. See the conclusions for more discussion of such rocks.

Neither do I discuss the neutrino geophysics operating in an "X-ray" transmission mode (Lindner et al., 2003). In this case, tomography is simpler than and easily combined with seismic studies. The paths are known and essentially straight with tiny known corrections for the aberration of light and general relativity from the mass of the Earth. The method is sensitive to discontinuities, in particular to the absolute depth of the core-mantle boundary.

2. Antineutrino Detectors

Detectors like KamLAND are sensitive to antineutrinos from the U and Th decay chains. Such detectors are nondirectional. Their signal is formally an integral over the Earth's volume

$$A(\vec{r}_{\mathrm{m}}) = \int_{\mathrm{vol}} \frac{(C_{\mathrm{U}} f_{\mathrm{U}} + C_{\mathrm{Th}} f_{\mathrm{Th}})\rho(\vec{r}_{\mathrm{s}})}{|\vec{r}_{\mathrm{s}} - \vec{r}_{\mathrm{m}}|^2} w(|\vec{r}_{\mathrm{s}} - \vec{r}_{\mathrm{m}}|) d\mathrm{vol}, \tag{1}$$

where the vector $\vec{r}$ represents position, the script m indicates the measurement point, the subscript s indicates the source point, C is concentration, f is proportionality constant between the element concentration and the number of antineutrinos detected, the subscripts U and Th relate to the decay chains for these elements, ρ is density, and w is a weak function of distance representing changes between antineutrino types. It may be defined to be of the order of unity without the loss of generality. Seismic studies and the Earth's mass and moment of inertia tightly constrain density over large regions of the Earth (e.g., Masters and Gubbins, 2003). Density in the mantle is mainly a function of depth and pressure. Lateral variations of the acceleration of gravity and seismic data constrain lateral variations in density. Even the largest expected lateral temperature variations (< 1000 K) affect it by only a few percent. Thus, the concentrations of U and Th are the only unknowns in (1).

It is well known that neutrino geophysics is not homologous to gravity. However, the nonuniqueness is similar. The acceleration of gravity at a point is

$$\vec{g}(\vec{r}_{\mathrm{m}}) = \int_{\mathrm{vol}} \frac{G\rho(\vec{r}_{\mathrm{s}})}{|\vec{r}_{\mathrm{s}} - \vec{r}_{\mathrm{m}}|^3} (\vec{r}_{\mathrm{s}} - \vec{r}_{\mathrm{m}}) d\mathrm{vol}, \tag{2}$$

where G is the gravitational constant and the attraction on the detector is in the direction of the source. There are two major differences between the antineutrino detector and a gravity meter. Gauss's law applies to gravity. Specifically, the total flux of the gravity vector through the Earth's surface gives the mass of the Earth without knowing the internal distribution of that mass. We cannot, however, get the total flux of antineutrinos from the Earth as the response of a concentrated antineutrino source (at say the Earth's center) differs somewhat from that for an equal shell-like sheet source spread out over the Earth's surface. Second, the gravimeter has vector properties. If we place equal masses above and below it, their attractions cancel. They add for an antineutrino detector.

We have good sampling of the Earth's gravity field. Still resolution is limited, crudely to length scales comparable to the depth. The mass distribution at depth is not unique. From gravity alone we cannot tell a variable thin massive sheet at the surface from a more geologically reasonable distribution of masses at depth. Similar nonuniqueness exists for neutrino detectors. In addition, we will have very poor spatial coverage in the reasonable future. With the detectors alone, we cannot resolve features finer than the spacing of our stations.

A simple assumption for interpreting the results of antineutrino experiments is that the Earth, except for the continental crust, is radially symmetric. The integral (1) then become

$$A(r_{\mathrm{m}}) = \int_{r_{\mathrm{s}}} (C_{\mathrm{U}} f_{\mathrm{U}} + C_{\mathrm{Th}} f_{\mathrm{Th}}) \rho(r_{\mathrm{s}}) W(r_{\mathrm{s}}) dr_{\mathrm{s}} \tag{3}$$

where the function W includes the net effects of the geometry of a spherical shell and changes in antineutrino type. We cannot resolve radial structure with the antineutrino data alone. We cannot even tell if the source is all at the Earth's center or all in a thin sheet at the surface as an extreme case.

Geophysicists have confronted the nonuniqueness of gravity data since local measurements became practical circa 1850. They use other available information including the outcrop distribution and measured densities of rocks at the surface and seismic data on the rocks at depth. They ask simple questions of gravity. For example, is a dome-shaped reflector at depth on a seismic section rock salt (which is less dense than the surrounding sediments) or the volcanic rock basalt (which is more dense)? This strategy carries through to antineutrino data. I give some real examples starting with KamLAND, a single available measurement, which has already been discussed extensively in the literature.

2.1. KamLAND

KamLAND provided the first detection of terrestrial antineutrinos (Araki et al., 2005). The site is within Japan, an island arc built on continental crust. The surface geology around the detector is complex; the local crust is depleted in U and Th with respect to average continental crust (Enomoto et al., 2005). A simple productive assumption is to divide the Earth into 3 domains, continental crust, mantle, and core with a correction for local geology (Figure 1). We suspect that the Th:U ratio is everywhere ~3.9 and use this assumption when we do not have enough data to resolve the two decay chains. Seismology gives that the core-mantle boundary is ~2900 km deep. Scientists typically but not universally suspect that it holds negligible amounts of U and Th. In any case, we cannot separate radially symmetric

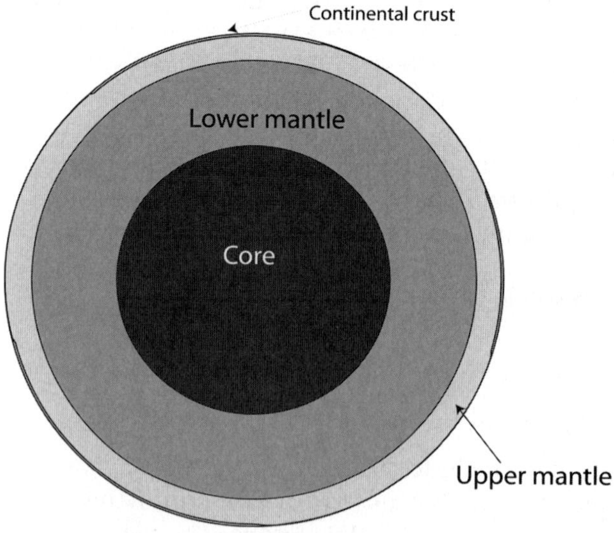

Figure 1. Schematic cross section of the Earth drawn to scale. The continental crust is ~40 km thick and contains about 1/3 of the Earths radioactive heat production. Plate tectonics and plume mix the upper mantle and the lower mantle over geological time. The deep lower mantle, however, may be chemically dense and have different concentrations of radioactive elements than the rest of the mantle.

sources in the core from those in the mantle. We do know where continents and continental crust are. We also have reasonable constraints that about 1/3 of the Earth's U and Th are in the continental crust. We assume either this fraction of the Earth's radioactivity that is in the crust or that an absolute amount is in the crust. The published calculations on KamLAND proceeded in these ways (Araki et al., 2005; Fiorentini et al., 2005; Enomoto et al., 2005). They garnered the available information on the Earth.

2.2. Hawaii

I now add a second detector near Hawaii in the middle of the Pacific Ocean (Enomoto et al., 2006, Submitted). This detector is far from continents and their crust. Hence we have geologically relevant spatial resolution when we combine its data with the KamLAND data. The continental signal is small and varies slowly across the ocean basin (Fiorentini et al., 2005).

For completeness, the oceanic crust is ~6 km thick in this region if we get away from the islands. To the first order, one does not have to take the existence of the oceanic crust into account (Figure 2). Ocean crust originally formed at a midoceanic ridge (e.g., Klein, 2003). The basaltic magma that formed the crust came from the underlying ~50 km of mantle. The melt carried essentially all the available U and Th with it but the composition of a column down to ~56 km depth remained unchanged. The analyst can ignore this stratification or include it as a small correction. Conversely, antineutrino data constrains the U and Th in the mantle source region, giving a hard constraint on models of basalt generation.

With a marine deployment and KamLAND, we have two equations with two unknowns (the concentrations of radioactive elements in the continental crust and in the mantle) and two measurements. The problem is well posed

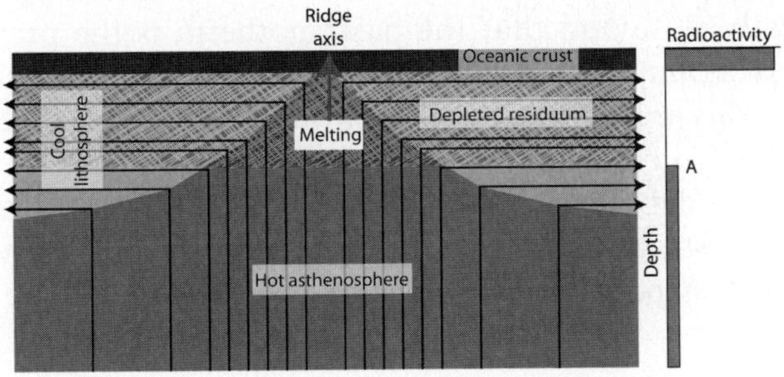

Figure 2. Schematic drawing of a mid-ocean ridge axis. Asthenospheric material upwells and melts above depth A, ~56 km. The U, Th, and K enter the melt and ascend to form the oceanic crust. Old cool oceanic lithosphere consists of ~6-km crust, ~50 km of depleted residuum, and material unaffected by melting beneath ~56 km depth. The average concentration of radioactivity above 56 km is that in the source composition mantle below 56 km.

but minimally determined. Subsequent deployments elsewhere in the Pacific basin will be quite valuable to detect lateral heterogeneity in the mantle. The second marine deployment is likely to find lateral heterogeneity if it exists. Three deployments without lateral variation constitute strong evidence against lateral heterogeneity.

2.3. Stable Continental Site

Mines of ~3-km depth in South Africa, India, the USA and Canada are attractive sites for an antineutrino detector. It turns out that work related to diamond interests makes data from these mines interpretable as one has independent constraints on the local concentration of radioactivity in the continental crust.

Diamonds are the stable phase of carbon ~200 km down in the Earth. Water- and CO_2-rich magmas, loosely called kimberlites carried diamonds from their zone of stability to the surface. These magmas ascended rapidly, reaching the surface on the order of hours to days. Many diamonds quenched quickly, rather than turning into graphite. The magmas also ripped rocks off the walls of their conduits. Many of these rocks, called xenoliths, reached the surface and quenched quickly, retaining the mineral composition that they had at depth. Petrologists examine the solid-solution chemistry of coexisting minerals (for example garnet and clinopyroxene) within these rocks and deduce the pressure and temperature of their equilibration (e.g., Kopylova et al., 1999; Rudnick and Nyblade, 1999). The pressure is simply the weight of a column of the overlying rocks. They convert pressure to depth, as the density and the acceleration of gravity are known.

This exercise on a suite of xenoliths yields the geotherm (temperature as a function of depth) beneath the diamond pipe, strictly speaking at the time the rocks ascended (Figure 3). South Africa, the Canadian Shield, and India are geologically stable. There is some justification for assuming that the geotherm is in steady state and that the past geotherm is the present one. One can modify this assumption to make small corrections for suspected geological events in this area (for example mantle plumes (e.g., Bell et al., 2003)). It is inappropriate to consider Japan or Italy to be stable.

To continue, the depths are shallow enough that one can treat the Earth as flat in a quick illustration. The geotherm gives the heat flow through the mantle (the heat loss per time per area) by conduction

$$q_{\mathrm{M}} = k\frac{\Delta T}{\Delta Z}, \tag{4}$$

where k is thermal conductivity, ΔT is the temperature change, and ΔZ is the depth interval. The xenoliths provide samples of the mantle rock for determining conductivity.

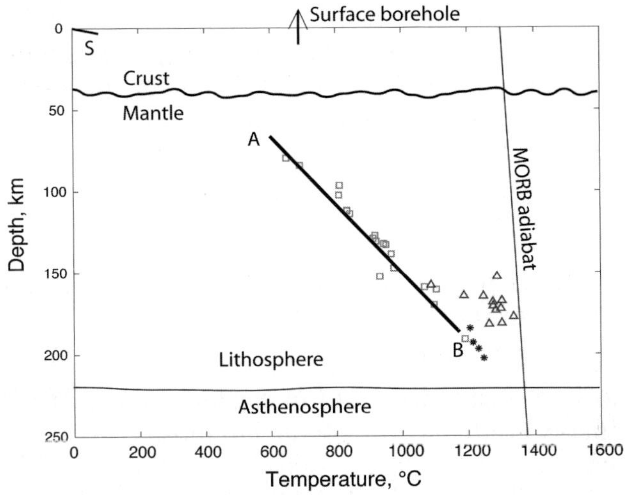

Figure 3. A xenolith geotherm drawn on a schematic cross section of a stable continent. In practice, one measures the surface thermal gradient and heat flow in a shallow borehole or mine (S) and the conductive gradient (AB) in the mantle lithosphere. Data from Jericho pipe in northern Canada are from Kopylova et al. (1999) who used the Brey and Köhler (1990) method to obtain pressure and temperature. Depth calculation from pressure by Sleep (2003). Symbols indicate various rock types: green squares represent coarse xenoliths that equilibrate over geological time; red triangles represent sheared xenoliths, and blue asterisks represent megacrystalline xenoliths.

The surface heat flow is directly measurable from the thermal gradient and the conductivity in a mine or borehole. Once this is done, one has enough information to estimate the radioactivity in the continental crust. That is, at steady state the radioactive heat production per area in the continental crust equals the difference between the heat flow at the surface and the heat flow in the mantle. For antineutrino studies, one must make assumptions on the ratios of U, Th, and K. Surface rocks as well as xenoliths derived from the crust help. The ratio of U to Th does not change much over resolvable spatial scales and K is typically a subordinate heat source. Atomic K:U is ~10,000. This procedure yields an estimate of the U and Th concentration in the mantle.

Conversely, if antineutrino studies have already determined mantle U and Th concentration, an antineutrino measurement on stable crust provides a strong constraint on the radioactive heat production within the crust. Then, one can use the surface heat flow and these data to estimate the geotherm where there are no xenolith data (e.g., Rudnick and Nyblade, 1999)

3. Application of Elemental Results

I now assume that we have analyzed data from a few antineutrino detectors. I will on occasion assume that we also have ^{40}K detectors. Treatment of K data is homologous to the treatment of U and Th data. I summarize geological arguments that constrain U, Th, and K with the intent that they will be run backwards once data are in hand, bringing their assumptions into scrutiny. I start with topics that merely require the bulk Earth composition and then proceed to topics that require some spatial resolution. This approach puts the topics in the order that they arose in the Earth's history.

3.1. Accretion of the Earth and Other Planets

The Earth accreted from a dust and gas nebula at the time that the Sun formed. The gas and dust started out in a giant molecular cloud at ~10 K. Gravitational energy contributed to solar luminosity. Its peak was 10 times the present value. In a few million years, luminosity declined to ~0.7 the present value (e.g., Sackmann et al., 1993). The dust condensed into small rocky bodies, called planetesimals, that eventually accreted into rocky planets, like the Earth (e.g., Davis, 2004).

The planet-forming process thus involved vaporization of dust grains as the nebula heated up and condensation as it cooled down. This occurred within an H_2-rich environment. Rocky bodies collided at cosmic velocities. They partly vaporized and condensed in H_2-poor environments. Primitive meteorites, called chondrites, contain presolar grains that never vaporized as well as high-temperature and low-temperature condensates. It is productive to view the Earth as an evolved product of the accretion of such a mixture. We have direct samples of our crust and the upper ~250 km of our mantle from xenoliths and outcrop. We have indirect samples from a similar depth range from volcanic rocks that equilibrated with the mantle in their source regions.

We need to be concerned with processes that differentially affected U and Th from the other rocky components of the Earth, rather than bulk loss of rock-forming material into the Sun or to space. Fractionation associated with volatility is a traditional starting point. For purposes of discussion, I divide the elements into 3 classes. Magnesium silicates and metal iron make up the bulk of rocky planets with $MgSiO_3$ condensing before Mg_2SiO_4. Fe metal condenses in the same temperature range. U and Th are among the first "refractory" elements to condense along with Ca, Ti, and Al. K in moderately volatile, condensing at somewhat lower temperatures than magnesium silicates. The behavior of U, Th and K, does not depend much on the presence of H_2.

One expects that all the available U and Th ended up in grains in the zone of the Earth's accretion, having either never vaporized or as a first condensate. The subsequent Mg silicates and iron acted as inert diluents. The ratio of diluents to U and Th, however, is not tightly constrained. The Mg:Si ratio in the Earth (1.27) is higher than the ratio (1.05) in CI chondrites. Two processes may explain this observation (e.g., Ringwood, 1990). The Earth is depleted in SiO_2 relative to the nebula because this more volatile component did not fully condense in the Earth's zone of the nebula. Alternatively, SiO_2 that failed to condense in the Earth's zone migrated outward locally enriching the zone where it condensed as a component of CI meteorites. One's estimate of total U and Th in the Earth varies by 20% depending on whether one normalizes with respect to Mg or Si. Potassium is more volatile than SiO_2. However, once it is within a rocky object it behaves similarly to U and Th. It is thus productive to consider U, K, and Th ratios. It is obvious that the part of the Earth that we can sample is depleted in K relative to meteorites and the nebula. The ratio K:U is ~10,000 within the Earth and ~80,000 in primitive meteorites.

Condensation of material at a single temperature and pressure that leaves 7/8 of the K in the vapor and 1/8 in the solid grains does not explain the rest of the elements in the Earth. In this case, the grains would have the nebular composition for elements somewhat less volatile than K and be quantitatively depleted in elements somewhat more volatile. In fact, the Earth contains significant quantities of more volatile elements. A useful approximation is that the Earth formed from ~85% of material that condensed with Mg silicates and ~15% of material with the nebular composition of moderately volatile elements. Potassium isotopes provide additional information (Richter, 1990). In addition to the radioactive isotope 40, potassium has two stable isotopes, 39 and 41. The isotopes fractionate significantly on condensation and on vaporization when kinetics limit these processes, but little at equilibrium at the K condensation temperature. Solar system material shows little fractionation of K isotopes. This implies that slow equilibrium processes dominate. It is consistent with (but does not require) that the Earth formed as an admixture of material that retained all the K (and hence no fractionation) with material quantitatively depleted in K.

Hard collisions partly vaporized bodies like the Earth's moon. Subsequent gravitational collapse occurred when both solid and gas were present. The mixture was dense, but had the compressibility of a gas. This assured that collapse occurred when there was still significant SiO_2, the most volatile major component, in the gas. Elements more volatile than silica, like K, stayed in the gas and were quantitatively depleted in the solid. The Moon shows evidence of this process and minor admixture of undepleted material. It is strongly depleted in K relative to the Earth but the K isotopes are not fractionated.

3.2. Fractionation within Rocky Bodies

Partial melting within nonsurviving bodies in the early solar system may have affected the final amounts of U and Th that ended up in the Earth. Basaltic melts formed in the first few million years of the accretion process. Samples of such melts now arrive at the Earth as meteorites derived, for example, from the asteroid Vesta. Iron meteorites represent the differentiated cores of scattered bodies (e.g., Kleine et al., 2004; Petaev and Jacobsen, 2004).

U, Th, and K behave similarly during partial melting of mantle materials nearly quantitatively entering the basaltic melt as at modern mid-oceanic ridges (Figure 2). This process leaves depleted mantle residuum at depth. U and Th are unlikely to enter iron cores, but K may as noted below.

Segregation is most likely to have had significant effect once a significant fraction of the Earth had accreted. By then the relative velocities of objects were high and collisions violent. Material ejected into space did not all re-accrete. U-, Th-, K-bearing basaltic rock on the surface was geometrically most likely to be ejected. This process depleted the Earth in these elements relative to the starting grain composition.

Variations in Pb-isotope composition of mantle-derived volcanic rocks show that the Earth's interior has been heterogeneous for billions of years (McDonough and Sun, 1995; Bennett, 2004). Lu–Hf systematics on 4.01–4.37 Ga detrital zircon crystals in an ancient sandstone indicate that isotopically distinct domains including continental crust formed just after the moon-forming impact, that is ~4.5 Ga (Harrison et al., 2005). These data, however, do not directly constrain the geometry of the heterogeneities. Two end-member hypotheses have been proposed. (1) The heterogeneities are small domains produced by plate tectonic processes, such as melting at midoceanic ridges and island arcs and subduction of crustal material including sediments (e.g., Meibom et al., 2005). That is, the mantle is homogeneous on the scale that one can hope to resolve with neutrino data. (2) The heterogeneities are large domains within the mantle that might be resolvable. Hybrid hypotheses exist. For example slabs have subducted beneath continents, but not beneath the Pacific Ocean for the last few hundred million years. One would expect that slab-rich regions of the deepest mantle might be enriched relative to average mantle in radioactive elements carried down in subducted sediments.

Helium isotopes were until recently thought to provide strong support for primordial regions in the Earth's mantle that had not degassed. The common isotope ^{4}He comes mainly from U- and Th-decay chains. The rare isotope ^{3}He in the mantle is mainly primordial from the accretion of the Earth. However, He does not preferentially enter mantle melts to the extent that U and Th do (Parman et al., 2005). The persistence of ^{3}He in the mantle is an indication that helium does not readily escape from the mantle during melting, rather than that pristine regions persist within the mantle.

The Earth became an essentially closed system after the moon-forming impact. None of the processes that I have discussed to this point segregate U from Th. If the traditional geologic reasoning is correct, antineutrino measurements will show that the bulk Earth Th:U is ~3.9. If not, Earth scientists have missed a major process.

3.3. Heat Budget and Convection in the Mantle

Radioactivity and secular cooling supply comparable amounts of heat from the Earth's interior (e.g., McDonough and Sun, 1995). Globally the total heat flow from the Earth is

$$4\pi r_E^2 q = \frac{4\pi r_{\mathrm{E}}^3}{3}\rho H - \frac{4\pi r_{\mathrm{E}}^3}{3}\rho C\frac{\partial T}{\partial t}, \tag{5}$$

where r_{E} is the Earth's radius, q is heat flow, ρ is density, H is radioactive heat generation per mass, C is specific heat, T is temperature, and t is time. It is convenient for geodynamics to consider the contributions of radioactivity in the continental crust separate as this heat escapes by conduction from a shallow depth (Figure 3). The heat flow is the contribution coming from the mantle supplied by radioactive heat generation and cooling in the deep interior. It is straightforward to treat the core and mantle as separate domains.

Earth scientists have direct but fuzzy constraints on the 3 terms in Equation (5). Heat flow from the mantle is a result of seafloor spreading (Stein and Stein, 1994). The local heat flow is inversely proportional to the square root of plate age. The global heat flow is equivalently inversely proportional to the square root of average plate age or directly proportional to square root of rate of formation of new area of seafloor. The current rate is approximately 3 $\mathrm{km}^2\ \mathrm{yr}^{-1}$ with the best estimate being 3.3 $\mathrm{km}^2\ \mathrm{yr}^{-1}$ (White et al., 1992). This is enough to replace the global area of the ocean basins in ~100 m.y. Geophysicists obtain this rate by studying magnetic lineations produced by seafloor spreading. The modern rate is well constrained, but the rate even 100 m.y. ago is poorly constrained because much crust of that age has already been subducted back into the deep mantle.

Geochemists constrain the cooling rate of the Earth's interior by inferring the source region temperature of magmas of various ages. Current estimates are around 50K/B.Y. with considerable uncertainty (Abbott et al., 1994; Galer and Mezger, 1998). As discussed above, geochemists have constrained the radioactivity in the mantle. Knowing this number gives the instantaneous cooling rate of the Earth's interior as well as constraining its thermal and tectonic history.

3.4. Hidden Reservoirs and Deep Lateral Heterogeneity within the Earth

Chemically dense regions may lurk near the base of the Earth's mantle. The properties of these regions if they exist are unclear. Seismologists have suggested thicknesses between ~100 and ~1000 km. Geochemistry and geodynamics provide few hard constraints. Fe-rich dregs might have settled out early in the history of the Earth. Al-rich basaltic rocks may separate from subducted lithosphere and accumulate at the base of the mantle.

Antineutrino studies are very relevant here as the dregs may be enriched (or even conceivably depleted) in radioactive elements. A thin dregs layer gets dragged into a cusp beneath the source of mantle plumes (Figure 4). A dregs layer ascends buoyantly upward, like a lava lamp (e.g., Davaille, 1999; Kellogg et al., 1999). Cool sinking slabs displace it downward.

We have a fuzzy idea of the geometry of these processes from seismology. Montelli et al. (2004) resolve plume conduits in the deep mantle. This result, however, is controversial (van der Hilst and de Hoop, 2005). The existence of lateral heterogeneity in the bottom several hundred kilometers of the mantle

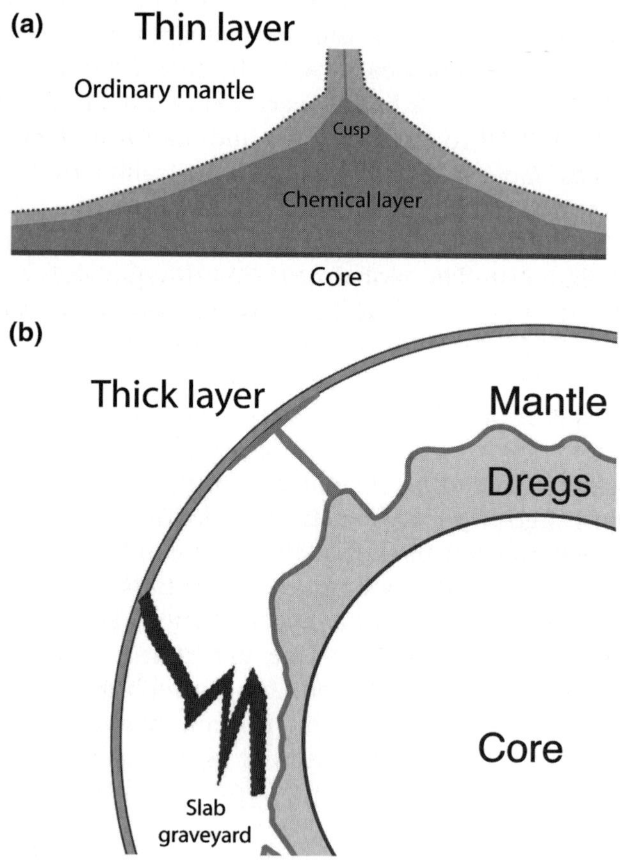

Figure 4. Schematic diagram of the geometry of a chemically dense layer at the base of the mantle. (a) Mantle plumes entrain a thin layer into cusps. (b) A thick layer deforms like a lava lamp. It is thin where slabs have gone down. The tops of some upwellings are the source plumes.

is not in doubt (Romanowicz, 2003). Broad regions of low seismic velocity called "superplumes" may be thick upwelled dregs. Lateral heterogeneity in the bottom ~100 km is also well established (e.g., Garnero and Lay, 2003), but may be associated with a thin chemical dense region or with a thermal boundary layer heated from below.

Seismic data give tenuous direct constraints on lateral variations of density in the deep mantle. The basic effects are simple, but the analysis is very mathematical in practice. Perturbations of density interfaces by seismic waves provide a restoring force analogous to water waves, which modestly augments the restoring force from elasticity. Seismometers are basically accelerometers that detect the variation of gravitational attraction from displacement of masses within the Earth by seismic waves as well as the motion from the waves themselves. These gravitational effects provide poor spatial resolution. In addition, spatial gradients of density reflect energy. Ishii and Tromp (2004) inverted their normal mode eigenfrequency data for density as well as seismic velocity. Taken at face value, their work indicates that the superplume regions are chemically denser than their surroundings. However, this result is controversial, as the resolution of lateral variations in density in this region of the mantle using normal mode data is poor (e.g., Kuo and Romanowicz, 2002).

Deployment of marine antineutrino detectors eventually to a ~1000 km spacing would be warranted if the initial marine deployments find lateral variation. Fluid dynamic calculations to track dregs within the mantle flow associated with surface plates and sinking slabs is feasible by simply adding a buoyant region to existing models (Steinberger et al., 2004).

Going deeper, K in the core or even the basal dregs layer constitutes a second important hidden reservoir (Rama Murthy, this volume). The heat from the decay of ^{40}K in the core would power convection within the core and the magnetic dynamo (Anderson, 2002; Buffett, 2002; Labrosse, 2003, Nimmo et al., 2004). It along with ^{40}K in the dregs layer would power mantle plumes. As the cosmic abundance of K is ~8 times that we see in the accessible Earth, the measurements need not be very precise to be quite useful. The measurements need to be more precise to sense lateral variation and hence distinguish a homogeneous core source from a dreg source.

4. Conclusions and Implications to Biology

The common radioactive elements U, Th, and K have played a central role in the Earth sciences since the advent of modern physics. Geochemists attempt to constrain their concentration in the mantle and the crust using samples including their decay products, mainly Pb and He. The inaccessibility of the deep Earth has brought numerous indirect arguments into play, especially cosmochemical ones on the formation of the planets. Reliable methods for studying the decay of uncommon and short-lived isotopes continue to come on line. Geodynamics is basically the study of the Earth as a heat engine. Radioactive decay is a major energy source. The assumptions embodied in this work are clear and quantitative. The geological arguments can be turned around, appraised, refined, or discarded as antineutrino data become available.

Geoneutrino physicists have amassed the key information that is already known about the Earth as well as the key issues. A few good measurements of geoneutrino flux will resolve key points, including the absolute concentration of U and Th in the mantle and whether the mantle composition varies laterally on a scale of 1000s of kilometers. Once data are available it will be evident whether one should concentrate on crustal or mantle processes. Potassium detection will constrain the heat budget of the core and how the Earth drives its magnetic field.

These issues relate to the Earth as a habitable planet. Ward and Brownlee (2000) contend that the Earth is special so that complex life is rare in the universe. On the other hand, hypothesizers on the formation and evolution of the Earth avoid special circumstances to get the standard form of the nebular hypothesis and the bulk silicate Earth. Our own existence helps little, we had to evolve and survive to be here to speculate.

The present (and post-accretion) U and Th concentration in the Earth is strong evidence as to whether the Earth picked up available nonvolatile material or whether it got much less or much more of its expected allotment by processes that we do not yet understand. The radioactive heat generation preordains how long our planet stays active, providing fresh rock with nutrients and dry land to biota. We will see whether a dregs layer exists in the deep mantle. It may sequester large quantities of biologically important elements including water.

Finally antineutrino data may provide a biomarker in addition to how our arena of life, the Earth, formed and performed over geological time. Subducted organic-rich sediments may have partitioned U but not Th into the dregs. Chemically, uranium has the +4 oxidation state in common igneous rocks and behaves similarly to Th. Weathering in oxidizing conditions produces the +6 oxidation state which is somewhat soluble in water. Uranium precipitates from solution in the reducing conditions within organic-rich sediments. Pb-isotopes in highly metamorphosed 3.8-Ga sediments indicate such segregation of U from Th (Rosing and Frei, 2004). This is well before the rise of otherwise geologically detectable (few ppm) O_2 in the air at 2.45 Ga from mass independent fractionation of sulfur (Farquhar et al., 2002). Mass independent sulfur fractionation in diamond inclusions also provides the earliest hard evidence of sediment subduction in the mantle at 2.9 Ga (Farquhar and Wing, 2003).

Acknowledgements

I thank NSF grant NSF EAR-0406658 (NHS) for support. This work was performed as part of collaboration with the NASA Astrobiology Institute Virtual Planetary Laboratory Lead Team. Barbara Romanowicz provided helpful comments.

References

Abbott, D. L., Burgess, L., Longhi, J., and Smith, W. H. F.: 1994, *J. Geophys. Res.* **99**, 13835–13850.

Anderson, O. L.: 2002, *Earth Planetary Interiors* **131**, 1–17.

Araki, T. and many others: 2005, *Nature* **436**, 499–503.

Bell, D. R., Schmitz, M. D., and Janney, P. E.: 2003, *Lithos* **71**, 273–287.

Bennett, V. C.: 2004, in H. D.Holland and K. K.Turekian (eds.), *The Mantle and Core, Vol, 2, Treatise on Geochemistry*, Elsevier, Amsterdam, pp. 493–519.

Brey, G. P. and Köhler, T.: 1990, *J. Petrol.* **31**, 1353–1378.

Buffett, B. A.: 2002, *Res. Lett.* **29**(12) art. 1566.

Davaille, A.: 1999, *J. Fluid Mech.* **379**, 223–253.

Davis, A. M. (ed.): 2004, in H. D. Holland and K. K. Turekian (eds.), *Meteroites Comets, and Planets, Vol. 1, Treatise on Geochemistry*, Elsevier, Amsterdam, 737 pp.

Enomoto, S., Ohtani, E., Inoue, K., and Suzuki, A.: 2005 "Neutrino geophysics with KamLAND and future prospects", arXiv:hep-ph/0508049.

Farquhar, J. and Wing, B. A.: 2003, *Earth Planet Sci. Lett.* **213**, 1–13.

Farquhar, J., Wing, B. A., McKeegan, K. D., Harris, J. W, Cartigny, P., and Thiemens, M. H.: 2002, *Science* **298**, 2369–2372.

Fiorentini, G., Lissia, M., Mantovani, F., and Vannucci, R.: 2005, *Earth Planet. Sci. Lett.* **238**, 235–247.

Galer, S. J. G. and Mezger, K.: 1998, *Precambrian Res.* **92**, 387–412.

Garnero, E. J. and Lay, T.: 2003, *Phys. Earth Plan. Inter.* **140**, 219–242.

Harrison, T. M., Blichert-Toft, J., Muller, W., Albarede, F., Holden, P., and Mojzsis, S. J.: 2005, *Science* **310**, 1947–1950.

Herndon, J. M.: 2003, *Proc. Nat. Acad. Sci. USA* **100**, 3047–3050.

Ishii, M. and Tromp, J.: 2004, *Phys. Earth Plan. Inter.* **146**, 113–124.

Kellogg, L. H., Hager, B. H., and van der Hilst, R. D.: 1999, *Science* **283**, 1881–1884.

Klein E. M.: 2003, in H. D. Holland and K. K. Turekian (eds.), *Treatise on Geochemistry*, Vol. 3., edited by R. L. Rudnick, Elsevier Amsterdam, pp. 433–463.

Kleine, T., Mezger, K., Munker, C., Palme, H., and Bischoff, A.: 2004, *Cosmochim. Acta* **68**, 2935–2946.

Kopylova, M. G., Russell, J. K., and Cookenboo, H.: 1999, *J. Petrol.* **40**, 79–104.

Kuo, C. and Romanowicz, B.: 2002, *Geophys. J. Inter.* **150**, 162–179.

Labrosse, S.: 2003, *Phys. Earth Plan. Inter.* **140**, 127–143.

Lindner, M., Ohlsson, T., Tomas, R., and Winter, W.: 2003, *Astropart. Phys.* **19**, 755–770.

Masters, G. and Gubbins, D.: 2003, *Phys. Earth Plan. Inter.* **140**, 159–167.

McDonough, W. F. and Sun, S. S.: 1995, *Chem. Geol.* **120**, 223–253.

Meibom, A., Sleep, N. H., Zahnle, K. and Anderson, D.L.: (2005), in G. R. Foulger, J. H. Natland, D. C. Presnall and D. L. Anderson (eds.), Plates, plumes, and paradigms: Geological Society of America Special Paper 388, pp. 347–363.

Montelli, R., Nolet, G., Dahlen, F. A., Masters, G., Engdahl, E. R., and Hung, S. -H.: 2004, *Science* **303**, 338–343.
Nimmo, F., Price, G. D., Brodholt, J., and Gubbins, D.: 2004, *Geophys. J. Int.* **156**, 363–376.
Parman, S. W., Kurz, M. D., Hart, S. R., and Grove, T. L.: 2005, *Nature* **437**, 1140–1143.
Petaev, M. I. and Jacobsen, S. B.: 2004, *Meteor. Planet. Sci.* **39**, 1685–97.
Rama Murthy, this volume.
Richter, F. M.: 1990, *Geochim. Cosmochim. Acta* **68**, 4971–4992.
Ringwood, A. E.: 1990, in H. E.Newsom and J. H.Jones (eds.), *Origin of the Earth*, Oford University Press, New York, pp. 101–134.
Romanowicz, B.: 2003, *Ann. Rev. Earth Planet. Sci.* **31**, 303–328.
Rosing, M. T. and Frei, R.: 2004, *Earth Planet. Sci. Lett.* **217**, 237–244.
Rudnick, R. L. and Nyblade, A. A.: 1999. in Y. Fei, C. M. Bertka and B. O. Mysen (eds.), *Mantle Petrology: Field Observations and High Pressure Experimentation: A Tribute to Francis R. (Joe) Boyd*, Geological Society, Special Publication 6, pp. 3–12.
Sackmann, L. -J., Boothroyd, A. I., and Kraemer, K. E.: 1993, *Astrophys. J.* **418**, 457–468.
Schuiling, this volume.
Sleep, N. H.: 2003, *Geochem. Geophys. Geosyst.* **4**, 1079, doi:10.1029/2003GC000511.
Stein, C. A. and Stein, S.: 1994, *J. Geophys. Res.* **99**, 3081–3095.
Steinberger, B., Sutherland, R., and O'Connell, R. J.: 2004, *Nature* **430**, 167–173.
van der Hilst, R. D. and de Hoop, M. V.: 2005, *Geophys. J. Int.* **163**, 956–961.
Ward, P. D. and Brownlee, D.: 2000, *Rare Earth: Why Complex Life is Uncommon in the Universe*, Copernicus, New York, 333 pp.
White, R. S., McKenzie, D., and O'Nions, R. K.: 1992, *J. Geophys. Res.* **97**, 19683–19715.

Earth, Moon, and Planets (2006) 99:359–368
DOI 10.1007/s11038-006-9131-5

Physics in Next Geoneutrino Detectors

ATSUTO SUZUKI
Research Center for Neutrino Science, Tohoku University, Aramaki, Aoba, Sendai 980-8578, Japan
(E-mail: atsuto.suzuki@kek.jp)

(Received 2 October 2006; Accepted 27 October 2006)

Abstract. The KamLAND liquid scintillator detector demonstrated the detection of antineutrinos produced by natural radioactivities in the Earth, so-called geoneutrinos. Although this first result of geoneutrinos is consistent with current geophysical models, more accurate measurements are essential to provide a new window for exploring the inside of the Earth. In this article I would like to discuss the future prospects of KamLAND geoneutrino detection, and the possibility of directional measurement of incoming geoneutrinos. It is interesting to consider the application of geoneutrino detectors to measurements of other neutrino signals. The possibility of detecting the solar 7Be, pep and CNO neutrinos is discussed. A new type detector concept is proposed not only to explore the precise measurement of reactor neutrino oscillations but also to enable us to realize the neutrino tomography inside the Earth.

Keywords: Geoneutrino, solar neutrino, reactor neutrino

1. Introduction

The KamLAND liquid scintillator detector demonstrated the detection of antineutrinos produced by natural radioactivities in the Earth, so-called geoneutrinos. Although this first result of geoneutrinos is consistent with current geophysical models, more accurate measurements are essential to provide a new window for exploring the inside of the Earth. In this article I would like to discuss the future prospects of KamLAND geoneutrino detection, and the possibility of directional measurement of incoming geoneutrinos. It is interesting to consider the application of geoneutrino detectors to measurements of other neutrino signals. The possibility of detecting the solar ^{7}Be, pep and CNO neutrinos is discussed. A new type detector concept is proposed not only to explore the precise measurement of reactor neutrino oscillations but also to enable us to realize the neutrino tomography inside the Earth.

2. Geoneutrino Detection After KamLAND First Result

A total of 152 candidate events are observed in the KamLAND first geoneutrino result based on a total detector live-time of 749.1 ± 0.5 days (Araki et al., 2005a). These events are heavily polluted by background. The number of total background events is estimated to be 127 ± 13, where 82.3 ± 7.2 events are from the reactor neutrinos, 42 ± 11 from the $^{13}C(\alpha, n)^{16}O$ reactions and 2.38 ± 0.01 from the accidental coincidences. Thus 25^{+19}_{-18} events are the geoneutrino signals produced from the ^{238}U and ^{232}Th decay chains. This result is consistent with the 19 events predicted by our reference Earth model (Enomoto et al., hep-ph/0508049). The energy spectra of $\bar{\nu}_e$ candidate events including the associated background events are shown in Figure 1.

Improving the signal to noise ratio of $\bar{\nu}_e$ candidate events is the next priority in KamLAND. Currently the KamLAND detector performance is investigated deploying radioactive sources only along the vertical axis and using events induced by cosmic-ray mucons. An expected reduction of the detector systematic error to 4% from the present 6.5% is achieved by the off-axis calibration system (Figure 2) which is under construction. This off-axis system makes it possible to place the calibration sources at every location inside the detector fiducial volume. In addition to upgrading the signal to noise ratio for geoneutrinos, the reduction of detector systematic error also provides the precise determination of the reactor neutrino energy spectrum in the geoneutrino energy range. Thus, more accurate subtraction of reactor events from the energy spectrum of geoneutrino candidate events can be

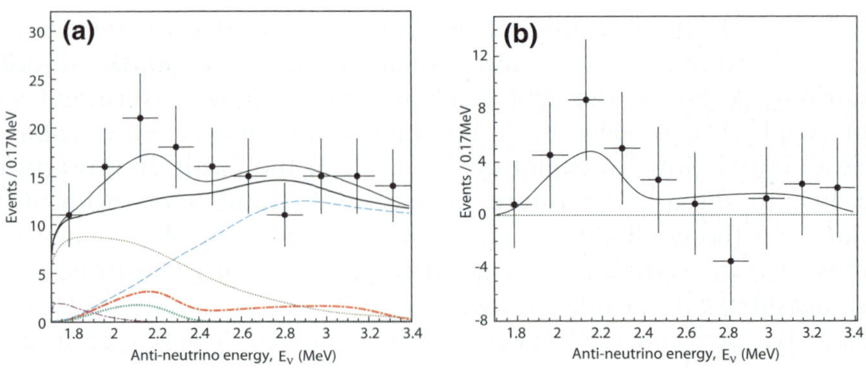

Figure 1. (a) $\bar{\nu}_e$ Energy spectra of the candidate events (data), the total expectation (thin solid black line), the total background (thick solid black line), the expected ^{238}U signals (dot-dashed red line), the expected ^{232}Th signals (dotted green line), and the background due to reactor $\bar{\nu}_e$ (dashed blue line), $^{13}C(\alpha, n)^{16}O$ reactions (dotted brown line) and random coincidences (dot-dashed blue line). (b) $\bar{\nu}_e$ energy spectrum of the candidate events subtracted by the total background.

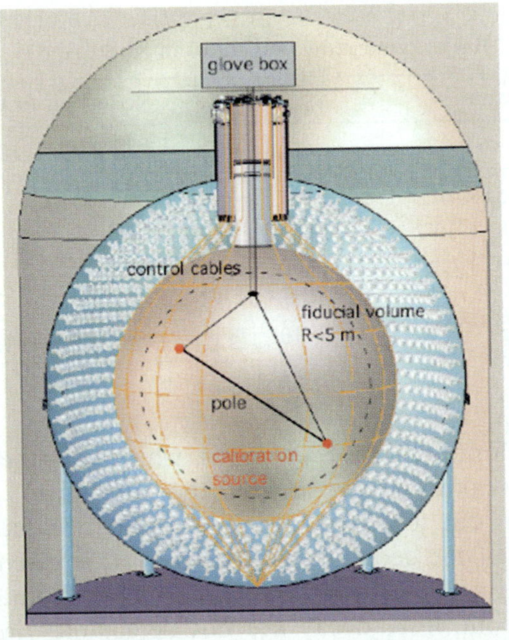

Figure 2. Off-axis detector calibration system.

done. Detector calibration using this system is performed beginning the summer of 2006.

KamLAND has a plan to upgrade the liquid scintillator purification system in 2006. New distillation and nitrogen purge systems are supposed to eliminate the serious background sources for the solar ^{7}Be neutrino detection, such as ^{210}Pb, ^{40}K, ^{222}Rn and ^{85}Kr. This new system is expected to reduce the backgrounds mimicking $\nu_e - e^-$ scattering events by a factor of 10^6 in the energy region of the solar ^{7}Be neutrino events (see Table 1). This reduction factor is far beyond the requirement of 10^2 for upgrading the signal to noise ratio of geoneutrinos, permitting detection of geoneutrinos free from the accidental and the $^{13}C(\alpha, n)^{16}O$ background events. Data-taking in KamLAND after calibrating the detector performance by the off-axis system and reducing background events by the new purification system will start in 2007, the beginning of KamLAND-II.

A major physics target in KamLAND-II is to determine the total geoneutrino flux produced from the ^{238}U and ^{232}Th decay chains, $N_U + N_{Th}$ and the flux ratio of N_U/N_{Th} with high significance. Assuming the best-fit values of $N_U + N_{Th}$ and $(N_U - N_{Th})/(N_U + N_{Th})$ derived from the first KamLAND geoneutrino result (Araki et al., 2005a), the data sample based on two years of KamLAND-II runs gives more severe constraints as shown in Figure 3(a). However taking the value predicted by the Bulk Silicate Earth

TABLE 1
Background reduction status in KamLAND

Impurity	Concentration		Reduction factor		
	Present	Goal	Required	Distillation	Purge
^{238}U	$(3.5 \pm 0.5) \times 10^{-18}$ g/g	10^{-16} g/g	OK		
^{232}Th	$(5.2 \pm 0.8) \times 10^{-17}$ g/g	10^{-16} g/g	OK		
^{40}K	$< 2.7 \times 10^{-16}$ g/g	10^{-18} g/g	10^{-2}	$< 10^{-2}$	
^{85}Kr	~1 Bq/m^3	~1μ Bq/m^3	10^{-6}	$< 10^{-5}$	$< 10^{-5}$
^{210}Pb	~10^{-20} g/g	~10^{-25} g/g	10^{-5}	$< 10^{-4}$	

(BSE) model (McDonough and Sun, 1995), which uses the Th/U mass ratio of 3.9, gives a weaker constraint on the flux ratio of N_U/N_{Th} (see Figure 3(b)). This is due to a large amount of reactor neutrino events in the higher energy part of the geoneutrino spectrum. It is essential to build next generation geoneutrino detectors to be farther away from reactors.

Antineutrinos from 40 K decay are not detectable via the inverse beta reaction because their maximum energy is below threshold. It is difficult to observe the ^{40}K geoneutrinos through neutrino-electron scatterings, because there are a huge number of recoiling electrons produced by the solar ^{7}Be, pep, CNO and ^{8}B neutrinos. As M. Chen pointed out in this workshop, going beyond the Krauss et al. paper (1984) is requisite to overcome this difficulty. New ideas are indispensable.

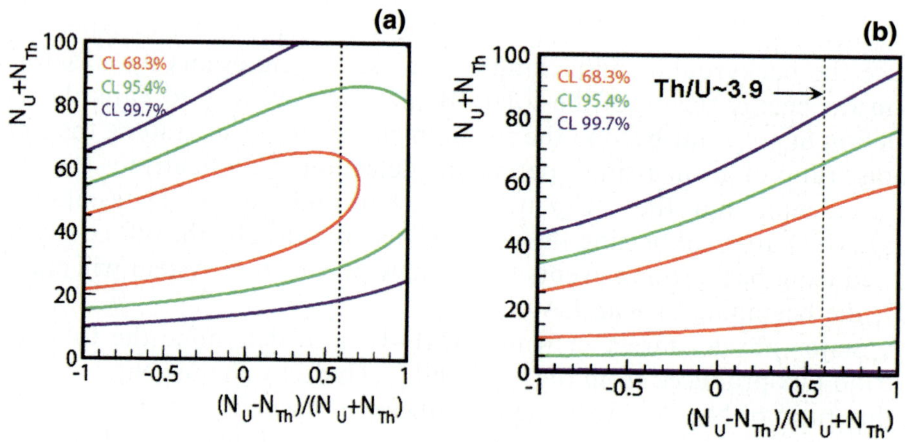

Figure 3. Confidence intervals in the total flux of $N_U + N_{Th}$ and the flux ratio of $(N_U - N_{Th})/(N_U + N_{Th})$ expected with 2 years exposure time in KamLAND-II. The same best fit value of the KamLAND first result (Araki et al., 2005a) is used in (a). The BSE model prediction is used in (b).

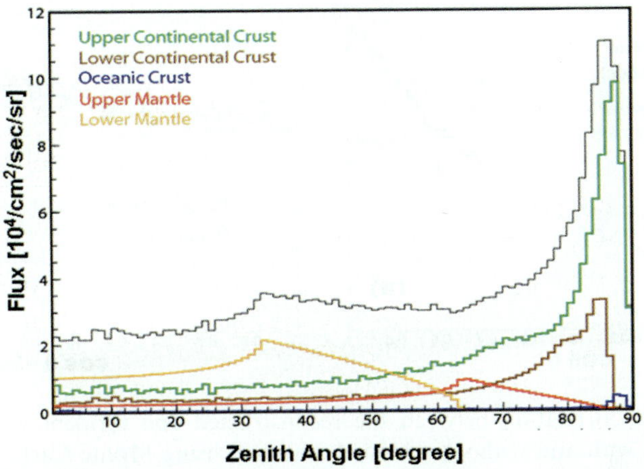

Figure 4. Zenith angle distributions of the ^{238}U geoneutrinos at Kamioka produced in the upper/lower continental crust, oceanic crust, and upper/lower mantle.

Directional measurement of incoming geoneutrinos would provide much information on the heat generation inside the Earth. Figure 4 show the zenith angle distribution of the ^{238}U geoneutrinos at Kamioka calculated by our reference model (Enomoto et al., hep-ph/0508049) (see also Fields and Hochmuth, hep-ph/0406001.). It can be seen that the zenith angle distribution gives the Crust and Mantle components of geoneutrinos. It would be interesting also to verify a null contribution of the ^{238}U and ^{232}Th geoneutrinos in the Earth Core which is the basic assumption of the BSE model (McDonough and Sun, 1995).

In detecting $\bar{\nu}_e$s via neutron inverse β-decay, $\bar{\nu}_e + p \rightarrow e^+ + n$, the incident $\bar{\nu}_e$ direction is approximated to the direction determined by two vertexes of the prompt e^+ and the delayed γ-ray produced by the thermal neutron captured on a proton or on a material loaded inside liquid scintillator. If the γ-ray production position is well identified, the measured direction shows a useful correlation to the $\bar{\nu}_e$ direction as shown in Figure 5(a). However precise measurement of the delayed γ-ray production position is difficult due to multiple scatterings. The direction obtained by reconstructed vertexes of the e^+ and γ-ray show less correlation to that of $\bar{\nu}_e$ (Figure 5(b)). One possibility to solve this problem is to develop a material loaded liquid scintillator which provides delayed α-particles and/or β-rays instead of γ-rays after capturing thermal neutrons by the loaded materials. The $\bar{\nu}_e$ direction measurement is the most urgent task in future geoneutrino experiments.

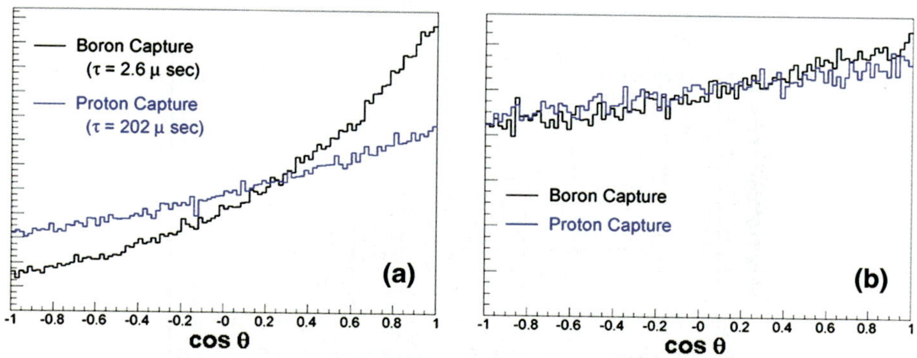

Figure 5. Angular correlation between the reconstructed and incident $\bar{\nu}_e$ direction for the liquid scintillator with and without a loaded material, using Monte Carlo generated events. The direction is obtained by the reconstructed vertex of e^+s and the simulated position of γ-rays (a), and by the reconstructed vertexes of e^+s and γ-rays (b).

3. Detection of Solar ^{7}Be, pep and CNO Neutrinos

Thanks to solving the solar neutrino problem, now is the time to realize Ray Davis's dream of exploring the solar interior by neutrinos. It is essential to measure the solar ^{7}Be, pep and CNO neutrinos for studying the early stage of stellar evolution. In the present KamLAND the main background sources for these neutrinos are ^{238}U, ^{232}Th, ^{40}K, ^{85}Kr and ^{210}Pb dissolved into the liquid scintillator. Table 1 is a list of the current concentration of radioactivity inside liquid scintillator, the goal value for detecting the solar ^{7}Be, pep and CNO neutrinos, the required reduction factors, and the reduction factor of the newly developed distillation and nitrogen purge systems.

The expected energy spectrum of the ^{7}Be neutrino candidate events in KamLAND-II is shown in Figure 6. After the successful purification, it can be seen that the ^{7}Be neutrino window is opened. The solar pep and CNO neutrinos are dominated in the energy region between 0.8 MeV and 1.5 MeV, being above the broad bump of ^{11}C background events produced by cosmic-ray mucons (see Figure 7). Going deep underground is necessary to reduce this background. In KamLAND-II it is crucial to identify the ^{11}C events. The 3-fold coincidence combining the incident cosmic-ray mucon, subsequently produced neutrons and ^{11}C decay is applied. The identified ^{11}C events are eliminated from the candidates, and the energy distribution of ^{11}C events obtained in the energy of $E > 1.5$ MeV is used to evaluate the distribution of the energy of $E < 1.5$ MeV. From Figure 7 it appears that KamLAND-II has a chance to observe the direct evidence of the ^{7}Be production and the CNO fusion process inside the Sun.

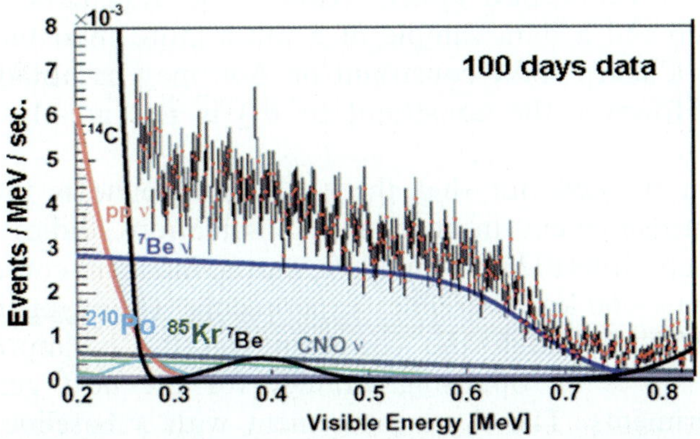

Figure 6. Visible energy distribution in (0.2–0.8) MeV region expected by the 100 days data of KamLAND-II. The pp neutrino signals are completely masked by the ^{14}C events. The spectra of the ^{210}Po and the ^{85}Kr are subtracted by identifying the ^{210}Po and the ^{85}Kr decay chains.

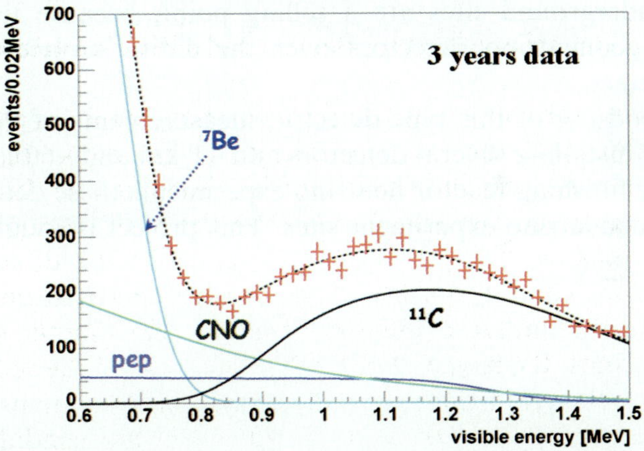

Figure 7. Visible energy distribution in (0.6–1.5) MeV region expected by the 3 years data of KamLAND-II. The ^{11}C events are reduced by applying the 3-fold coincidence, and its energy distribution in $E < 1.5$ MeV is possible to subtract by using the measured spectrum in $E > 1.5$ MeV.

4. Precise Measurement of Reactor Neutrino Oscillations

The reactor antineutrino disappearance was reconfirmed by the second KamLAND reactor result with 99.998% significance level (Araki et al., 2005b). The ratio of the observed inverse β-decay events to the expected number without $\bar{\nu}_e$ disappearance is $0.658 \pm 0.044(\text{stat}) \pm 0.047(\text{syst})$. It can be seen that the systematic error is already comparable to the statistical error. For further study of reactor neutrino oscillations, it is essential to reduce the systematic error. As mentioned in Sect. 2, the off-axis detector calibration system reduces the systematic error to 4%. With 4% error and a data sample of 5 times more than the KamLAND second result, a sharp spike constraint on Δm^2 may be obtained as shown in Figure 8. However the constraint of θ_{12} is not greatly improved by KamLAND-II.

It has been pointed out that the ~60 km baseline is the best choice for the θ_{12} measurement in reactor $\bar{\nu}_e$s oscillation studies (Bandyopadhyay et al., hep-ph/0410283). Figure 9 shows the 3 σ constraint on θ_{12} expected from ~60 km baseline experiments (Bandyopadhyay et al., hep-ph/0410283). Although the constraint on θ_{12} is improved, getting close to reactors is an opposing strategy for the next generation geoneutrino experiments. The θ_{13} measurement with a baseline of ~1 km is in a similar situation. How we realize experiments detecting reactor neutrinos and geoneutrinos together is an interesting challenge for us. Our proposal is to develop a neutrino detector of several tons with the capability to shield and reject cosmic-rays. Neutrino detectors functioning in shallow underground sites are a selling point, because they enable us to distribute geoneutrino detectors over the entire surface of the Earth like seismometers.

Making good use of this type detector, measurements of θ_{12} and θ_{13} are carried out by installing several detectors at a ~1 km and ~60 km away from reactors. After finishing reactor neutrino experiments these detectors may be delivered to geoneutrino experiment sites. This project is roughly illustrated in Figure 10.

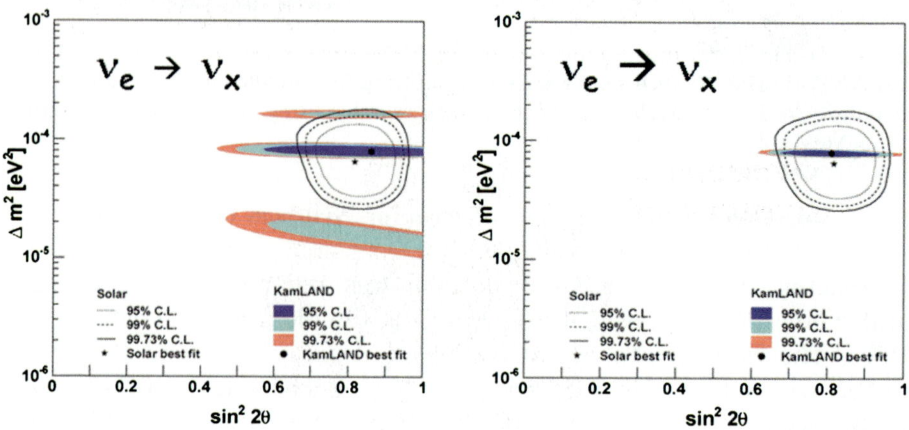

Figure 8. Contour plots of Δm^2– $\sin^2 2\theta$ plane. (a) is the present KamLND result, and (b) is the expectation of KamLAND-II discussed in the text.

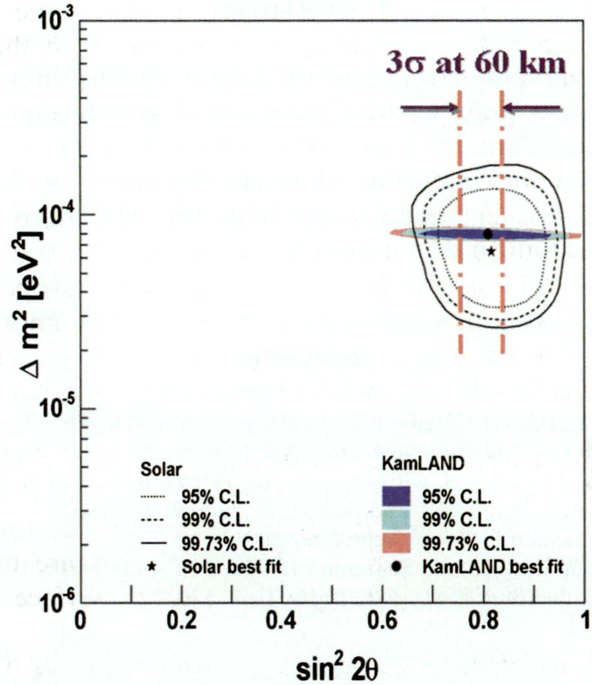

Figure 9. Sensitivity of 3σ range of allowed values for $\sin^2 2\theta$ in the case of $\Delta m^2_{12} = 8 \times 10^{-5}\,\mathrm{eV}^2$ (Bandyopadhyay et al., hep-ph/0410283).

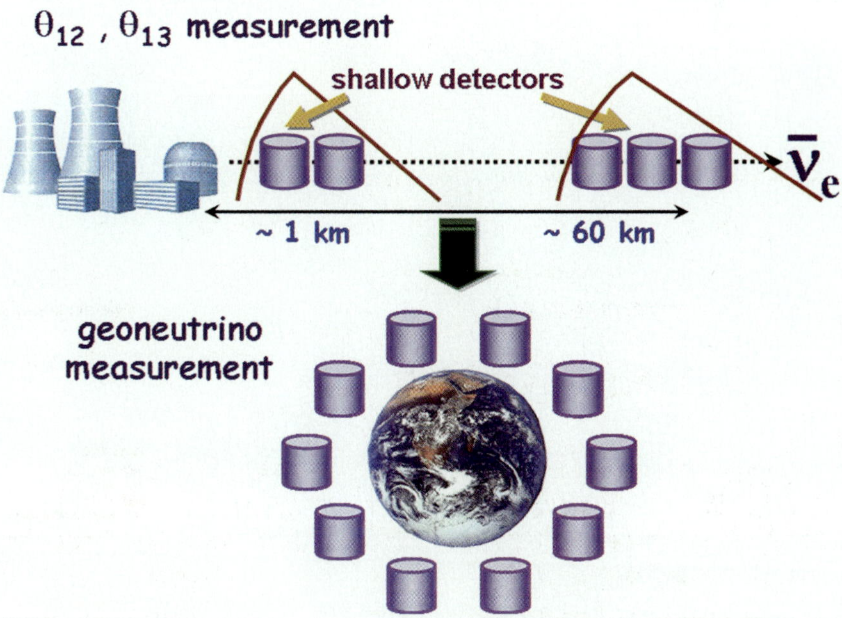

Figure 10. Module-type detector for reactor neutrinos and geoneutrinos.

5. Conclusions

Although the method to detect geoneutrinos was established by Kam LAND, this is only the first step. Separate detection of geoneutrinos coming from Crust, Mantle and Core, the geoneutrino measurements from K, U and Th, and the multi-site measurements of geoneutrinos are issues for the next generation detectors. A modular-type detector which yields a direction measurement is an interesting candidate.

References

Araki, T., et al. (KamLAND Collaboration).: 2005a, *Nature* **436**, 499.
Araki, T., et al. (KamLAND Collaboration).: 2005b, *Phys. Rev. Lett.* **94**, 081801.
Bandyopadhyay, A., Choubey, S., Goswami, S. and Petcov, S. T.: hep-ph/0410283.
Enomoto, S., Ohtani, E., Inoue, K. and Suzuki, A.: hep-ph/0508049.
Fields, B. D. and Hochmuth, K. A.: hep-ph/0406001.
Krauss, L. M., Glashow, S. L. and Schramm, D. N.: 1984, *Nature* **310**, 191.
McDonough, W. F. and Sun, S.-s.: 1995, *Chem. Geol.* **120**, 223.